T5-CVD-944

Thin Liquid Film Phenomena

W.B. Krantz, D. T. Wasan and R. K. Jain, editors

P. V. Nerad, associate editor

Zeev Dagan
A.S. Dimitrov
Stig E. Friberg
D. Gallez
P.A. Gauglitz
Kevin J. Gleason
Pil-Soo Hahn
I.B. Ivanov
Rakesh K. Jain
J.G.H. Joosten
S. Kalachandra
William B. Krantz
P.A. Kralchevsky
Hironobu Kunieda
Charles Maldarelli
A.K. Malhotra
C.A. Miller
P. Neogi
Bruce A. Nerad
A.D. Nikolov
Dilva Padovan
M. Prevost
C.J. Radke
Roderick J. Ray
Eli Ruckenstein
D.O. Shah
Ashutosh Sharma
John C. Slattery
T.T. Traykov
B.G. Volintine
Darsh T. Wasan
P.C. Wayner, Jr.
Donald R. Woods

AIChE Symposium Series

Number 252 — 1986 — Volume 82

Published by

American Institute of Chemical Engineers

345 East 47 Street — New York, New York 10017

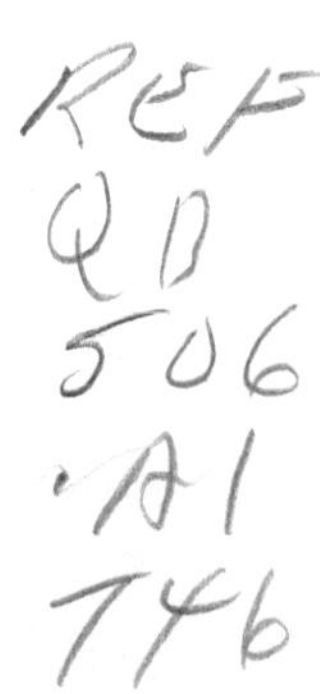

American Institute of Chemical Engineers
345 East 47 Street, New York, N.Y. 10017

AIChE shall not be responsible for statements or opinions advanced in papers or printed in its publications.

Library of Congress Cataloging-in-Publication Data
Main entry under title:

Thin liquid film phenomena.

(AIChE symposium series ; no. 252, v. 82)
Proceedings of a symposium organized by AIChE's Interfacial Phenomena Technical Program Committee in cooperation with the ACS Division of Colloid and Surface Chemistry at the Annual Fall Meeting of the American Chemical Society in Chicago, Ill., on Sept. 8-13, 1985.
Includes index.
1. Liquid films—Congresses. 2. Surface chemistry—Congresses.
I. Krantz, William B., 1939- II. Wasan, D. T. III. Jain, Rakesh K. IV. AIChE's Interfacial Phenomena Technical Program Committee. V. American Chemical Society. Division of Colloaid and Surface Chemistry. VI. Series: AIChE symposium series ; no. 252.
QD506.A1T46 1986 541.3'453 86-32069
ISBN 0-8169-0394-8

Printed in the United States of America by
Twin Production & Design

FOREWORD

The thin liquid films of interest here refer to liquid films or regions of liquid layers whose thickness is less than that of the long range attractive London-dispersion intermolecular forces, that is, less than approximately 100 nm. Increased attention is being focused on thin liquid film phenomena occurring in areas of engineering interest such as foam drainage, drop and bubble coalescence, multiphase flow through porous media, coating processes, biological thin films, and fabrication processes involved in the microelectronics industry. The recognition of the importance of thin liquid film phenomena in current as well as emerging technologies motivated AIChE's Interfacial Phenomena Technical Program Committee in cooperation with the ACS Division of Colloid and Surface Chemistry to organize the symposium on "Thin Liquid Film Phenomena" at the Annual Fall Meeting of the American Chemical Society in Chicago, Illinois, on September 8-13, 1985. The intent of this symposium volume, which constitutes the proceedings of this ACS symposium, is to provide an overview of the fundamentals of thin liquid film phenomena, while including the latest developments in this field by leading researchers from throughout the world. A particular thrust of this symposium volume is to explore the new directions in thin liquid film research which have emerged in response to the problems posed in our chemical and materials processing industries. As such, this symposium volume represents the first research monograph on the subject of thin liquid film phenomena published in the U.S.

William B. Krantz, *editor*
Professor
Department of Chemical Engineering
University of Colorado
Boulder, CO

Darsh T. Wasan, *editor*
Professor and Chairman
Department of Chemical Engineering
Illinois Institute of Technology
Chicago, IL

Rakesh K. Jain, *editor*
Professor
Department of Chemical Engineering
Carnegie-Mellon University
Pittsburgh, PA

CONTENTS

THIN LIQUID FILM PHENOMENA —INTRODUCTION

William B. Krantz ■ Department of Chemical Engineering, University of Colorado, Boulder, CO
Darsh T. Wasan ■ Department of Chemical Engineering, Illinois Institute of Technology, Chicago, IL
Rakesh K. Jain ■ Department of Chemical Engineering, Carnegie-Mellon University, Pittsburgh, PA

This paper serves to introduce the subject of this symposium volume, thin liquid films, to the reader. It provides a brief chronology on the development of thin liquid film research. The role of thin liquid films in current as well as new and emerging technologies is indicated. The organization and scope of this symposium volume then are given.

This AIChE Symposium Series volume emanated from the ''Symposium on Thin Liquid Film Phenomena'' held at the Annual Fall Meeting of the American Chemical Society (ACS) in Chicago, Illinois, on September 8-13, 1985. This Symposium was sponsored by the ACS Division of Colloid and Surface Chemistry in cooperation with AIChE's Interfacial Phenomena Technical Program Committee. This volume represents the first research monograph on the timely subject of thin liquid films published in the United States.

The thin liquid films of interest here refer to liquid films or regions of liquid layers whose thickness is less than that of the long range attractive London-dispersion intermolecular forces, that is, less than approximately 100 nm. Derjaguin is generally credited with initiating the study of these thin liquid films. Derjaguin (1) sought to explain why the pressure in the thin liquid film between two flat solid surfaces immersed in a nonwetting liquid, was higher than in the bulk liquid surrounding the plates. Derjaguin demonstrated that this excess pressure, which is referred to as the ''disjoining pressure'', arose because the intermolecular potential energy field at any point in a thin liquid film is different from that which would prevail in the same thickness of liquid if it were underlaid by a very thick bulk phase of the same liquid.

The pressure in a thin film need not always be greater than that of the liquid in the bulk phase. For example, the pressure in the thin film between two flat solid surfaces immersed in a wetting liquid will be less than that in the bulk liquid. This negative pressure then is referred to as the ''conjoining pressure''.

Derjaguin (2, 3) developed a simple gas-bubble pressure technique to measure this disjoining pressure which he found to be typically of the order 100 Pa. The ensuing years have led to more sophisticated and accurate methods for measuring the disjoining and conjoining pressures, some of which will be discussed in this symposium volume.

For nearly 30 years, thin liquid film research focused on quiescent liquid films in the absence of any dynamic effects. Scheludko (4) and his coworkers initiated the interest in the instability of thin liquid films. The instability of a thin liquid film refers to its propensity to develop corrugations when subjected to small spontaneous perturbations in its shape. Clearly, the shape of a thin liquid film can be perturbed only if one or more of its bounding surfaces is deformable. The particular problem which appears to have spurred the Bulgarian school of thin film investigators to explain thin film instabilities, was that of the coalescence of two drops or bubbles suspended in an immiscible liquid phase. That is, they sought to resolve the anomaly that two drops or bubbles cannot in theory coalesce by mere draining of the thin liquid film separating them, since the ratio of the viscous to inertia forces becomes unbounded as the film thickness becomes infinitesimally thin. Of course, this paradox is resolved if the thin film separating the drops or bubbles can rupture, for example, by the growth of a spontaneous perturbation in its shape.

Scheludko and Ivanov's early studies of the instability of thin liquid films excited considerable theoretical analyses of thin films. Two seemingly distinct mathematical approaches evolved for incorporating the driving forces causing the instability into the equations of motion. The first, which is referred to as the ''disjoining pressure approach'', was pioneered by Vrij (5), Ivanov (6), and others. In this approach, the driving force for the instability is expressed as the gradient of the disjoining pressure. The second mathematical approach, which is referred to as the ''body force approach'', was developed by Felderhof (7), Ruckenstein and Jain (8), and others. In this approach, the driving force for the instability is included as a body force in the equations of motion and expressed in terms of a gradient of a potential function which is referred to as the ''excess intermolecular potential energy''. The interrelationship between these two approaches has been established only very recently by Maldarelli et al. (9). This symposium volume will focus considerable attention on recent developments in the study of thin liquid film instability.

More recent thin film studies revealed the formation of stepwise transitions or stratification phenomena in films formed from concentrated surfactant solutions. Recent data of Wasan and his co-workers (10) indicate that the thickness of the transition is constant and equal to the micellar-lamella thickness. These data suggest a new mechanism of stability due to the stabilizing force of the micellar periodic structures in the film.

A recent development concerns ''pseudothin'' liquid film phenomena. This refers to physicochemical effects within a very thin region of a relatively thicker fluid layer, for which the intermolecular force field becomes important because of very steep gradients. That is, if a very steep temperature or concentration gradient is impressed at the deformable surface of a liquid, the thin region defined by the thickness of the zone over which the gradient is steep, takes on thin-film-like characteristics. It has been demonstrated recently by Ray et al. (11) that very steep concentration or temperature gradients normal to an interface can induce convective instabilities via a driving force expressed as the gradient of the excess intermolecular potential energy. As such, these gradient-driven instabilities are clearly related to thin liquid film instabilities. However, they are distinct from Marangoni-driven instabilities because the driving force for the gradient-driven instability mechanism appears either in the equations of motion or in the surface normal stress boundary condition, whereas for the Marangoni mechanism it appears in the tangential stress boundary condition. This symposium volume will review this recent work on pseudothin liquid film phenomena as well.

The Symposium on Thin Liquid Film Phenomena was organized because of the ever-increasing importance which this subject has for both current as well as new and emerging technologies. The importance of thin liquid films in the coalescence of drops and bubbles has already been mentioned. The draining of thin films interlacing foams is critical to foam stability. Tertiary oil recovery and other processes concerned with multiphase flow have focused attention on the dynamics of the thin films which form as bubbles and droplets squeeze through porous media. The spreading of liquids on solid surfaces involved in coating and other similar processes involve moving contact lines where three phases meet. Clearly, these moving contact lines always involve thin film phenomena within at least some distance from the contact line. A subject of continually increasing interest is biological thin films such as deformable cell walls and lipid membranes. Finally, the rapidly emerging microelectronics industry employs a variety of deformable films to selectively etch, form, and protect chips, microsensors, and other types of microcircuitry. A particular thrust of this symposium volume is to explore the new directions in thin liquid film research which have emerged in response to the problems posed in our chemical and materials processing industries.

The authors of the papers constituting this symposium volume include several of the pioneers in the field of thin liquid films as well as relatively new researchers in this field whose novel work is charting new directions for thin film research. The breadth and wide appeal of thin liquid film research is evidenced by the range of scientific disciplines covered by the contributors to this volume. In addition to its interdisciplinary flavor, the coverage in this volume is truly international, drawing from leading thin liquid film research groups from throughout the world.

The organization of this AIChE Symposium Volume will parallel that of the ACS Symposium on Thin Liquid Film Phenomena. First, three overview papers will summarize the state-of-the-art concerning the fundamentals of liquid film thinning and stratification phenomena, thin film stability, and steep gradient induced pseudothin film instability. These review papers then are followed by 14 papers which present the state of the art in thin liquid film research and which outline new directions in this rapidly evolving research field.

ACKNOWLEDGMENTS

The co-editors gratefully acknowledge Ms. Patrice Nerad and Ms. Maura Mullen for their assistance in preparing this symposium volume, and NSF grant No. CBT-8506904 for financial support. Finally, the co-editors express their appreciation to ACS's Division of Colloid and Surface Chemistry for sponsoring this symposium and to the American Institute of Chemical Engineers for publishing this symposium volume.

LITERATURE CITED

1. Derjaguin, B., Koll. Z., 69, 155 (1934).

2. Derjaguin, B., Acta Physicochimica U.S.S.R., 5, 1 (1936).

3. Derjaguin, B., Acta Physicochimica U.S.S.R., 10, 333 (1937).

4. Scheludko, A., Proc. Kon. Ned. Akad. Wet., B65, 87 (1962).

5. Vrij, A. and J. Th. G. Overbeek, J. Am. Chem. Soc., 90, 3074 (1968).

6. Ivanov, I., B. Radoev, E. Manev, and A. Scheludko, Trans. Faraday Soc., 66, 1262 (1970).

7. Felderhof, B. U., J. Chem. Phys., 49, 44 (1968).

8. Ruckenstein, E and R. K. Jain, Faraday Trans. II, 70, 132 (1974).

9. Maldarelli, C., R. K. Jain, I. B. Ivanov and E. Ruckenstein, J. Colloid and Int. Sci., 78, 118 (1980).

10. Wasan, D. T. and A. D. Nikolov, Paper presented at the Sixth Int. Symposium on Surfactants in Solutions, New Dehli, Aug. 18-22, 1986. See also: Society of Petroleum Engineers Preprint, SPE 15433, paper presented at the SPE 61st Annual Technical Conference, New Orleans, Oct. 5-8, 1986.

11. Ray, R. J., W. B. Krantz and R. L. Sani, J. Membrane Sci., 23, 155 (1985).

THIN LIQUID SURFACTANT FILM DRAINAGE PHENOMENA—A REVIEW

D. T. Wasan and A. K. Malhotra* ■ Department of Chemical Engineering, Illinois Institute of Technology, Chicago, IL 60616

Both experimental and theoretical investigations of the dynamic behavior of thin liquid surfactant films, such as those existing between two approaching droplets/bubbles or a drop coalescing at a bulk interface, are reviewed. The phenomenon of layered structuring in thin films arising because of surfactant micelle interactions, and the influence of these microstructures on the drainage process, are presented. Results of our more recent theoretical model to predict the rate of drainage of surfactant/polymer stabilized foam and emulsion films from the knowledge of bulk and interfacial properties are discussed. These results indicate that the rate of drainage of thin films is strongly dependent upon the surfactant adsorption kinetics, selective surfactant solubility, and interfacial rheological properties such as elasticity and interfacial viscosity.

*Presently with Polaroid Corporation, New Bedford, MA 02745

INTRODUCTION

Thin liquid fioms have gained large scientific and industrial applications. Free liquid films are considered as models for foams, films of one liquid in another as a model for emulsion, a film of liquid between solids as a model for colloidal suspensions, and a thin film on a solid substrate as a model for polymolecular adsorption. In the present review, only foam and emulsion films are considered.

The extent and rate of drainage of liquid from the interior of the lamella are two of the most important factors determining the stability of foam and emulsion films.

The main stages in the formation and evolution of the liquid film (between two approaching drops or bubbles) as shown in Figure 1 are:

(a) Two drops approach each other resulting in their hydrodynamic interaction;
(b) Deformation of the drops leading to a bell shaped formation, which is called a 'dimple';
(c) The dimple gradually disappears and a plane-parallel film of radius, R, is formed. The film draines under the combined action of suction at Plateau borders and the disjoining pressure;
(d) When the disjoining pressure is negative, it favors the growth of corrugations at the film surfaces and at a critical thickness, h_{cr}, either the film ruptures or a jump transition occurs leading to a stable structure. This process of transition to stable state is known as 'black spot formation' since at these thicknesses the film appears grey or black (Sheludko (1));
(e) The black spots increase in size and cover the whole film;
(f) The formation of an equilibrium film whose lifetime can be virtually unlimited.

Hence if the colliding drops have axial symmetry, the process of coalescence can be split into (a) mutual approach of two droplets to form a plane-parallel film and (b) thinning of the film to such a thickness so that rupture can occur, followed by (c) rupture itself when a hole is formed. Stages (a) and (c) occur immeasureably fast so that the lifetime of the intervening film is essentially given by stage (b). Observations of MacKay and Mason (2) and Scheludko et al. (1) suggest that the most critical period is during drainage of the film over a thickness range of 200 nm to critical thickness. Thus, the lifetime of the intervening film is an important characteristic of dispersed systems, and is directly related to their stability.

The forces of interaction that govern the lifetime of thin liquid films are the capillary pressure (suction at Plateau borders) and the disjoining pressure. The thermodynamic properties of thin liquid films are different from those of the bulk solutions. These films possess an excess chemical potential that is manifested into an excess pressure. Deryaguin (1) coined the term 'disjoining pressure' to characterize this excess pressure. Disjoining pressure consists of the electrostatic repulsion forces between ions on the two surface layers, the attractive van der Waals forces among all the molecules of the film, and the steric forces due to steric hindrance in closely packed monolayers.

EXPERIMENTAL

In order to predict the lifetimes of thin films it is essential to know both the rate of drainage, as well as the critical thickness at which the film ruptures. The rate of film thinning and its thickness can be measured using several methods including optical methods. These methods are discussed in detail elsewhere (3).

Effect of Surfactant Concentration on Film Drainage

Measurements of the drainage time, critical thickness and the velocity of thinning have been reported using the interference microscopic technique proposed by Sheludko (1). Recently Rao et al. (4) employed this technique to study the drainage behavior of aqueous foam films (of radius, $R=9x10^{-2}$) formed from lauryl sulfate (an ionic surfactant), SLS solutions. Figure 2 shows the surface tension and the drainage time as a function of the bulk SLS concentration for an NaCl concentration of 0.25 kmol/m^3. The surface tension decreases with increasing SLS concentration and reaches a constant value. The abrupt change in the surface tension corresponds to the critical micelle concentration (CMC) for the system. The drainage time increases with increasing surfactant concentration, reaches a maximum value and then decreases. The concentration corresponding to the maximum drainage time is very close to the CMC.

Figure 3 shows the dependence of the critical thickness on the bulk SLS concentration. The critical thickness decreases with increasing concentration and reaches an almost constant value near the CMC of the system. At low surfactant concentration, the amplitude of surface waves, which causes the films to rupture or form black spots, is large. Low surfactant concentrations thus cause the films to rupture at a large critical thickness. As the surfactant concentration increases, the waves are damped, causing the films to drain to lower thicknesses before rupture or black spot formation is seen (1). The critical thickness also decreases as the radius of the film is decreased.

The dependence of the drainage time on SLS concentration is best understood if the two portions of the curve, below and above CMC, are considered separately. The increase in the drainage time below CMC is due to the decrease in both surface tension and the critical thickness. The drainage time depends on the capillary pressure, P_c which is equal to $2\sigma/R_c$, where σ is the surface tension and R_c is radius of the capillary. Hence, any decrease in the capillary pressure results in a decrease in the velocity of thinning and, consequently, an increase in the drainage time. As is shown in Figure 3, an increase in the bulk concentration causes the critical thickness of the film to decrease. Thus below the critical micelle concentration of the surfactant (CMC), the combined effects of the capillary pressure and the critical thickness increase the drainage time of the film.

At CMC, the surface of the film is completely saturated with the surfactant. Any expansion of the surface results in a lowering of the surface concentration at the surface, with a consequent rise in the surface tension. Above CMC, as both micelles and monomers are able to diffuse to the surface, equilibrium is restored more rapidly. A certain minimum surface tension gradient, however, has to be maintained during drainage. This can be achieved only by increasing the rate of thinning, which results in a decrease in the drainage time. Since the surface elasticity is a measure of the surface tension gradient, a change in the surface elasticity should influence the drainage time (4). Rao et al. (4) reported data on the effect of SLS concentration on the apparent surface elasticity at concentrations above CMC. Their data showed that elasticity decreased as the concentration increased and that the elasticity was maximum near CMC.

Recently, Manev et al. (5) investigated the drainage behavior of microscopic circular horizontal films formed from concentrated solutions of SLS in the range from 1 to 15 wt % (0.034 to 0.52 kmol/m^3). At a sufficiently high concentration of surfactant (SLS), formation of films possessing a layered or stratified structure has been observed. The drainage times of the films, together with the bulk viscosities of the solutions, are presented in Table 1. The drainage time of SLS film increases with an increase in surfactant concentration for two principal reasons: First, because of an increase in the bulk viscosity of the SLS solutions and second, because of the time required for the development of each consecutive stage of transition until the final state of a thin stable black film is formed. During its drainage, the stratifying film passes through several metastable states. The metastable equilibrium thicknesses of the stratifying foam and emulsion films measured as a function of sodium lauryl sulfate concentration are shown in Figure 4. The stratifying emulsion films were of oil (toluene)-in-water type. As can be seen from the figure, foam and emulsion films exhibit similar stratification patterns. The stratification phenomenon has also been observed with nonionic surfactants (6). Recently, Wasan and Nikolov (6) explained the process of stratification on the basis of a micelle-laticing structure model.

Effect of Film Size on Film Drainage

In our laboratory, Manev et al. (7) and Dinardo (8) have taken extensive data on drainage of aqueous foam and emulsion films as a function of film radius. The experiments were performed with films of radii 5 to 50 x 10^{-5}m, applying the microinterferometric technique of Scheludko. In Figure 5, the experimental drainage times for a constant range of thickness (from 200 to 50nm) and plotted on a log-log scale as a function of film radius. The plot yields a linear dependence with a slope of about 0.8 which is much smaller than the slope of 2.0 predicted by the Reynolds' equation.

A MODEL FOR FILM DRAINAGE

Reynolds (9) was the first to study the rate of approach between two surfaces separated by a draining thin film. His analysis assumed that the two surfaces were both flat and rigid. His result for the rate of drainage is given by:

$$V_{RE} = -\frac{dh}{dt} = \frac{8Fh^3}{3\pi\ \mu R^4} \qquad (1)$$

where

h = half film thickness

μ = film viscosity

R = film radius

F = force

t = time

As pointed out by Zapryanov et al. (10) and Oppenheim (11) the Reynolds' equation (1) represents a conservative prediction. Due to the mobility of the interfaces, the rate of film thinning can be several times greater than predicted by this equation.

Much work has been done since then to model film drainage of both flat and dimpled liquid surfactant films. Manev et al. (7) have shown that the dimple formation in foam and emulsion films is dependent on the radius of the film. At film radii below 10^{-4}m, dimple can't be detected at any thickness. However, at larger film radii, the dimple is spontaneously ejected at a thickness of about 2000 Å and an almost plane-parallel film is formed. Hence, the assumption of a plane-parallel film is justified for film thicknesses smaller than 2000 Å.

In what follows, we will review the predictions of recent generalized model (see Figure 6) for the drainage of surfactant/polymer stabilized plane-parallel foam and emulsion films (12). The model accounts for flows in both film and dispersed phases, the partitioning of surfactant, interfacial viscosities and their gradients, interfacial elasticity and mass transfer involving both the bulk and interfacial diffusion. Surfactant transport from the bulk to the interface is described by a two step mechanism: (i) diffusion of the surfactant from the bulk to the sublayer (which lies immediately below the interface), followed by (ii) the adsorption of the surfactant from the sublayer to the interface. The second step is represented by a kinetic expression (Miller and Kretzschmar (13); Borwankar and Wasan, (14)). The effect of surfactant partitioning on the rate of film thinning is examined for a) surfactant soluble in both the film and the dispersed phases, b) surfactant soluble in the film phase only and c) surfactant soluble in dispersed phase only. The present model employs the plane-parallel film model to describe thinning of a liquid film (between two drops/bubbles) in terms of dimensionless numbers and is an extension and supplement to earlier analyses (Good, (15); Traykov and Ivanov (16); Zapryanov et al., (10). Details of this model development are presented elsewhere (12). However, the importance of the bulk and interfacial properties on the rate of drainage of thin liquid films are highlighted below.

Figure 7 shows the variation of interfacial mobility as a function of dimensionless film thickness, h, for different values of interfacial viscosity number, E_S. For the present model the interfacial mobility is given by

$$\frac{V}{V_{RE}} = 1 - \frac{3\pi}{h^2 F} \sum_{n=1}^{N} A_n \frac{e^{-\lambda_n h}}{1+\lambda_n h} J_0(\lambda_n). \qquad (2)$$

The effect of elasticity number, E_s, on dimensionless drainage time is depicted in Figure 8. Liquid movement along the film surface results in a local variation of interfacial surfactant concentration which produces an interfacial stress opposing the liquid flow (Marangoni-Gibbs effect). At the same time a concentration gradient is established between the bulk and the interface. This causes surfactant transport to the interface in an attempt to restore the original condition. The elasticity number, E_s, is a measure of the relative importance of these two competing effects. At high values of elasticity number ($E_s > 10^6$), bulk and interfacial diffusion can not counterbalance the interfacial tension gradient (the Marangoni-Gibbs effect) and hence the velocity of thinning is given by the Reynolds equation (1). However, for small values of E_s, even at a moderate interfacial viscosity number ($V_s=10^2$), the drainage velocity is several times greater than the Reynolds velocity. An increase in interfacial viscosity number, V_s, results in decreased interfacial mobility and hence higher drainage time. For low values of interfacial viscosity number ($V_s < 1$), the interfacial viscosity V_s has very little contribution and hence does not appreciably affect the thinning velocity. At high values of interfacial viscosity number ($V_s > 10^5$), the interface is rendered immobile and hence interfacial mobility and drainage time are independent of interfacial elasticity number and interfacial viscosity number.

Figure 9 demonstrates the effect of viscosity ratio on drainage time. As is evident from this figure, the viscosity ratio ($V_I = \tilde{\mu}'/\mu'$) has no effect on drainage time at high values of elasticity number ($E_s \geq 10^6$). Therefore, emulsion films behave like foam films at high values of E_s. However, at lower values of E_s, an increase in the viscosity ratio, V_I, lowers interfacial mobility and hence increases drainage time, due to the increased bulk stress from the drop phase which opposes the outward flow of film liquid. An increase in the viscosity ratio has the same effect as an increase in elasticity number but the influence of elasticity number on both interfacial mobility as well as drainage times is more pronounced.

As the film thins, the surfactant is swept outwards from the center of the flattened area, thus perturbing the equilibrium surfactant concentration on the interface. The perturbation in interfacial concentration results in surfactant transfer (interfacial and bulk diffusion) which tends to restore the equilibrium distribution on the interface. Interfacial diffusion eliminates interfacial tension gradients, so that an increase in diffusivity ratio ($D_f = D'_s/D'$) results in increased interfacial velocity and hence faster drainage rate (see Figure 10). However, for high values of elasticity number ($E_s \geq 10^7$), the drainage time has little dependence on diffusivity ratio, D_f.

The forward rate constant for the film phase, K_1, was found to have very little or no effect on the velocity of thinning. Hence, the surfactant transport from the film onto the interface is diffusion controlled. This is in agreement with the qualitative argument proposed by Ivanov (17).

The forward rate constant for the drop phase, $\tilde{K}_1$, was found to have an appreciable effect on drainage time (see Figure 11). Higher values of $\tilde{K}_1$ imply lesser resistance to adsorption and hence results in increased velocity of thinning. For high value of elasticity number ($E_s \geq 10^6$), the drop phase forward rate constant, $\tilde{K}_1$ was found to have no effect on drainage time.

If the bulk concentration of the surfactant is high and its adsorption relatively weak (i.e. small A_d), the bulk and interfacial diffusion can counterbalance the perturbation in interfacial concentration due to the expansion of the interface. This results in high velocity of thinning and hence lower drainage time. Figure 12 shows the effect of adsorption number on drainage time. As the adsorption number, A_d, increases the drainage time increases. At high values of elasticity number ($E_s \geq 10^6$), the adsorption number has almost no effect on the drainage time.

So far we have considered the surfactant to be soluble in both the film and the drop phases. We now present results when the surfactant is soluble in the film or in the drop phase only.

Figure 13 compares the effect of selective surfactant solubility on interfacial mobility and drainage time for the cases where (a) the surfactant is soluble in only the film phase, and (b) the surfactant is soluble in the drop phase. It is noted that lower interfacial mobility and higher drainage times are obtained in the first case, while higher interfacial mobility and lower drainage times are obtained in the second case. As pointed out by Lee and Hodgson (18) and Ivanov (17), when the surfactant is soluble in the film phase it has to diffuse a long way through the film perimeter. Since the driving force is the gradient of surfactant concentration along the interface, the diffusion can not eliminate the interfacial tension gradient, which opposes thinning. When the surfactant is soluble in the drop phase, the surfactant flux counterbalances the perturbation in interfacial concentration, Γ, caused by the convective flux, resulting in a higher velocity of thinning. The emulsions with surfactant soluble in the film phase are thus more stable than those with surfactant soluble in the drop phase. This conclusion is in agreement with Bancroft's rule and has been discussed previously by Davies(19), Sherman (20) and more recently by Traykov and Ivanov (16).

CONCLUDING REMARKS

The results of our generalized flat film model indicate that the rate of film drainage depends on interfacial viscosity number between 10' and 10^4. Outside this range, interfacial viscosity number has no effect. The following variables decrease interfacial mobility for all values of interfacial viscosity number between 10^1 and 10^4: (1) high value of elasticity number; (2) high viscosity ratio; (3) low diffusivity ratio; (4) low drop phase forward rate constants; (5) high adsorption number; (6) low distribution number; and (7) surfactant solubility in the continuous (film) phase only.

A comparison of theoretical predictions with experimented results on the rate of drainage of thin films requires data on bulk and interfacial properties (e.g. bulk and interfacial diffusion coefficients, both interfacial shear and dilational viscosities, interfacial tension, bulk and interfacial concentrations) for the system under consideration. The values for most of these can vary several orders of magnitude from one system to another. Several separate experiments need to be performed to obtain accurate data on these properties. In view of this, a comparison has been made to the limited available data on drainage of thin liquid films. This comparison has been presented by us elsewhere (10,12).

ACKNOWLEDGMENTS

This study was supported partly by the National Science Foundation and partly by the Department of Energy.

NOTATION

A_d = Adsorption number = $\Gamma_o'/C_o'R'$
A_n = the unknown constants
C_o' = equilibrium surfactant concentration
D' = bulk diffusion coefficient
D_F = diffusivity ratio, D_s'/D'
D'_s = interfacial diffusion coefficient
E_o = Gibbs elasticity
E_s = elasticity number = $E_o R'/\mu'D'$
h' = half-film thickness
h = dimensionless film thickness = h'/R
$\tilde{K}_1$ = forward rate constant for the dispersed phase
p'_m = pressure in the meniscus
r' = distance along the r axis
R = film radius
V = rate of thinning of film
V'_r = velocity component in the radial direction
V_{RE} = Reynolds velocity
V_s = interfacial viscosity number = $\frac{\mu_s'}{\mu'R'}$
Z' = distance along the z axis

Greek Letters

Γ' = interfacial concentration
λ_n = nth root of $J_\cdot(\lambda) = 0$

$\tilde{\mu}'$ = viscosity of dispersed phase
μ' = viscosity of continuous (film) phase
μ^{s}, = interfacial viscosity

LITERATURE CITED

1. Sheludko, A. D., in "Advances in Colloid and Interface Science," Vol. 1, p. 391 (1967).

2. MacKay, G. D. M., and Mason, S. G., Can. J. Chem. Engng. 41, 204 (1963).

3. Sonntag, H., and Strenge, K., "Coagulation and Stability of Disperse Systems", Halstead Press, New York (1972).

4. Rao, A. A., Wasan, D. T., and Manev, E. D., Chem. Engng. Commun. 15, 63 (1982).

5. Manev, E. D., Sazdanova, S. V., and Wasan, D. T., J. Dispersion Sci. & Tech. 3,435 (1982).

6. Wasan, D. T. and Nikolov, A. D., Paper to be presented at the 6th International Symposium on Surfactants in Solution, New Delhi, August 18-22, 1986.

7. Manev, E. D., Sazdanova, S. V., and Wasan, D. T., J. Colloid Interface Sci. 97, 591 (1984).

8. DiNardo, P. E., M. S. Thesis, Illinois Institute of Technology, Chicago, Illinois, 1984.

9. Reynolds, O., Phil. Trans. Roy. Soc. (London) A177, 157 (1886).

10. Zapryanov, Z., Malhotra, A. K., Aderango, N., and Wasan, D. T., Int. J. Multiphase Flow 9, 105 (1983).

11. Oppenheim, J. P., Ph. D. Thesis University of Houston, Houston, Texas, 1983.

12. Malhotra, A. K., and Wasan, D. T., Manuscript submitted to Chem. Eng. Commun., July, 1986.

13. Miller, R., and Kretzschmar, G., Coll. Polym. Sci. 258, 85 (1980).

14. Borwankar, R. P., and Wasan, D. T., Chem. Engng. Sci. 38, 1637 (1983).

15. Good, P. A., Ph.D. Thesis, University of Minnesota, Minneapolis, 1974.

16. Traykov, T. T., and Ivanov, I. B., Int. J. Multiphase Flow, 3, 471 (1977).

17. Ivanov, I. B., Pure Appl. Chem. 52, 1241 (1980).

18. Lee, J. C., and Hodgson, T. D., Chem. Engng. Sci. 23, 1375 (1968).

19. Davies, J. T., Proc. 2nd Int. Congr. Surface Activity (J. C. Schulman, Ed.), Vol. 1, p 426 (1957).

20. Sherman, P., "Emulsion Science", Academic Press, New York (1968).

Table 1. Effect of surfactant concentration on drainage of SLS films. Temperature 25° C.

Δt is drainage time from 300 to 100 nm.

SLS conc., kmol/m^3	refr. index	viscosity, cp	Δt, sec R=0.1 mm	Δt, sec R=0.2 mm
0.017	1.3335	0.91	10.0	17.5
0.024	1.3337	0.92	10.5	18.0
0.035	1.3341	0.94	10.5	18.0
0.069	1.3352	1.01	12.0	21.5
0.138	1.3375	1.14	16.5	27.0
0.277	1.3422	1.64	19.5	33.0
0.520	1.3540	2.38	28.0	49.0

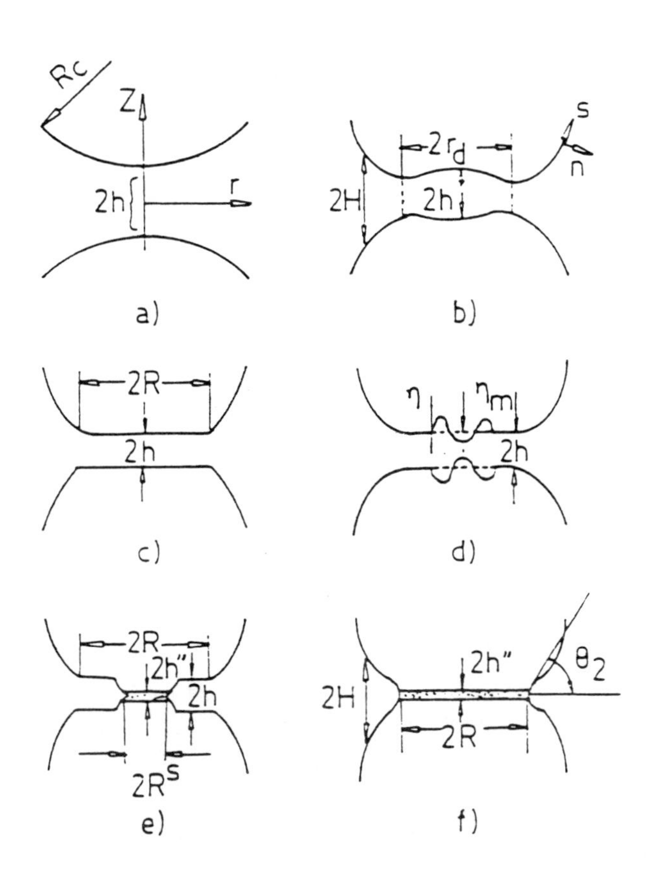

Figure 1. Main stages in the evolution of a thin film.

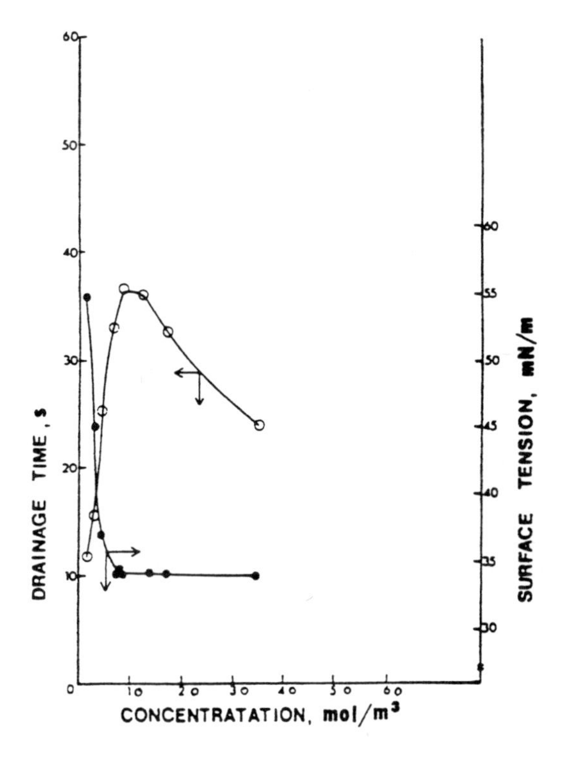

Figure 2. Drainage time and surface tension as a function of SLS bulk concentration. $'K' = \infty$, $C_{el} = 0.25$ kmol/m^3 NaCl, $R = 9 \times 10^{-2}$mm. [Ref. *4*]

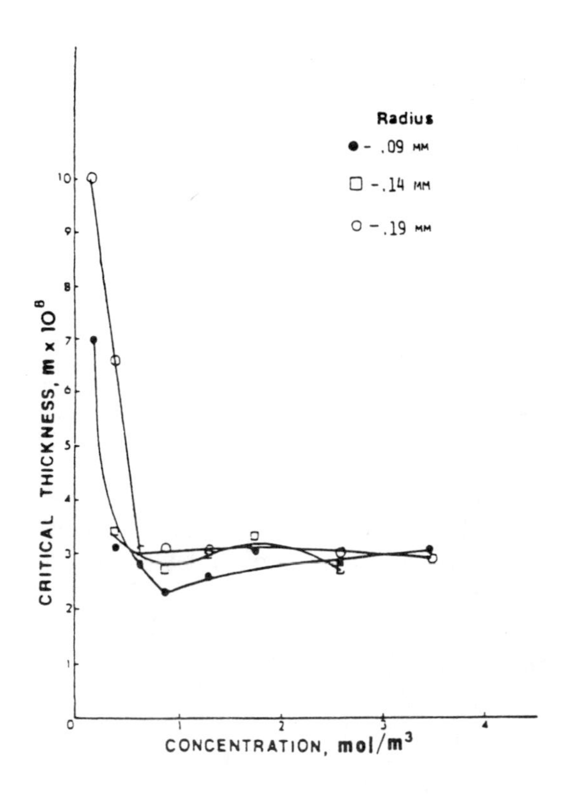

Figure 3. Critical thickness as a function of SLS bulk concentration. $'K' = \infty$, $C_{el} = 0.1$ kmol/m^3 NaCl. [Ref. *4*]

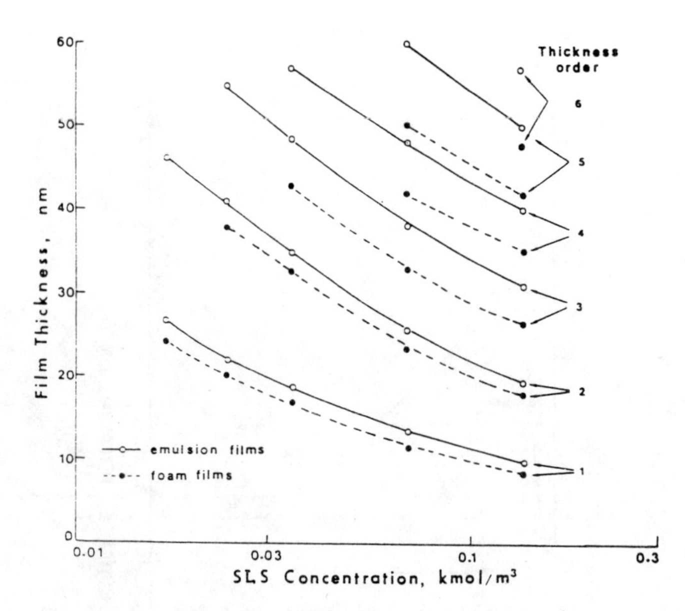

Figure 4. Thickness of stratifying SLS films of different orders as a function of SLS bulk concentration. [Ref. *5*]

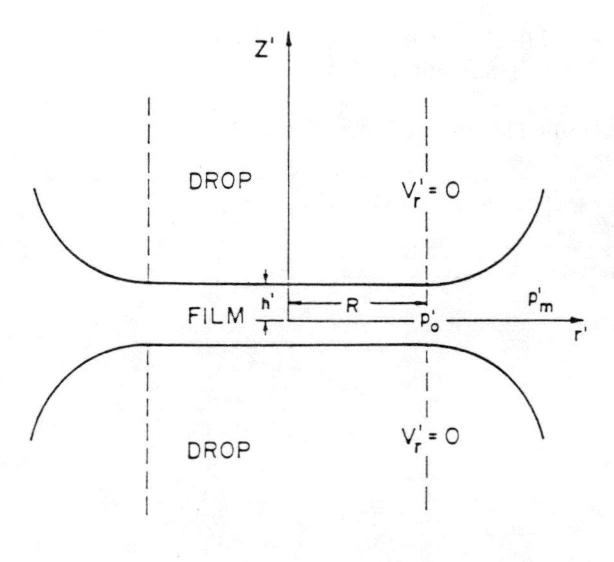

Figure 6. Film configuration and coordinate system [Ref. 12]

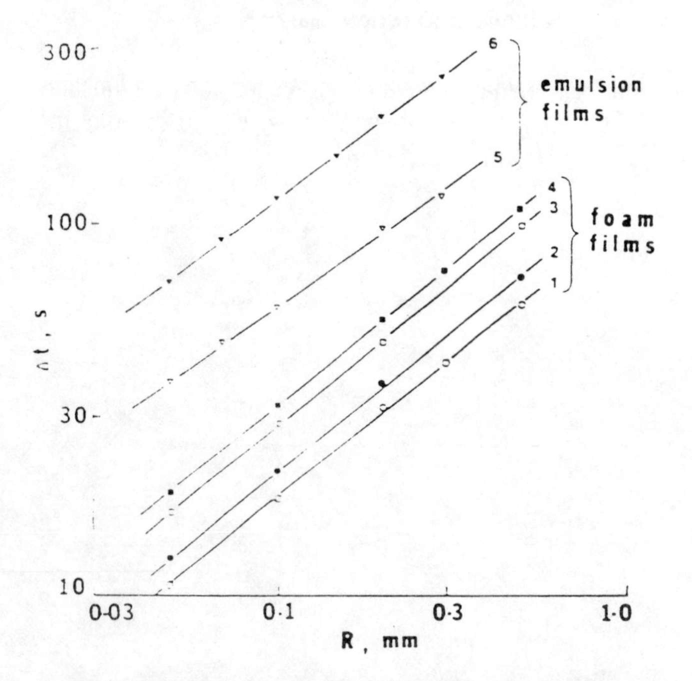

Figure 5. Drainage time Δt from a thickness h_0 = 200 nm to a constant thickness h = 50 nm for aqueous foam and emulsion films as a function of film radius R. Foam films: (1) 4.3×10^{-4} kmol/m^3 SLS + 0.1 kmol/m^3 NaCl. (2) 4.3×10^{-4} kmol/m^3 SLS + 0.25 kmol/m^3 NaCl. (3) 3.5×10^{-3} kmol/m^3 SLS + 0.25 kmol/m^3 NaCl. (4) 8.7×10^{-4} kmol/m^3 SLS + 0.25 kmol/m^3 NaCl. Emulsion films: (5) 5.6×10^{-3} kmol/m^3 SLS + 0.3 kmol/m^3 NaCl. (6) 4.3×10^{-4} kmol/m^3 SLS + 0.1 kmol/m^3 NaCl. [Ref. *7*]

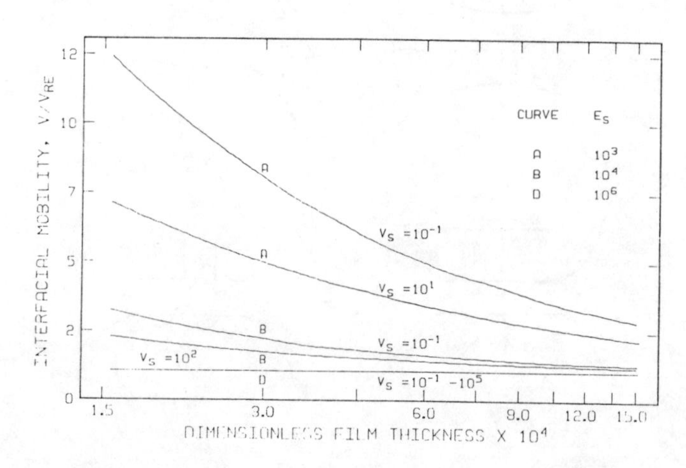

Figure 7. Effect of film thickness on interfacial mobility for various values of elasticity number, E_s, and interfacial viscosity number, V_s.

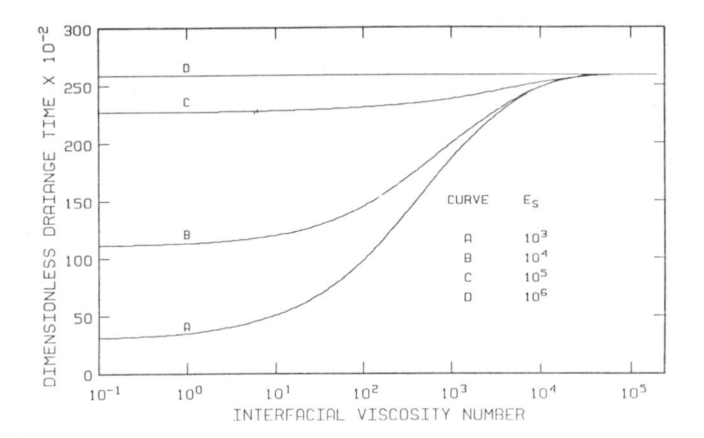

Figure 8. Effect of interfacial viscosity number, V_S, on drainage time, t, for various values of elasticity number, E_S.

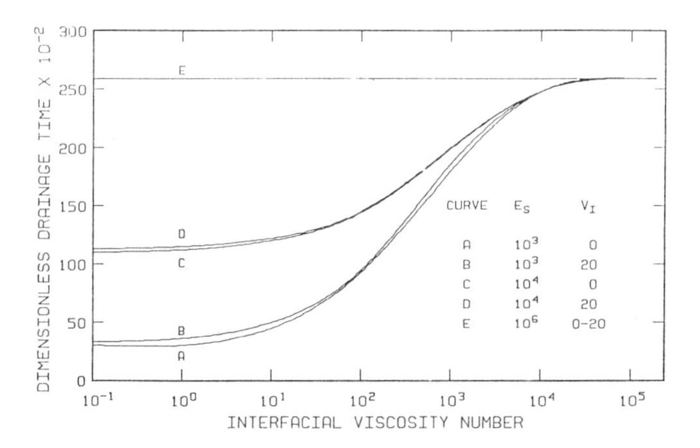

Figure 9. Effect of interfacial viscosity number, V_S, on drainage time, t, for various values of elasticity number, E_S, and viscosity ratio, V_i.

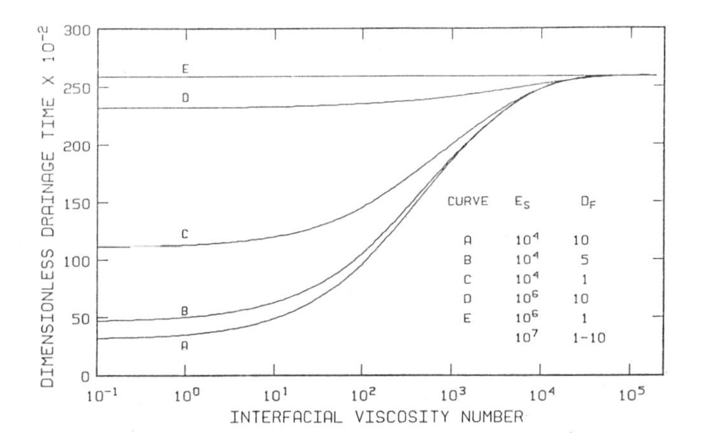

Figure 10. Effect of interfacial viscosity number, V_S, on drainage time, t, for various values of elasticity number, E_S, and diffusivitv ratio, D_f.

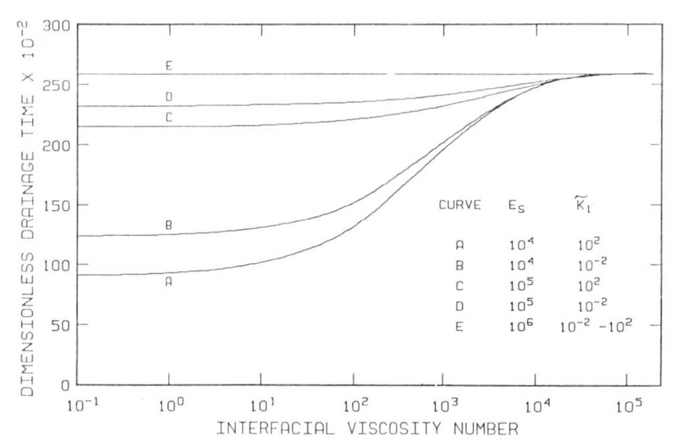

Figure 11. Effect of interfacial viscosity number, V_S, on drainage time, t, for various values of elasticity number, E_S, and forward rate constant for the drop phase, k_1.

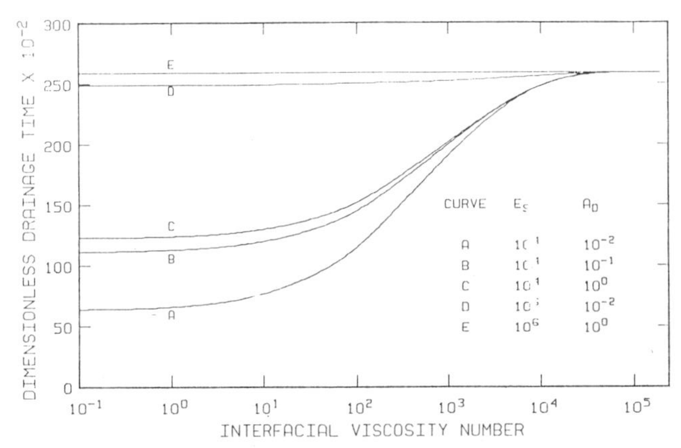

Figure 12. Effect of interfacial viscosity number, V_S, on drainage time, t, for various values of elasticity number, E_S, and adsorption number, A_d.

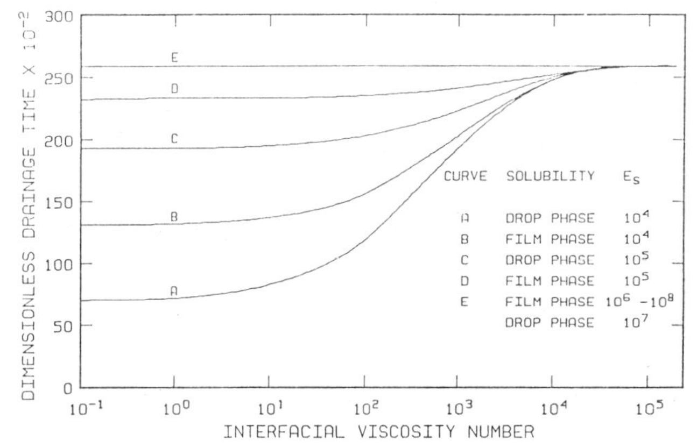

Figure 13. Effect of selective surfactant solubility on drainage time, t, for different values of E_S and V_S.

THE INFLUENCE OF THE LONG RANGE INTERMOLECULAR INTERACTIONS ON THE HYDRODYNAMIC STABILITY OF ULTRATHIN FILMS

Charles Maldarelli ■ Department of Chemical Engineering, Institute of Applied Chemical Physics, City College of New York, New York, NY 10031

At thicknesses of the order of 100 nm and below, the hydrodynamic stability of ultrathin films is influenced by the long range intermolecular interactions between molecules (i.e. the electrostatic forces between charged molecules and the van der Waals forces between uncharged molecules). Because of the importance of such very thin films in science and technology, a great deal of research has been undertaken to understand this influence. In this article this research is reviewed. Four major areas are studied. The first is the proper methodology for incorporating, from a continuum framework, the van der Waals forces among the molecules of a thin film system. The second area studies the nature of van der Waals induced instabilities in ultrathin lamellae. The third area concerns the elastic stabilization of van der Waals induced instabilities when the film exhibits an elastic rheology. The final area covers the influence of electrostatic forces. The stabilization of electrolyte films through double layer repulsion is discussed, and instabilities derived from potential differences applied across thin dielectric films is studied.

The long range intermolecular forces between molecules - i.e. the electrostatic interactions between charged molecules, and the van der Waals interactions between uncharged ones - critically influence the hydrodynamic stability of thin films when the film thicknesses are of the order of (or less than) the range of interaction of the long range forces. Inasmuch as this range is between 1 - 100 nm, the films whose hydrodynamic stability are affected are ultrathin. Nevertheless, the occurrence of such films in science and technology is widespread and, when present, their hydrodynamic stability is usually important. Some examples are given in the following paragraph.

In a foam or emulsion, particles of the dispersed phase may sometimes coalesce, leading ultimately to a breakdown of the dispersion. The stages which occur in coalescence involve the mutual approach of the dispersion particles, the formation of a radially bounded, thinning, near-planar film, and finally the rupture of the intervening lamella at thicknesses in the range of 10 - 100 nm. The rupture has been attributed to a hydrodynamic instability induced by van der Waals interactions. It is also well recognized that electrostatic forces within the film (double layer repulsion) can stabilize the film and therefore the dispersion. (For reviews, see Scheludko (1) and Aveyard and Vincent (2).) Nonuniformaties in film thicknesses caused by instabilities arising from long range forces also occur in the deposition of ultrathin lamellae on solid substrates as, for example, in the electostatic powder coating (Corbett (3) and Cross (4)) and semi-conductor deposition (Dunn (5)) technologies. Additional examples are related to biological phenomena: Theories of instabilities in thin films derived from long range interactions have been utilized to understand the physicochemical mechanisms underlying plasma membrane deformations of biological cells (Jain *et al*.(6)), lipid vesicles (Wendel *et al*.(7)) and the rupture of the ocular tear film in the human eye (Lin and Brenner (8, 9)).

This article is intended to provide a review of the research which has been undertaken on the influence of the long range interactions on the hydrodynamic stability of ultrathin lamellae. To detail the essentials of the subject matter, the scope of this review is restricted to nonthinning films and to linear instabilities. Nonlinear instabilities derived from van der Waals interactions have been studied by Williams and Davis (10) and Chen and Slattery (11) (See also Joosten's study of solitary waves and solitons in thin films (13).) The influence of film drainage on the development of van der Waals induced instabilities is also studied in (11) (and (12)), as well as in the studies of Ivanov and his coworkers (14-19). It is noted here that special attention in this review is focused on the influence of the film rheology on the hydrodynamic stability because in the practical applications of the theory, this rheology is diverse (e.g., the lamellae formed in dispersions are usually Newtonian, while those encountered in the deposition technologies or in biological applications are elastic or viscoelastic).

A brief outline of this review is as follows. The text is organized into three sections. In the first, a continuum formulation of the van der Waals forces among the molecules in a thin film system is described in detail. The need for a continuum formulation of the intermolecular force derives from the fact that the other forces balanced in the momentum equation for the film (e.g. electrostatic and hydrodynamic forces) are described from a continuum framework. The reason for the extended inclusion in this review is that the continuum van der Waals force has only been introduced ad hoc in several papers, and a detailed development which unifies and systematizes the treatments in the literature is necessary for a clear presentation. The second section details the mathematical framework of the linear theory of thin film stability. The third section describes separately the van der Waals and electrostatic induced linear instabilities.

CONTINUUM FORMULATION OF VAN DER WAALS FORCES IN A THIN FILM SYSTEM

The problem of the description and mathematical formulation of the van der Waals interactions in a thin film system is more difficult than the corresponding problem for a bulk interfacial system because of the presence, in a thin film, of two interfaces which are separated by a distance that is of the same order of magnitude as the range of the intermolecular interaction (O (1 - 10^2nm)). The usual hydrodynamic treatment of a fluid interface as a zero thickness boundary at which the bulk normal stresses are balanced by the Laplace pressure and the bulk tangential stresses are equated to the surface tension gradient (cf. Levich (20)) must be supplemented to account for the effect of the close proximity of the film surfaces. The methodology for accounting for this effect in the hydrodynamic field and boundary equations can be derived by utilizing a thermodynamic principle in which the film system is subject to an isothermal, equilibrium deformation and the variation of the system Helmholtz free energy is equated to the work done by external forces. To illustrate, in its simplest form, the principle and the type of equations which result, consider a film system consisting of a solute-free, unbounded, liquid sheet surrounded by its vapor (a free liquid film system). Only deformations which are incompressible as well as isothermal are considered. It is necessary to remark that, by utilizing such a thermodynamic approach, the field and boundary equations which result are only valid for equilibrium states of the film: However, since the relaxation time for intermolecular forces is extremely fast, these equilibrium equations can also be used for describing the intermolecular contributions to irreversible film processes such as viscous flow.

For a planar free film which undergoes a virtual, isothermal displacement history $\underline{u}$ (considered a function of the spatial coordinates z_1, z_2, z_3 and time (t)) and which is subject to an external body force (per unit mass) $\underline{f}(z_k)$, the conservation of the system Helmholtz free energy A may be expressed by the relation

$$\delta A = \int_V \rho(z_k, t)\underline{f}(z_k)\cdot\delta\underline{u}\ dV \qquad (1)$$

where $\delta\underline{u}$ is the virtual velocity, $\rho(z_k, t)$ is the mass density distribution in the film system, V is the system volume and δA is the variation in the total free energy. Two approaches have evolved in the literature in an attempt to formulate an expression for the Helmholtz free energy of a deformed film system. The first approach utilizes the rigorous thermodynamics of a planar film system to express the free energy of an interfacially perturbed film (Ivanov (16)). Since the final form for the surface stress balance in the procedure contains a disjoining pressure term, this method is referred to as the disjoining pressure approach. The second approach is based upon the procedure of expressing the system free energy in terms of pairwise intermolecular potential energy function integrals (Felderhof (21) and Maldarelli *et al*. (22)). These two approaches are described below.

Disjoining Pressure Approach

The thermodynamics of a free, precisely planar liquid film containing dissolved solutes has been studied extensively (cf. Toshev and Ivanov (23), de Feijter *et al*. (24) and Marmur (25) and the references cited in these articles). Here the formulation and results of Toshev and Ivanov (23) are used (see also Ivanov and Toshev (26)).

Utilizing as a framework the Gibbs treatment of a bulk interface (i.e. the interface between two semi-infinite fluids), Toshev and Ivanvov devised an idealized system wherein two planar dividing surfaces (separated by a distance h_p) partition the film system into three noninteracting regions which are regarded as homogeneous in intensive properties up to the dividing interfaces. (The three regions of the idealized system are denoted by the Roman numerals I, II and III; "II" refers to the film phase abd "I" and "III" denote, respectively, the bounding upper and lower vapor phases (see Fig. 1, below.) The uniform mass density of the gas phase is denoted by ρ^v and that of the liquid phase by ρ^ℓ. The volumes of the three regions of the idealized system are denoted by V^v for the gas phases and V^ℓ for the film phase.)

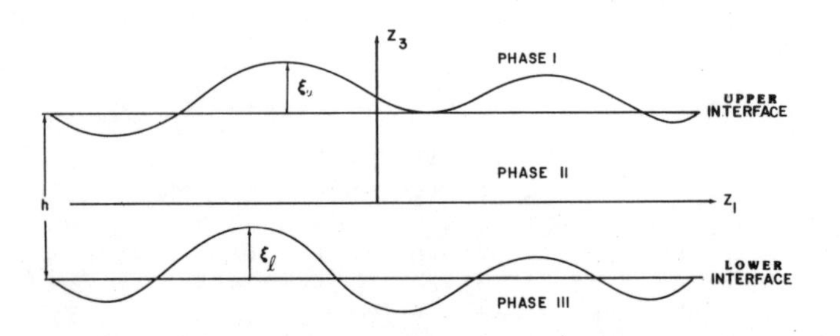

Figure 1. Schematic of the Gibbs construction of the idealized film system; the plane $z_3 = 0$ coincides with the midplane of the unperturbed film.

The free energy of the film system may then be expressed as a sum of the energy of the idealized system and an excess free energy; thus

$$A_p = a_p\tilde{f}_p + 2\int_{V^v} f^v(\rho^v)dV + \int_{V^\ell} f^\ell(\rho^\ell)dV \qquad (2)$$

where A_p is the Helmholtz free energy of the planar system, a_p is the area of one of the dividing surfaces, $\tilde{f}_p$ is the excess free energy (per unit area) and $f^\ell(\rho^\ell)$ and $f^v(\rho^v)$ are, respectively, the energies (per unit volume) of infinite reservoirs of liquid at a density ρ^ℓ and vapor at a density ρ^v. (Specifically, from thermodynamics, $f(\rho) = -p(\rho) + \rho\mu(\rho)$, where $p(\rho)$ is the pressure and $\mu(\rho)$ is the chemical potential per unit mass.) The excess free energy per unit area ($\tilde{f}_p$) can be shown to be a function of the film thickness (h_p) and the surface excess concentration of the film liquid (Γ_p) where

$$a_p\Gamma_p = \int_{V_p} \rho_p(z_k,t)dV - 2\int_{V^v} \rho^v dV - \int_{V^\ell} \rho^\ell dV \qquad (3)$$

In Eq. (3), V_p and $\rho_p(z_k,t)$ are the total volume and mass density distribution of the planar system. The dependence of $\tilde{f}_p$ on the variables Γ_p and h_p is given by the following derivatives:

$$\left.\frac{\partial\tilde{f}_p}{\partial\Gamma_p}\right|_{h_p} = \mu_{sp}(h_p,\Gamma_p) \qquad (4)$$

$$\left.\frac{\partial\tilde{f}_p}{\partial h_p}\right|_{\Gamma_p} = -\Pi(h_p,\Gamma_p) \qquad (5)$$

where $\Pi(h_p,\Gamma_\rho)$ is the disjoining pressure and $\mu_{sp}(h_p,\Gamma_p)$ is the surface chemical potential. The general requirement for equilibrium is that $\mu_{sp}(h_p,\Gamma_p) = \mu^v(\rho^v) = \mu^\ell(\rho^\ell)$.

In the disjoining pressure procedure, the Helmholtz free energy of an interfacially deformed film is expressed by utilizing the above outlined idealized system for each instantaneous shape of the film system. The shapes of the dividing surfaces at one particular time are determined by locating these surfaces in the interfacial zones and requiring that their normal vectors coincide with the direction of the local mass density gradient. For succeeding times, the shapes are determined from the condition that the dividing surfaces are material interfaces, i.e. they deform with the material particles which comprised the surface at the reference time. (The elevation of the film interfaces from planes perpendicular to the transverse direction of the film are denoted by $\frac{h}{2} + \xi_u$ and $-\frac{h}{2} + \xi_\ell$ (see Fig. 1; h denotes the mean thickness of the unperturbed film).)

The central approximation in the disjoining pressure procedure is the utilization of the planar function $\tilde{f}_p$ to express the excess Helmholtz free energy of the interfacially deformed film. Since only the thermodynamics of a symmetrical planar system has been studied, the disturbances considered here must be restricted to deformations in which the film is deformed in varicose squeezing modes (i.e., $\xi_u = -\xi_\ell = \bar{\xi}$) because only such disturbances are symmetric with respect to the midplane of the unperturbed film. For perturbations of this type, the system Helmholtz free energy is formulated as follows in the disjoining pressure approximation:

$$A = \int_{-\infty}^{\infty}\int_{-\infty}^{\infty} \tilde{\tilde{f}}_p(\Gamma_p=\Gamma, h_p=h+2\bar{\xi}(z_1,z_2,t))\sqrt{b}\, dz_1dz_2 + \int_{-\infty}^{\infty}\int_{-\infty}^{\infty}\Big(\int_{\frac{h}{2}+\bar{\xi}}^{\infty} f^v(\rho^v)dz_3 + \int_{-\infty}^{-\frac{h}{2}-\bar{\xi}} f^v(\rho^v)dz_3 + \int_{-\frac{h}{2}-\bar{\xi}}^{\frac{h}{2}+\bar{\xi}} f^\ell dz_3\Big)dz_1dz_2 \qquad (6)$$

where b is equal to the determinant of the surface metric

$$b = 1 + \left(\frac{\partial\bar{\xi}}{\partial z_1}\right)^2 + \left(\frac{\partial\bar{\xi}}{\partial z_2}\right)^2 \qquad (7)$$

and $\Gamma(z_1,z_2,t)$ is the surface excess concentration of liquid,

$$\int_{-\infty}^{\infty}\int_{-\infty}^{\infty} \Gamma(z_1,z_2,t)\sqrt{b}\, dz_1dz_2 = \int_{-\infty}^{\infty}\int_{-\infty}^{\infty}\Big(\int_{-\infty}^{\infty}\rho(z_k,t)dz_3 - \int_{\frac{h}{2}+\bar{\xi}}^{\infty}\rho^v dz_3 - \int_{-\frac{h}{2}-\bar{\xi}}^{\frac{h}{2}+\bar{\xi}}\rho^\ell dz_3 - \int_{-\infty}^{-\frac{h}{2}-\bar{\xi}}\rho^v dz_3\Big)dz_1dz_2 \qquad (8)$$

The procedure of utilizing $\tilde{f}_p$ to express the deformed excess free energy is obviously restricted to situations in which the interfacial perturbation wavelengths are much larger than the mean film thickness, since it is only in such "long-wavelength" situations that the film may be regarded as an assemblage of first elements.

Upon substituting the formulation for A (Eq. 6) into Eq. (1) and noting that the dividing surfaces are material interfaces, the following variational expression is obtained:

$$\int_{-\infty}^{\infty}\int_{-\infty}^{\infty}(\mu_{sp}\delta\Gamma - 2\Pi(h+2\bar{\xi})\delta\underline{u}\cdot\underline{k} + \tilde{f}_p\nabla_s\cdot\delta\underline{u})\sqrt{b}\, dz_1dz_2 - \int_V \rho(z_k,t)\underline{f}(z_k)\cdot\delta\underline{u}dV = 0 \qquad (9)$$

where $\underline{k}$ is the unit vector in the z_3 direction, $\delta\Gamma$ is the substantial variation in the surface excess concentration of liquid and $\nabla_s\cdot\delta\underline{u}$ is the surface divergence of the virtual velocity (for a definition of the surface divergence see McConnell (27)). The variables $\delta\Gamma$ and $\delta\underline{u}$ are not arbitrary but are constrained by the requirements of mass conservation (Eq. (8)) and incompressibility ($\nabla\cdot\delta\underline{u} = 0$). Therefore, in order to derive the equilibrium differential equations, the variation in Eq. (8) is adjoined to Eq. (9) by utilizing the Lagrange multiplier λ and the incompressibility constraint is adjoined to Eq. (9) by means of the Lagrange multiplier $-p^\alpha$(the undefined incompressibility pressure of region α (α = I, II or III) of the idealized system). After equating to zero the coefficients of $\delta\Gamma$ and $\delta\underline{u}$ in the adjoined expression and then self-consistently solving for λ, the following field equations and boundary conditions at the upper interface result:

$$-\nabla p^{\alpha} + \rho^{\alpha}\underline{f} = 0 \quad (10)$$

$$\mu_{sp}(\Gamma,\ h+2\bar{\xi}) = \text{constant} \quad (11a)$$

$$(-p^{II}+p^{I}) - 2H\sigma_p^f(\Gamma,h+2\bar{\xi}) -$$
$$\Pi(\Gamma,h+2\bar{\xi})(\underline{n}\cdot\underline{k}) = 0 \quad (11b)$$

where $\underline{n}$ is the unit normal of the upper interface (oriented in the direction of the upper gas phase), H is the mean curvature of the upper surface (McConnell (27)) and $\sigma_p^f(h_p,\Gamma_p)$ is the surface tension of the film defined by Toshev and Ivanov (23) as

$$2\sigma_p^f = \tilde{f}_p - \mu_{sp}\Gamma_p \quad (12)$$

Equation (11b) is of the typical form for the interfacial stress balance in physicochemical hydrodynamics (cf., for example, Levich (20)) except for the presence of the disjoining pressure term. For a planar, uniform film ($H = 0$, $\underline{n} = \underline{k}$), Eq. (11b) indicates that the gas pressure above the film is balanced by the pressure within the film and the disjoining pressure. A second significant feature of Eq. (11b) is the fact that the tension coefficient appearing in the Laplace pressure term is not the interfacial tension between a bulk phase of the film liquid and its vapor, but rather the surface tension of the film, a quantity related to the bulk tension (σ_∞) by the expression (23)

$$\sigma_p^f = \sigma_\infty + \frac{1}{2}\int_{h_p}^{\infty} \Pi\, d\tilde{h}_p \quad (13)$$

where the integration in (13) is at constant chemical potential. In practice, the difference between σ_p^f and σ_∞ is small; for $\Pi = \bar{A}/6\pi h_p^3$ (where $\bar{A}$ is the Hamaker constant) and for the typical values for $\bar{A}$ of 10^{-20}J and h of 10^{-8}m, $\sigma_p^f - \sigma_\infty$ is of the order of only 10^{-6}N/m.

The restriction of the above method to long wavelength disturbances is a serious one for systems in which hydrodynamically unstable disturbance wavelengths are of the order of the film thickness. The second approach for developing an expression for δA, the body force procedure, is free of the long wavelength restriction because the excess free energy in this procedure is formulated by integrating a potential energy function over the instantaneous, deformed shape of the film. This potential energy function is defined through a procedure in which the total system free energy is decomposed into two terms, one corresponding to the intermolecular interaction of a bulk interfacial system and the second related to the excess interaction resulting from the close proximity of the film interface. This decomposition and the resulting set of field and boundary relations in the body force procedure are detailed below.

Body Force Procedure

The free energy of a surface deformed bulk interfacial system in which semi-infinite volumes of the film liquid and its vapor are interfaced may be expressed in a long wavelength formulation. (The precise domain of deformation length scales for which a long-wavelength formulation of the free energy is valid for a bulk interfacial system is discussed after Eq. (17)). Consider two identical liquid-vapor bulk systems, one deformed interfacially in the shape of the upper interface (with the vapor above the liquid) and the other in the shape of the lower interface (with the liquid above the vapor); the free energies of these systems are denoted, respectively, by $A_{\infty\ell}$ and $A_{\infty u}$ and are represented by the following integral equations:

$$A_{\infty u} = \int_{-\infty}^{\infty}\int_{-\infty}^{\infty}\Big(\int_{\frac{h}{2}+\xi_u}^{\infty} f^v(\rho^v)dz_3$$
$$+ \int_{-\infty}^{\frac{h}{2}+\xi_u} f^\ell(\rho^\ell)dz_3\Big)dz_1dz_2$$
$$+ \int_{-\infty}^{\infty}\int_{-\infty}^{\infty} \sigma_\infty \sqrt{b_u}\, dz_1dz_2 \quad (14)$$

$$A_{\infty\ell} = \int_{-\infty}^{\infty}\int_{-\infty}^{\infty}\Big(\int_{-\frac{h}{2}+\xi_\ell}^{\infty} f^\ell(\rho^\ell)dz_3$$
$$+ \int_{-\infty}^{-\frac{h}{2}+\xi_\ell} f^v(\rho^v)dz_3\Big)dz_1dz_2$$

$$+ \int_{-\infty}^{\infty} \int_{-\infty}^{\infty} \sigma_{\infty} \sqrt{b_{\ell}}\, dz_1 dz_2 \qquad (15)$$

where σ_{∞} is the bulk surface tension of the film liquid interfaced with its vapor, ρ^{v} and ρ^{ℓ} are, respectively, as before, the uniform vapor and liquid mass densities in the idealized system and b_u and b_{ℓ} are the surface metrics of the upper and lower interfaces, respectively, and are defined below.

$$b_u = 1 + \left(\frac{\partial \xi_u}{\partial z_1}\right)^2 + \left(\frac{\partial \xi_u}{\partial z_2}\right)^2 \qquad (16)$$

$$b_{\ell} = 1 + \left(\frac{\partial \xi_{\ell}}{\partial z_1}\right)^2 + \left(\frac{\partial \xi_{\ell}}{\partial z_2}\right)^2 \qquad (17)$$

The procedure of utilizing the planar surface tension σ_{∞} to evaluate the energy of a perturbed bulk interfacial system has been reviewed in the literature; detailed calculations on the interfaction tension of small spherical droplets (Ono and Kondo (28)) have indicated that for surface disturbances with wavelengths on the order of or greater than 10nm, the variation in the tension is less than 20 percent. Consequently, application of Eqs. (14) and (15) to the extreme case of disturbances on the order of the film thickness (i.e. 10-100nm) is self-consistent.

Upon summing Equations (14) and (15), the following relation is obtained:

$$A_{\infty u} + A_{\infty \ell} - A_{\ell} =$$

$$\int_{-\infty}^{\infty} \int_{-\infty}^{\infty} \Big(\int_{\frac{h}{2}+\xi_u}^{\infty} f^{v}(\rho^{v}) dz_3 + \int_{-\frac{h}{2}+\xi_{\ell}}^{\frac{h}{2}+\xi_u} f^{\ell}(\rho^{\ell}) dz_3$$

$$+ \int_{-\infty}^{-\frac{h}{2}+\xi_{\ell}} f^{v}(\rho^{v}) dz_3 \Big) dz_1 dz_2$$

$$+ \int_{-\infty}^{\infty} \int_{-\infty}^{\infty} \sigma_{\infty} \sqrt{b_u}\, dz_1 dz_2 + \int_{-\infty}^{\infty} \int_{-\infty}^{\infty} \sigma_{\infty} \sqrt{b_{\ell}}\, dz_1 dz_2 \qquad (18)$$

where A_{ℓ} is the free energy of an infinite reservoir of liquid at a density ρ^{ℓ}. Explicit expressions for the free energies A, $A_{\infty u}$ and $A_{\infty \ell}$ may be formally expressed in terms of the volume integral of a pairwise potential energy function; thus

$$A_u = \frac{1}{2} \int_{-\infty}^{\infty} \int_{-\infty}^{\infty} \int_{-\infty}^{\infty} \rho_u(z_k,t) W_u(z_k,t) dz_3 dz_1 dz_2 \qquad (19)$$

$$A_{\infty u} = \frac{1}{2} \int_{-\infty}^{\infty} \int_{-\infty}^{\infty} \int_{-\infty}^{\infty} \rho_{\infty u}(z_k,t) W_{\infty u}(z_k,t) dz_3 dz_1 dz_2 \qquad (20)$$

$$A_{\infty \ell} = \frac{1}{2} \int_{-\infty}^{\infty} \int_{-\infty}^{\infty} \int_{-\infty}^{\infty} \rho_{\infty \ell}(z_k,t) W_{\infty \ell}(z_k,t) dz_3 dz_1 dz_2 \qquad (21)$$

where $\rho_{\infty_u}(z_k,t)$ and $\rho_{\infty_{\ell}}(z_k,t)$ are the density distributions in the bulk interfacial systems.

The potential functions $W(z_k,t)$, $W_{\infty u}(z_k,t)$ and $W_{\infty \ell}(z_k,t)$ appearing in the above set of equations are defined as

$$W(z_k,t) =$$

$$\int_{-\infty}^{\infty} \int_{-\infty}^{\infty} \int_{-\infty}^{\infty} \rho(z_k',t) w(|\underline{r}(z_k - z_k')|) dz_3' dz_2' dz_1' \qquad (22)$$

$$W_{\infty u}(z_k,t) =$$

$$\int_{-\infty}^{\infty} \int_{-\infty}^{\infty} \int_{-\infty}^{\infty} \rho_{\infty u}(z_k',t) w(|\underline{r}(z_k - z_k')|) dz_3' dz_2' dz_1' \qquad (23)$$

$$W_{\infty \ell}(z_k,t) =$$

$$\int_{-\infty}^{\infty} \int_{-\infty}^{\infty} \int_{-\infty}^{\infty} \rho_{\infty \ell}(z_k',t)\; w(|\underline{r}(z_k - z_k')|) dz_3' dz_2' dz_1' \qquad (24)$$

where $\underline{r}(z_k - z_k')$ is the vector extending from the source point with coordinates z_k' to the field point with coordinates z_k and w(r) is a pairwise molecular potential energy function (per unit mass). Upon evaluating the variation in A, $A_{\infty u}$, and $A_{\infty \ell}$ resulting from the virtual velocity induced redistribution of mass in the system, the following equation is obtained:

$$\delta A - \delta A_{\infty u} - \delta A_{\infty \ell} =$$

$$\int_{-\infty}^{\infty}\int_{-\infty}^{\infty}\int_{-\infty}^{\infty}(\rho\nabla W-\rho_{\infty u}W_{\infty u}-\rho_{\infty\ell}\nabla W_{\infty\ell})\cdot\delta\underline{u}\,dz_3dz_2dz_1 \quad (25)$$

The variation in the sum $A_{\infty u}+A_{\infty\ell}$ may be expressed in terms of surface integrals by differentiating Eq. (18) with respect to time; thus, from Eqs. (18) and (25), the following relation results:

$$\delta A = \int_{-\infty}^{\infty}\int_{-\infty}^{\infty}\sigma_\infty(\nabla_s\cdot\delta\underline{u})\sqrt{b_u}\,dz_1dz_2 +$$

$$\int_{-\infty}^{\infty}\int_{-\infty}^{\infty}\int_{-\infty}^{\infty}(\rho\nabla W-\rho_{\infty u}\nabla W_{\infty u}-\rho_{\infty\ell}\nabla W_{\infty\ell})\cdot\delta\underline{u}\,dz_3dz_2dz_1$$

$$+\int_{-\infty}^{\infty}\int_{-\infty}^{\infty}\sigma_\infty(\nabla_s\cdot\delta\underline{u})\sqrt{b_\ell}\,dz_1dz_2 \quad (26)$$

Adjoining the incompressibility ($\nabla\cdot\delta\underline{u}=0$) to the variatonal equation (Eq. (26)) via a Lagrange multipliers and then equating to zero the coefficient of $\delta\underline{u}$ yields the following field equations and boundary relation for the upper interface:

$$-\nabla p^\alpha+\rho^\alpha\underline{f}-(\rho\nabla W-\rho_{\infty u}\nabla W_{\infty u}-\rho_{\infty\ell}\nabla W_{\infty\ell})=0 \quad (27)$$

$$(-P^{II}+P^{I})\underline{n}-2H\underline{n}\sigma_\infty=0 \quad (28)$$

It is useful to note that the above derivation of Eqs. (27) and (28) does not require that the interfacial disturbances be symmetric.

The final form of relations (27) and (28) indicates that the influence of the intermolecular forces on the film mechanics is expressed, in contrast to the disjoining pressure approach, in terms of both field and boundary quantities: The field contribution $(\rho\nabla W - \rho_{\infty u}\nabla W_{\infty u} - \rho_{\infty\ell}\nabla W_{\infty\ell})$ is termed the excess van der Waals body force per unit volume and its appearance in the equation of mechanical equilibrium in the form of a body force motivates the terminology "body force method" for the above procedure.

A significant simplification in the expression for the excess van der Waals body force may be achieved if it is assumed that the mass density distribution in the film system is sufficiently discontinuous so that the gas phase profiles may be approximated by the constant value ρ^v and the liquid phase profile by the value ρ^ℓ. With this assumption, it can easily be demonstrated that the excess van der Waals body force in the film phase is identically equal to zero and that the excess force in the gas phases are given by the (negative) gradients of the following scalar functions:

$$\tilde{W}^{I}(z_k,t) = \int_{z_3+\frac{h}{2}}^{\infty}\int_u^{\infty}2\pi r\tilde{w}(r)\,dr\,du +$$

$$\int_{-\infty}^{\infty}\int_{-\infty}^{\infty}\int_0^{\xi_\ell(\beta,\gamma,t)}\tilde{w}(\sqrt{((z_3+\tfrac{h}{2}-\alpha)^2+(z_1-\beta)^2+(z_2-\gamma)^2)})\,d\alpha\,d\beta\,d\gamma \quad (29)$$

$$\tilde{W}^{III}(z_k,t) = \int_{\frac{h}{2}-z_3}^{\infty}\int_u^{\infty}2\pi r\tilde{w}(r)\,dr\,du -$$

$$\int_{-\infty}^{\infty}\int_{-\infty}^{\infty}\int_0^{\xi_u(\beta,\gamma,t)}\tilde{w}(\sqrt{((\tfrac{h}{2}-z_3+\alpha)^2+(z_1-\beta)^2+(z_2-\gamma)^2)})\,d\alpha\,d\beta\,d\gamma \quad (30)$$

where $\tilde{W}^{I}$ and $\tilde{W}^{III}(z_k,t)$ denote the excess van der Waals potentials of the gas phases of the film system and the potential $\tilde{w}(r)$ is given by

$$\tilde{w}(r) = (\rho^\ell-\rho^v)^2 C(r) \quad (31)$$

The function C(r) is the intermolecular potential energy (per unit mass) of two molecules (of the type which comprise the film system) separated by a distance r. For film systems composed of three different solvents (the upper and lower phases composed, respectively, of solvents "I" and "III" and the film phase composed of solvent "II"), an analogous derivation yields the above distributions for $\tilde{W}(z_k)$ with the following relation for $\tilde{w}(r)$:

$$\tilde{w}(r) = \rho^{I}\rho^{III}w_{I,III}(r) + \rho^{II}\rho^{II}w_{II,II}(r) - \rho^{I}\rho^{II}w_{I,II}(r) - \rho^{II}\rho^{III}w_{II,III}(r) \quad (32)$$

where $w_{i,j}(r)$ is the potential energy between two molecules from solvents i and j.

In all studies which have utilized the body force procedure, the above simplification has been employed and hence intermolecular interactions in these studies have been described by the scalar field $\tilde{W}(z_k,t)$. Since the pressure in an incompressible medium remains undefined, the excess potential and pressure terms in the equation of motion can be reformulated as the gradient of the sum of the pressure and the van der Waals fields. The boundary conditions may then be recast in terms of the combined scalar field $(p + \tilde{W})$ and the values of the excess potential at the boundaries. Consequently, even in the body force procedure, the influence of the intermolecular interactions can be described completely through the boundary conditions. Recall, however, that only the body force method is valid for short wavelength disturbances. An exact congruence between the disjoining pressure and body force methods for long wavelength disturbances cannot be achieved since the former approach is not formulated in a precise asymptotic manner (Eq. (6) is not an exact asymptotic statement) and could not therefore be matched with an asymptotic expansion, for long waves, of the body force equations. Nevertheless, for describing the linear, long wavelength stability of squeezing modes both approaches yield essentially the same result when σ_∞ is not in the millidyne/cm range (cf. the results in Maldarelli *et al*. (22). The disjoining pressure approach has the primary advantage that it is valid for all types of intermolecular interactions while the body force procedure is restricted to those interactions which may be described by pairwise additivity. The advantage of the body force method is that it is applicable to the complete spectrum of disturbance wavelengths and may easily be applied to unsymmetrical film systems and different modes of vibration of the film.

THE MATHEMATICAL FRAMEWORK OF LINEAR INSTABILITIES IN THIN FILMS

In this section is detailed the eigenvalue equations which are obtained from a linear stability analysis of a thin film system. The model system chosen is a planar film bounded by semi-infinite Newtonian phases. As mentioned in the introduction, in order to describe simply the influence of the long range forces on the hydrodynamic stability, only a nonthinning (zero flow) base state is considered. Finally, in order to study the influence of the film rheology, an isotropic viscoelastic model for the film is adopted in which the stress tensor is formulated as a convolution memory integral of the product of the rate of strain tensor and a relaxation function. The adoption of such a linear rheology does not prove restrictive since the present analysis is only concerned with describing the linear dynamics from a zero motion base state, and Joseph (29) has shown that under these circumstances, the memory integral representation is sufficient for the Noll "simple" fluids.

The Eigenvalue Equations

For a nonthinning, unbounded, planar film (of thickness h and surrounded by incompressible Newtonian fluids) the eigenvalue equations may be formulated solely in terms of the variables $u(z_3,k_1,k_2,\omega)$, $v^I(z_3,k_1,k_2,\omega)$ and $v^{III}(z_3,k_1,k_2\omega)$, the Fourier-Laplace transformed functions of, respectively, the z_3 components of the film displacement vector and the upper and lower semi-infinite fluid velocity vectors. In this form, the eigenvalue equations are detailed below. (For derivations see Lucassen *et al*. (30) and Vrij *et al*. (31) for Newtonian fluid films and Maldarelli and Jain (32, 33) and Steinchen *et al*. (34) for the treatment of viscoelastic films.)

$$(D^4 - (2k^2 + \rho^i\omega/\mu^i)D^2 + k^4 + k^2\rho^i\omega/\mu^i)v^\alpha = 0 \qquad (\alpha = \mathrm{I,III}) \qquad (33)$$

$$(D^4 - (2k^2 + \rho^{II}\omega/\mu(\omega))D^2 + k^4 + k^2\rho^{II}\omega/\mu(\omega))u = 0 \qquad (34)$$

(equations of motion)

$$v^I = \omega u \qquad Dv^I = \omega Du \qquad (z_3 = \frac{h}{2})$$

(continuity of velocity) (35)

$$\omega\mu(\omega)(D^2 + k^2)u - \mu^I(D^2 + k^2)v^I - k^2\gamma_1^1\, Dv^I(\frac{h}{2}) - k^2\gamma_2^1 Dv^{III}(-\frac{h}{2}) = 0 \qquad (z_3 = \frac{h}{2})$$

(tangential stress balance) (36)

$$k^4\Lambda_1^1 u(\frac{h}{2}) + k^4\Lambda_2^1 u(-\frac{h}{2}) + \mu^I D^3 v^I - (3\mu^I k^2 + \omega\rho^I)Dv^I - \omega\mu(\omega)D^3 u + (3\omega k^2\mu(\omega) + \omega^2\rho^{II})Du = 0 \qquad (z_3 = \frac{h}{2})$$

(normal stress balance) (37)

$$v^{III} = \omega u \quad Dv^{III} = \omega Du \qquad (z_3 = -\frac{h}{2}) \tag{38}$$

$$\mu^{III}(D^2+k^2)v^{III} - \omega\mu(\omega)(D^2+k^2)u - k^2\gamma_1^2 Dv^I(\frac{h}{2}) - k^2\gamma_2^2 Dv^{III}(-\frac{h}{2}) = 0 \qquad (z_3 = -\frac{h}{2}) \tag{39}$$

$$k^4\Lambda_1^2 u(\frac{h}{2}) + k^4\Lambda_2^2 u(-\frac{h}{2}) + \omega\mu(\omega)D^3 u - (3\omega k^2\mu(\omega)+\omega^2\rho^{II})Du - \mu^{III}D^3v^{III} + (3\mu^{III}k^2+\omega\rho^{III})Dv^{III} = 0 \qquad (z_3 = -\frac{h}{2}) \tag{40}$$

In the equations listed above, the notation "$D^n\theta$" indicates the nth derivative of θ with respect to z_3 and the variables μ^I, μ^{III} and $\mu(\omega)$ are, respectively, the Newtonian viscosities of the upper and lower bounding phases and the Laplace transform of the relaxation function. Definitions of the rest of the unidentified variables in the eigenvalue equations are:

$$k^2 = k_1^2 + k_2^2 \tag{41}$$

$$\Lambda_1^1 = \sigma_{\infty u} - \frac{1}{k^2}\Pi'(h) \tag{42}$$

$$\Lambda_2^1 = \Lambda_1^2 = -\frac{1}{k^2}I(k,h) \tag{43}$$

$$\Lambda_2^2 = \sigma_{\infty \ell} - \frac{1}{k^2}\Pi'(h) \tag{44}$$

$$\Pi(h) = 2\pi \int_h^\infty \int_s^\infty r\tilde{w}(r)drds \tag{45}$$

$$I(k,h) = \int_{-\infty}^{\infty}\int_{-\infty}^{\infty}\cos(k\eta)\tilde{w}(\sqrt{(h^2+\eta^2+\zeta^2)})d\zeta d\eta \tag{46}$$

In the above, the notation $\Pi'(h)$ denotes the derivative of $\Pi(h)$ with respect to h. In formulating the eigenvalue relations, the body force method has been utilized; thus $\sigma_{\infty u}$ and $\sigma_{\infty\ell}$ are the bulk interfacial tensions between phases I and II and II and III, respectively, and $\tilde{w}(r)$ is the composite potential energy function defined by Eq. (32). The variables $\gamma_j^i(\omega,k)$ ($i=1,2$; $j=1,2$) describe viscoelastic interfacial stresses derived from either Marangoni tensions caused by surfactant gradients along the surface or interfacial rheological forces as, for example, viscous shear tractions. Explicit expressions for these variables are obtained by specifying a particular surface rheology and solving the surface convective diffusion equation in order to evaluate the Maragoni force arising from surfactant gradients. For example, for a Newtonian surface containing a single, bulk insoluble, chemically inert surfactant, γ_2^1 and γ_1^2 are equal to zero, and γ_1^1 is of the form

$$\gamma_1^1(\omega,k) = -\mu_s + \left(\frac{\partial\sigma_{\infty u}}{\partial\Gamma}\right)_{(o)}\Gamma_{(o)}(\omega+k^2D_s)^{-1} \tag{47}$$

where D_s and μ_s are, respectively, the surface diffusion coefficient and surface shear viscosity and $\Gamma_{(o)}$ is the spatially uniform, base state surfactant concentration (all quantities defined with reference to the upper interface). (For an introduction to the incorporation of Marangoni and surface shear forces in a hydrodynamic stability analysis see Sorensen _et al_. (_35_) and for a complete treatment see the articles by Dalle Vedove and Sanfeld (_36_-_38_). The Newtonian interface is discussed by Scriven (_39_) and Slattery (_40_).)

Although the above eigenvalue set does not incorporate the influence, on the stability, of the electrostatic forces derived from double layers of charge, this influence may also be included. Terms corresponding to these forces renormalize the tension terms Λ_j^i and the surface stress coefficients γ_j^i. In addition, electrostatic forces give rise to the terms $\alpha_1^1 Dv^I(\frac{h}{2})$ and $\alpha_2^1 Dv^{III}(-\frac{h}{2})$ in the normal stress balance at $z_3 = \frac{h}{2}$ and the terms $\beta_1^1 u(\frac{h}{2})$ and $\beta_2^1 u(-\frac{h}{2})$ in the tangental balance at the upper interface. (Similar terms $\alpha_1^2 Dv^I(\frac{h}{2})$ and $\alpha_2^2 Dv^{III}(-\frac{h}{2})$ and $\beta_1^2 u(\frac{h}{2})$ and $\beta_2^2 u(-\frac{h}{2})$ exist in, respectively, the normal and tangential balances of the lower interface). The beta coefficients and the electrostatic terms of the renormalized gamma coefficients account for the tangential tractions arising from the action of the first order tangential electric field on the zero order surface charge density. The alpha coefficients and the electrical terms of the renormalized γ_j^i coefficients describe the influence of surface ion convection and diffusion on the electrostatic stresses acting on the film interfaces. Explicit expressions for the α_j^i and β_j^i coefficients and the Λ_j^i, γ_j^i renormalizations are dependent upon the manner in which the charge density is redistributed along the surface with the flow and the interfacial deformation. Particularly simple expressions are obtained when it is assumed that the surface ions remain in equilibrium with the bulk adjoining ones. Thus the surface charge is potentially determined, the coefficients α_j^i are identically zero and the parameters γ_j^i are not electrostatically renormalized. Detailed below are expressions for Λ_j^i in this latter regime for the case in which the film is bounded by identical phases. Two different electrostatic systems are considered, (1) a dielectric film bounded by electrolytes and (2) an electrolyte film surrounded by dielectric phases. The former system has been studied in the article by Maldarelli and Jain (<u>32</u>) wherein a complete derivation can be found of the equations detailed here for the regime of potential determination of the surface charge density. This system has also been examined by Wendel <u>et</u> <u>al</u>. ((<u>41</u>) and (<u>7</u>)), Bisch <u>et</u> <u>al</u>. (<u>42</u>) and Gallez <u>et</u> <u>al</u>. (<u>43</u>) for the case in which the surface charge density is determined by the interfacial fluid convection. Electrolyte films surrounded by dielectrics were initially studied by Felderhof (<u>21</u>) and later by Sche (<u>44</u>) and Sche and Fijnaut (<u>45</u>) for both convective and potential determination of the surface charge density. Derivations of the equations detailed here for electrolyte films in which the surface charge density is potentially determined can be found in the article by Felderhof (<u>21</u>).

(1) Dielectric Film Bounded by Identical Electrolytes:

(i) Base state conditions: Surface potentials (relative to bulk phases) are equal and zero potential drop exists across the film (Fig. 2a).

$$\Lambda_1^1 - \Lambda_2^1 = \sigma_\infty - \frac{1}{k^2}(\Pi'(h) - I(k,h)) - \frac{\epsilon^I \kappa \psi_s^2}{4\pi(1+\bar{\alpha})} \times \frac{\epsilon^I \kappa + (\epsilon^{II} k \tanh(kh/2) + \bar{K})}{\epsilon^I \kappa \bar{\alpha} + (\epsilon^{II} k \tanh(kh/2) + \bar{K})} \tag{48}$$

$$\Lambda_1^1 + \Lambda_2^1 = \sigma_\infty - \frac{1}{k^2}(\Pi'(h) + I(k,h) - \frac{\epsilon^I \kappa \psi_s^2}{4\pi(1+\bar{\alpha})} \times \frac{\epsilon^I \kappa + (\epsilon^{II} k \coth(kh/2 + \bar{K})}{\epsilon^I \kappa \bar{\alpha} + (\epsilon^{II} k \coth(kh/2) + \bar{K})} \tag{49}$$

(ii) Base state conditions: Surface potentials (relative to the bulk) equal in absolute value but of opposite sign; potential difference exists across the film of the same sign as (Fig. 2b-1) or of opposite sign as Fig. 2b-2) the surface potential of phase I.

$$\Lambda_1^1 - \Lambda_2^1 = \sigma_\infty - \frac{1}{k^2}(\Pi'(h) - I(k,h)) - \frac{\epsilon^I \kappa \psi_s^2}{4\pi(1+\bar{\alpha})} \times \frac{\epsilon^I \kappa + (\epsilon^{II} k \coth(kh/2) + \bar{K})}{\epsilon^I \kappa \bar{\alpha} + (\epsilon^{II} k \coth(kh/2) + \bar{K})} - \frac{\epsilon^I \psi_s}{4\pi(1+\bar{\alpha})}\left(\frac{\Delta\Phi}{h}\right)\frac{2\epsilon^{II} k \coth(kh/2)}{\epsilon^I k \bar{\alpha} + (\epsilon^{II} k \coth(kh/2) + \bar{K})} - \frac{\epsilon^{II}}{4\pi k}\left(\frac{\Delta\Phi}{h}\right)^2 \frac{(\epsilon^I \kappa \bar{\alpha} + \bar{K}) \coth(kh/2)}{\epsilon^I \kappa \bar{\alpha} + (\epsilon^{II} k \coth(\frac{kh}{2}) + \bar{K})} \tag{50}$$

$$\Lambda_1^1 + \Lambda_2^1 = \sigma_\infty - \frac{1}{k^2}(\Pi'(h) + I(k,h)) - \frac{\epsilon^I \kappa \psi_s^2}{4\pi(1+\bar{\alpha})} \times \frac{\epsilon^I \kappa + (\epsilon^{II} k \tanh(kh/2) + \bar{K})}{\epsilon^I \kappa \bar{\alpha} + (\epsilon^{II} k \tanh(kh/2) + \bar{K})} - \frac{\epsilon^I \psi_s}{4\pi(1+\bar{\alpha})}\left(\frac{\Delta\Phi}{h}\right)\frac{2\epsilon^{II} k \tanh(kh/2)}{\epsilon^I k \bar{\alpha} + (\epsilon^{II} k \tanh(kh/2) + \bar{K})} - \frac{\epsilon^{II}}{4\pi k}\left(\frac{\Delta\Phi}{h}\right)^2 \frac{(\epsilon^I \kappa \bar{\alpha} + \bar{K}) \tanh(kh/2)}{\epsilon^I \kappa \bar{\alpha} + (\epsilon^{II} k \tanh(\frac{kh}{2}) + \bar{K})} \tag{51}$$

(2) Electrolyte Film Surrounded by Identical Dielectrics:

Base conditions: Zero potential difference across the film and the film system overall electrically neutral (Fig. 2c).

$$\Lambda_1^1 - \Lambda_2^1 = \sigma_\infty - \frac{1}{k^2}(\Pi'(h) - I(k,h)) + \frac{\epsilon^{II}\Psi^2\kappa^3(1-\bar{\alpha}\tanh(\kappa h/2)\tanh(\kappa\bar{\alpha}h/2))}{4\pi k^2\tanh(\kappa\bar{\alpha}h/2)} \times \frac{(\epsilon^I k + \bar{K})\tanh(\kappa h/2) + \epsilon^{II}\kappa)}{(\epsilon^I k + \bar{K})\coth(\kappa\bar{\alpha}h/2) + \epsilon^{II}\kappa\bar{\alpha}} \quad (52)$$

$$\Lambda_1^1 + \Lambda_2^1 = \sigma_\infty - \frac{1}{k^2}(\Pi'(h) + I(k,h)) + \frac{\epsilon^{II}\Psi^2\kappa^3(1-\bar{\alpha}\tanh(\kappa h/2)\coth(\kappa\bar{\alpha}h/2))}{4\pi k^2\coth(\kappa\bar{\alpha}h/2)} \times \frac{(\epsilon^I k + \bar{K})\tanh(\kappa h/2) + \epsilon^{II}\kappa)}{(\epsilon^I k + \bar{K})\tanh(\kappa\bar{\alpha}h/2) + \epsilon^{II}\kappa\bar{\alpha}} \quad (53)$$

(Note that for both cases (1) and (2), $\Lambda_1^1 = \Lambda_2^2$ and $\Lambda_2^1 = \Lambda_1^2$). In deriving the above equations, the Debye-Huckel approximation is utilized to simplify the Poisson-Boltzmann equations. The variables ψ_s and $\Delta\Phi$ denote, respectively, the surface potential of the upper surface (relative to the potential in phase I far from the interface) and the potential of this surface less than that of the opposite interface. The variable Ψ in Eqs. (52) - (53) is the potential of the upper interface relative to the potential of the electrolyte reservoir from which the film is drawn. The variables ϵ^{II} and ϵ^{I} denote the dielectric constants of the film and surrounding phases, respectively, and κ represents the inverse Debye length of the electrolyte. Finally $\bar{\alpha}$ denotes the reduced wavenumber $\sqrt{(1+(k/\kappa)^2}$ and $-\bar{K}/4\pi$ denotes the differential variation in the surface charge density of the upper surface with a change in the potential of that interface, evaluated at the base state value of the surface potential. (In order to obtain the simplified expressions of Eqs. (48) - (53), the derivative of the surface charge of the second interface with respect to that interface's surface potential (evaluated at the base value) is also assumed to be equal to $-\bar{K}/4\pi$.)

It can easily be shown that $\bar{K} > 0$ for the case in which the equilibrium surface concentration of adsorbed ions increases monotonically with the bulk concentration of the ions. This parameter is also usually greater than zero when disassociation equilibrium determines primarily the surface charge density (cf., for example, the expressions developed by Ruckenstein and Prieve (46) for cell membrane ionic disassociation phenomena).

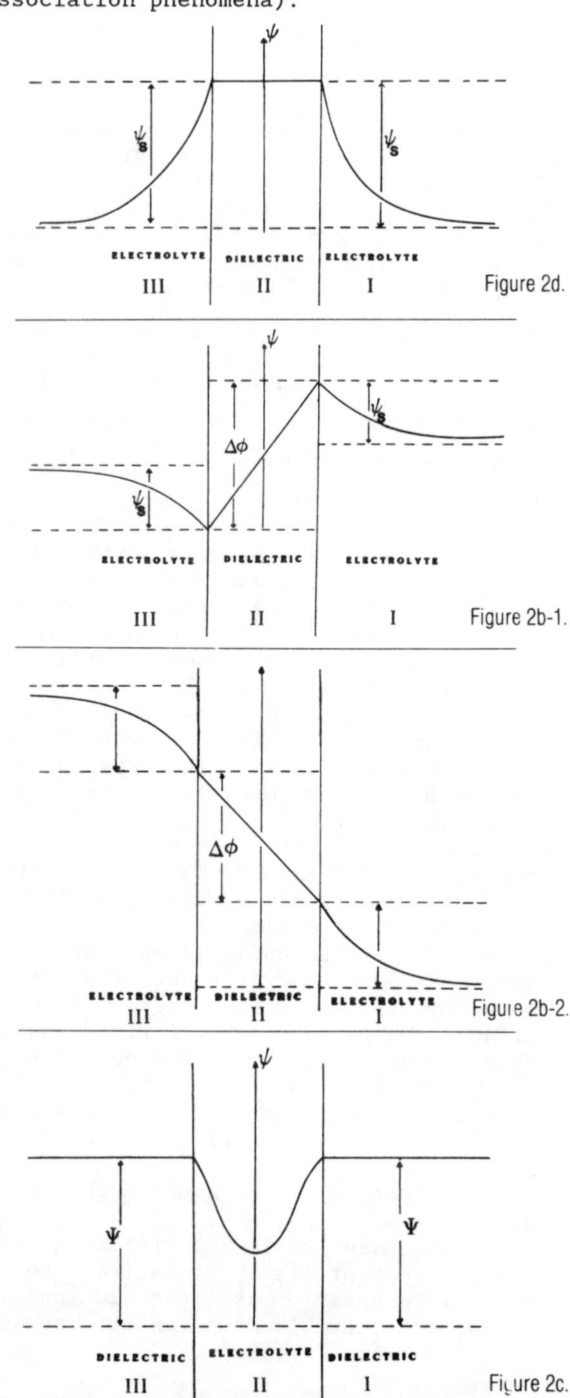

Figure 2. Base state electrostatic potential distributions considered for film systems bounded by identical phases. The diagrams describe the cases when the surface potential of the upper interface is positive relative to the bulk (i.e., $\psi_s > 0$); when $\psi_s < 0$, simply invert the figures.

General Characteristics of the Eigenvalue Equations

The above set of eigenvalue equations exhibit several significant features. Consider first the intermolecular interaction terms in the variables Λ_j^i; these terms arise from the presence of the excess van der Waals potential functions in the normal stress boundary conditions. Note, in particular, the term I(k,h) which represents the first order change in the potential of the upper surface as a result of the perturbation in the lower interface. This latter factor arises because the value of the excess potential of the upper phase is a function of the shape of the lower interface (cf. Eq. (29)). Thus, importantly, for the thin film normal stress balances, the condition at the upper (lower) surface involves the displacement function of the lower (upper) interface. Note, however, that even for an unsymmetrical film system the coupling in the normal stress balances is symmetric (note Eqs. (37), (40) and (43)); this fact is an obvious consequence of the symmetry in the expressions for $\tilde{W}^I(z_k,t)$ and $\tilde{W}^{III}(z_k,t)$ (cf. Eqs. (29) and (30)). Eigenvalue equations describing the stability of a film on a solid support can be obtained from Eqs. (33) - (40) by neglecting the terms $Dv^{III}(-h/2)$ in Eq. (36) and $u(-h/2)$ in Eq. (37) and replacing relations (38) - (40) by the equations $u(-h/2) = Du(-h/2) = 0$. (This problem is treated in detail by Ruckenstein and Jain (47) and Jain and Ruckenstein (48).)

A second significant characterstic of the eigenvalue relations is that they possess both symmetric and antisymmetric solution sets for the functions $v^I(z_3)$, $u(z_3)$ and $v^{III}(z_3)$ when the film system is symmetric, i.e., when the bounding phases are identical both mechanically and electrically (for this case, $\Lambda_1^1 = \Lambda_2^2$, $\Lambda_2^1 = \Lambda_1^2$, etc.). In the symmetric solution, $u(z_3)$ is an odd function of z_3 and the functions $v^I(z_3)$ and $v^{III}(z_3)$ are related by the equation $v^I(z_3) = -v^{III}(-z_3)$, $z_3 \geq \frac{h}{2}$; in the antisymmetric set, $u(z_3)$ is an even function of z_3 and $v^I(z_3)$ and $v^{III}(z_3)$ are equal with respect to the midplane of the unperturbed film, i.e., $v^I(z_3) = v^{III}(-z_3)$, $z_3 \geq \frac{h}{2}$. From the definitions of the antisymmetric and symmetric solutions it is evident that in the antisymmetric set the film thickness is preserved (to first order) while in the symmetric set the thickness fluctuates. Consequently (after Felderhof (21)) the symmetric solution set is termed the squeezing (SQ) eigenmode and the antisymmetric set is referred to as the stretching (ST) mode (see Fig. 3). Since the eigenvalue equations possess these two solution sets, the symmetrical system dispersion equation is factorable into two terms: one term, when equated to zero, represents the dispersion relation of the squeezing eigenmode and therefore defines the SQ mode's isolated singularities $\omega_i^{SQ}(k_1,k_2)$ and the other term, when equated to zero, corresponds to the stretching dispersion relation and hence defines the singularities $\omega_i^{ST}(k_1,k_2)$ of the ST mode,.

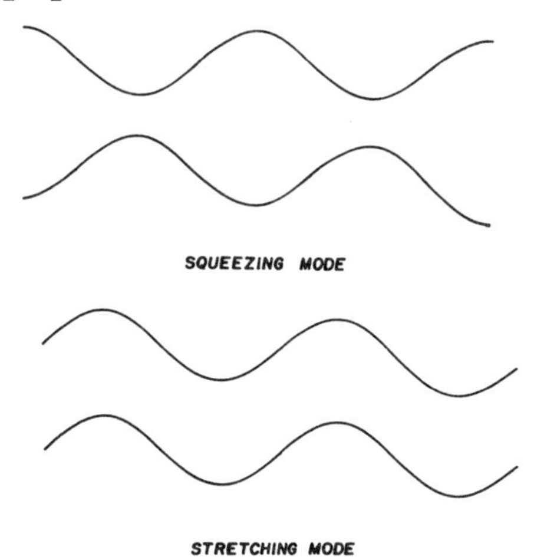

Figure 3. The eigenmodes of a symmetrical film system.

Adjoint Characteristics of the Eigenvalue Equations

The adjoint equations of the eigenvalue relations are derived in the usual manner by first multiplying the field equations by the complex conjugate of the dependent functions and integrating over the domains of definition, secondly reducing the differential order by integration by parts and finally simplifying and combining the integral equations for each domain by utilizing the boundary conditions (cf., for example, Chandrasekhar (52)). Detailed below are the real and imaginary parts of the expression obtained after these manipulations for a symmetrical film system for which the surface charge density is potentially determined ($\alpha_j^i = 0$, γ_j^i not electrically renormalized) and negligible in the base state ($\beta_j^i = 0$).

$$\begin{aligned}
&\mathrm{Re}(\omega)\Big\{2k^4(\Lambda_1^1 \mp \Lambda_2^1)|u(\tfrac{h}{2})|^2 - \\
&2k^2\mathrm{Re}(\gamma_1^1 \pm \gamma_2^1)|\omega|^2|Du(\tfrac{h}{2})|^2/\mathrm{Re}(\omega) \\
&+|\omega|^2\rho^{II}\int_{-h/2}^{h/2}(|Du|^2+k^2|u|^2)dz_3 + \\
&2\rho^I\int_0^\infty(|Dv^I|^2+k^2|v^I|^2\ )dz_3\Big\} \\
&+\ |\omega|^2\mathrm{Re}(\mu)\int_{-h/2}^{h/2}(|D^2u|^2+k^4|u|^2)dz_3 \\
&+\ k^2|\omega|^2\mathrm{Re}(2\mu)\int_{-h/2}^{h/2}|Du|^2dz_3 \\
&+2\mu^I\int_0^\infty(|D^2v^I|^2+2k^2|Dv^I|^2+k^4|v^I|^2)dz_3+ \\
&4k^2|\omega|^2(\mathrm{Re}(\mu)-\mu^I)\mathrm{Re}(D\bar{u}(\tfrac{h}{2})u(\tfrac{h}{2})) = 0 \quad (54)
\end{aligned}$$

$$\begin{aligned}
&-\mathrm{Im}(\omega)\{2k^4(\Lambda_1^1 \mp \Lambda_2^1)|u(\tfrac{h}{2})|^2 \\
&+2k^2\mathrm{Im}(\gamma_1^1 \pm \gamma_2^1)|\omega|^2|Du(\tfrac{h}{2})|^2/\mathrm{Im}(\omega) \\
&-|\omega|^2\rho^{II}\int_{-h/2}^{h/2}(|Du|^2+k^2|u|^2)dz_3 \\
&-2\rho^I\int_0^\infty(|Dv^I|^2+k^2|v^I|^2)dz_3\} \\
&+|\omega|^2\mathrm{Im}(\mu)\int_{-h/2}^{h/2}(|D^2u|^2+k^4|u|^2)dz_3 \\
&+k^2|\omega|^2\mathrm{Im}(2\mu)\int_{-h/2}^{h/2}|Du|^2dz_3 \\
&+4k^2|\omega|^2\mathrm{Im}(\mu)\mathrm{Re}(D\bar{u}(h/2)u(h/2)) = 0 \quad (55)
\end{aligned}$$

(The upper sign refers to the squeezing mode, while the lower sign refers to the stretching vibration. In the above, the overbar indicates the complex conjugate.)

For certain types of symmetrical film systems, the following thereoms can easily be established from the above two adjoint relations:

(i) Thereom 1: If $\mathrm{Re}(\omega) = 0$, then $\mathrm{Im}(\omega) = 0$ (a classical exchange of stabilities thereom)

(ii) Thereom 2: If $\mathrm{Re}(\omega) > 0$, then $\mathrm{Im}(\omega) = 0$ (instability is nonoscillatory)

(iii) Thereom 3: If $\mathrm{Im}(\omega) \neq 0$, the $\mathrm{Re}(\omega) < 0$ (oscillatory states decay)

Sufficient conditions for the validity of these thereoms are (for $\mu^I \neq 0$): (i) $\mathrm{Im}(\omega\mu)/\mathrm{Im}(\omega) \geq 0$ for all ω and (ii) either $\mathrm{Im}[\omega(\gamma_1^1 \pm \gamma_2)]/\mathrm{Im}[\omega] \leq 0$ and $\mu^I = \mathrm{Im}(\omega\mu)/\mathrm{Im}[\omega]$ for all ω or tangential immobility ($Du(\frac{h}{2})=0$). When $\mu^I = 0$, condition (i) must be strictly greater than zero for the thereoms to be valid. Condition (i) is satisfied for all isotropic elastic ($\mu(\omega)=G/\omega$, Newtonian ($\mu(\omega)=\mu$), Kelvin body ($\mu(\omega)= G/\omega+\mu$) and Maxwell fluid ($\mu(\omega)=G/(\omega+G/\mu)$) films. The inequality $\mathrm{Im}(\omega(\gamma_1^1 \pm \gamma_2^1))/\mathrm{Im}(\omega) \leq 0$ is satisfied for Newtonian interfaces which are clean or occupied by bulk insoluble (and in some cases soluble) surfactants which diffuse along (but do not react on) the surfaces (cf. the expression for γ_1^1 given by Eq. (47)).

Compatibility with the equality $\mu^I = \mathrm{Im}\ [\omega]$ is more restrictive. Satisfaction for the common case of (isotropic) Newtonian lamellae requires equality of the film and bounding phase viscosities.

MARGINAL STABILITY LOCI AND CONDITIONS FOR INSTABILITY

Examined in this section are the relations which characterize the marginal stability (i.e., $\mathrm{Re}(\omega) = 0$; $\frac{d}{dk}\mathrm{Re}(\omega) \neq 0$) of thin films in the presence of the long range intermolecular interactions described. Consideration here is restricted to symmetrical film systems as these admit the mathematically simple squeezing and streching vibration modes. Marginal stability criteria for asymmetrical systems are described by Maldarelli _et al_. (_22_) for van der Waals instabilities. In addition, only nonoscillatory, marginally stable states are evaluated here. Although these need not be the only states from which instability developes (even for the SQ and ST modes of symmetrical systems), it appears evident from the thin film stability literature that interfacial instabilities arising from nonoscillatory states are the more significant. This section is divided as follows: First, the marginal curves corresponding to van der Waals instabilities in uncharged fluid films is considered. Second, the stabilization of such instabilities as a consequence of the elastic interactions inherent in a viscoelastic solid film is evaluated. Third, the electrostatic stabilization owing to double layer replusion within electrolyte films is discussed and finally instabilities in electrolyte-bounded dielectric films arising form the negative electric film tension and the transmembrane potential are studied.

Van der Waals Instabilities in Fluid Films

It is not difficult to demonstrate from the eigenvalue relations (Eqs. (33) - (40)) that if the film behaves rheologically as a viscoelastic fluid (i.e., the relaxation function tends to zero at infinite time; in the Laplace domain $\lim_{\omega\to 0}(\omega\mu(\omega)) = 0$), the wavenumbers k_c which are (nonoscillatorily) marginally stable are given by the equation

$$\Lambda_1^1 \mp \Lambda_2^1 = 0, \text{ or, equivalently,}$$

$$\sigma_\infty - \frac{1}{k_c^2}(\Pi'(h) \mp I(k_c,h)) = 0 \quad (56)$$

In Eq. (56), the upper sign refers to the SQ mode and the lower sign corresponds to the ST vibration. Note, importantly, that the above criteria is independent of the limiting behavior of $\gamma_j^i(\omega,k)$ as $\omega\to 0$. This latter characteristic of fluid films is only strictly true as long as $\lim_{\omega\to 0} \gamma_j^i(\omega,k)$ is either zero, negative or unbounded. Thus fluid films which are either surfactant free ($\gamma_j^i = 0$) or effectively tangentially immobile ($\gamma_j^i(\omega,k) \to \infty$) are described by Eq. (56). An example where $\gamma_j^i(0,k)$ may become positive is when a chemically reacting surfactant is present on the interface; for this case a second marginal stability curve exist (cf. (36-38) and (48). Note also from Eq. (56) that the stability branches are independent of any rheological parameters. This latter property is a consequence of the fact that (1) the base state of the film system is assumed to be nonthinning and (2) the stability branches represent nonoscillatory marginal states. When the film exhibits elastic behavior in its deformation modes, elastic coefficients appear in the criteria (see the following subsection).

If London's Law ($w_{i,j}(r) = -C_{i,j}r^{-6}$) is utilized in evaluating the composite intermolecular potential $\tilde{w}(r)$ of Eq. (32), then

$$\Pi(h) = -\bar{A}/6\pi h^3 \quad (57)$$

$$I(k,h) = -\bar{A}k^2K_2(kh)/4\pi h^2 \quad (58)$$

where $K_2(x)$ is a modified Bessel function of the second kind of order 2 and $\bar{A}$ is the Hamaker constant of the film system, defined as follows for a symmetrical system

$$\bar{A} = \pi^2 \{(\rho^I)^2 C_{I,I} + (\rho^{II})^2 C_{II,II} - 2\rho^I\rho^{II}C_{I,II}\} \quad (59)$$

Using Eqs. (57) and (58), the stability branches may be written in the following nondimensional form

$$\alpha - (4/(k_ch)^2 - 2K_2(k_ch)) = 0 \quad \text{(ST)} \quad (60)$$

$$\alpha - (4/(k_ch)^2 + 2K_2(k_ch)) = 0 \quad \text{(SQ)} \quad (61)$$

where α is the nondimensional group $\sigma_\infty/(\bar{A}/8\pi h^2)$.

Consider first the ST mode. The term $4/(k_ch)^2 - 2K_2(k_ch)$ decreases monotonically from one to zero as the reduced critical wavenumber (k_ch) increases from zero to infinity. Thus only positive values of α will admit critical wavenumbers. Shown in Fig. 4 is a diagnostic plot of the values of α which correspond to a particular value of k_ch as given by Eq. (60). Consider first the case in which $\bar{A} > 0$ (Note that in these discussions, σ_∞ is always considered greater than or equal to zero). It can easily be shown (for both the SQ and ST modes) from the dispersion equations that instability arises (around the critical value) when $\Lambda_1^1 \mp \Lambda_2^2 < 0$; thus unstable points lie beneath the curve of Fig. 4. From the figure, it is evident that (1) a sufficient condition for instability of the ST mode is that $\alpha < 1$ ($\sigma_\infty < \bar{A}/(8\pi h^2)$) and (2) when instability does arise, the ST mode is unstable to all waves with wavelengths larger than the critical one. Since $\bar{A}$ is typically of the order of 10^{-19}-10^{-20} J, only very small values of σ_∞(<10^{-3}dyne/cm for h ≥ 10nm) are indeed less than the group $\bar{A}/(8\pi h^2)$ and consequently ST mode instabilities are only realizable in extremely low tension film systems.

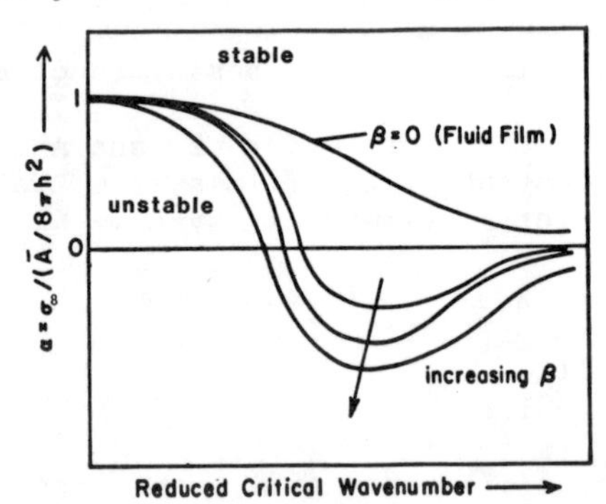

Figure 4. ST mode nonoscillatory marginal stability branches for a viscoelastic fluid film (demarcated $\beta = 0$) and viscoelastic solid films with a fluid interfacial rheology ($\lim_{\omega \to 0} (\omega \gamma_j^i (\omega, k)) = 0$).

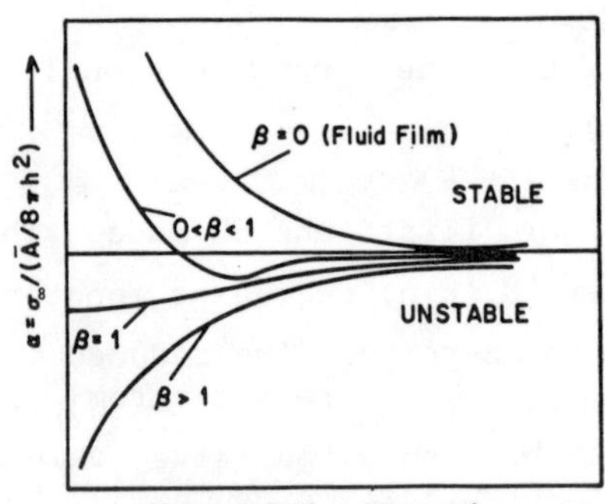

Figure 5. SQ mode nonoscillatory marginal stability branches for a viscoelastic fluid film (demarcated $\beta = 0$) and viscoelastic solid films with a fluid interfacial rheology ($\lim_{\omega \to 0} (\omega \gamma_j^i (\omega, k)) = 0$).

Markedly different behavior is found in the SQ mode's marginal stability curve (Fig. 5, above). Clearly, the fact that this curve diverges monotonically to infinity at long waves indicates that there exists a single critical wavenumber for every (positive) value of α. Hence, for $\bar{A}$>0, the SQ mode is always unstable, with instability again arising for all waves larger than the critical one. However, in physical systems, the film is not infinite in extent but is confined to a length scale ℓ, with wavelengths less than ℓ realizable. Hence the SQ mode is only unstable when the critical wavelength computed from Fig. 5 is less than the horizontal scale ℓ.

The above conclusions are valid for $\bar{A}$ > 0 (α>0). When $\bar{A}$ < 0 (α<0) both diagnostic plots indicate that no critical wavenumbers exist. In addition, for $\bar{A}$ < 0, it can easily be verified from the dispersion equations that all real parts of the ω roots are less than zero. Thus for $\bar{A}$ < 0, the film is stable.

<u>Elastic Stabilization of the van der Waals Instability</u>

When the film exhibits a solid viscoelastic rheology, i.e., when $\lim_{\omega \to 0}[\omega\mu(\omega)] \neq 0$, the nonoscillatory marginal stability criteria became a function of the zero frequency limiting behavior of the complex shear viscosity $\mu(\omega)$, and also the limiting behavior of $\omega\gamma_j^i(\omega,k)$ as $\omega \to 0$. Here, two different formulations for γ_j^i are considered: (1) any fluid form for which $\lim_{\omega \to 0}(\omega\gamma_j^i) = 0$ (cf. Eq. (47)) and (2) the form corresponding to surface tangential immobility, i.e., $\omega\gamma_j^i \to \infty$. For these conditions, the marginal stability branches are functions of α and the dimensionless group β, where β is defined as $Gh/(\bar{A}/8\pi h^2)$. The variable G is an elastic shear modulus defined as $\lim_{\omega \to 0}[\omega\mu(\omega)]$. The group β is a measure of the elastic restoring capability of the film in relation to the van der Waals interactions. In what follows below, $\bar{A}$ (and hence β and α) is regarded as positive; it may be verified that when $\bar{A}$ < 0, the film is stable in both the SQ and ST modes.

Consider first the ST mode for a fluid surface ($\lim_{\omega \to 0}[\omega\gamma_j^i] = 0$). The stability branch for varying β is plotted in Fig. 4 (above). The branch indicates that the influence of film elasticity is to shift the critical wavelength for instability to values larger than the fluid film value. However, the condition for instability established for a fluid film, i.e., $\sigma_\infty < \bar{A}/(8\pi h^2)$, still remains valid. For a surface tangentially immobile film this latter condition is altered by the film elasticity (Fig. 6, below). The long wavelength limit of the stability curves in Fig. 6 are equal to 1 - $\beta/2$; complete stabilization of the ST mode from the van der Waals instability occurs when $\beta \geq 2$ as for this value no critical wavenumbers exist and it may be shown that all real parts of ω are less than zero.

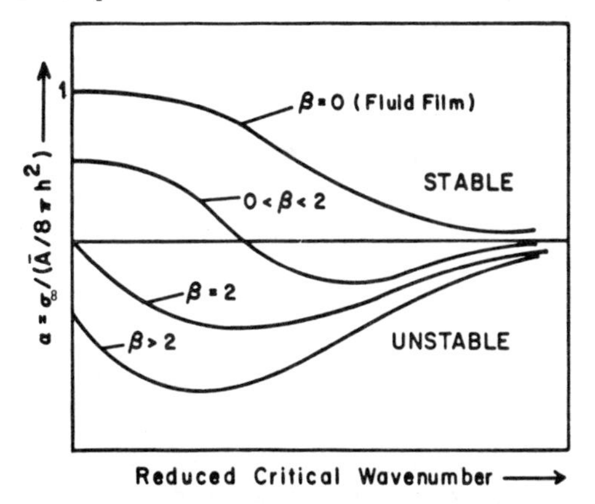

Figure 6. ST mode nonoscillatory marginal stability branches for a tangentially immobile, isotropic viscoelastic solid film.

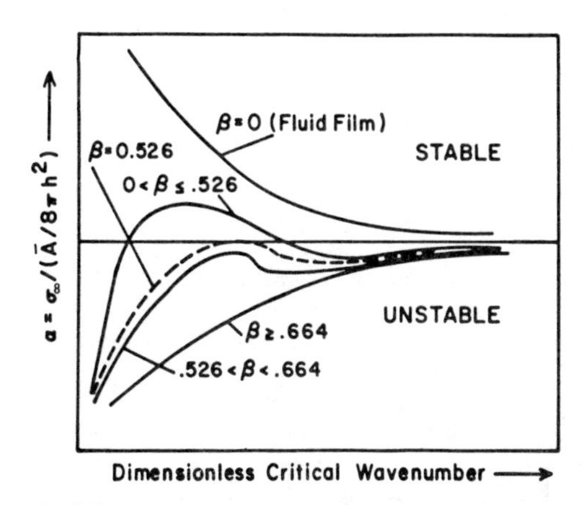

Figure 7. SQ mode nonoscillatory marginal stability branches for a tangentially immobile, isotropic viscoelastic solid film.

Complete stabilization of the SQ mode at a critical value of β is found for both surface conditions (Figs. 5 and 7). For the fluid surface (Fig. 5), the SQ mode is unstable for $0 \leq \beta < 1$ with the increasing values of β shifting the critical wavelength of instability to larger values. When $\beta \geq 1$, no critical wavenumbers exist and it may be shown from the dispersion equation that the SQ mode is stable. The tangentially immobile film (Fig. 7) exhibits a critical value of β equal to .526. There is, however, an interesting difference between the stability curves of the SQ mode for the two surface conditions. For the fluid surface, the SQ mode is unstable to all waves larger than the critical value (Fig. 5) while the immobile surface displays a finite band of unstable waves (Fig. 7).

The stabilization of the SQ mode by elastic restoring forces is significant as it may help in understanding the stability of many ultrathin elastic films such as biological membranes and vesicle lamellae. Such structures, if they behave rheologically as fluids, would be unstable to squeezing vibrations because of the van der Waals interactions..

<u>Electrostatic Double Layer Repulsion and Instabilities</u>

The essential characteristics of the influence of electrostatic double layers on the film stability may be illustrated by examining the relatively simple case of a tangentially immobile, Newtonian film. For such a film, the principle of exchange of stabilities is valid, and it can be shown that the (nonoscillatory) marginal stability curve is again constructed from the equation $\Lambda_1^1 \mp \Lambda_2^1 = 0$, where the Λ_j^i coefficients here are the electrostatically renormalized ones detailed by Eqs. (48) - (53). As before, instability arises (around the critical wavelength) for wavelengths for which $\Lambda_1^1 \mp \Lambda_2^1 < 0$.

Consider first the case of an electrolyte film (neutral overall) drawn from an infinite reservoir and surrounded by dielectrics. The film interfaces are assumed to be at the same potential (cf. Fig. 2c and Eqs. (52 and (53)). For the SQ mode, the term $1-\bar{\alpha}\tanh(\kappa h/2)\tanh(\kappa\bar{\alpha}h/2)$ in Eq. (52) varies monotonically from positive to negative values as k increases from zero. Thus, for this mode, the presence of the double layers within the film acts to stabilize the system from long wavelength disturbances (double layer repulsion), but has the opposite effect for short wavelength perturbations. In the limit as $k \to 0$ (extremely long waves), both the electrostatic and van der Waals terms in Eq. (52) diverge as $1/k^2$:

$$\Lambda_1^1 - \Lambda_2^1 = \frac{2}{k^2}\left\{-\frac{\bar{A}}{2\pi h^2} + \frac{\epsilon^{II}\Psi^2\kappa^3}{4\pi \sinh(\kappa h)}\left(\frac{\bar{K}\tanh(\kappa h/2) + \epsilon^{II}\kappa}{\bar{K}\coth(\kappa h/2) + \epsilon^{II}\kappa}\right)\right\} + O\left(\frac{1}{k}\right) \quad (k \to 0) \tag{62}$$

Consequently, for long waves, the double layer repulsion within the film stabilizes the system from the van der Waals instability when

$$\frac{\epsilon^{II}\psi^2\kappa^3}{4\pi \sinh(\kappa h)}\left\{\frac{\bar{K}\tanh(\kappa h/2)+\epsilon^{II}\kappa}{\bar{K}\coth(\kappa h/2)+\epsilon^{II}\kappa}\right\} > \frac{\bar{A}}{2\pi h^2} \tag{63}$$

By defining the negative electrostatic disjoining pressure $(-\Pi_{e\ell}(h))$ as the difference (in the base state) between the hydrostatic pressure in the film reservoir and the pressure in the dielectric at the film surface, it can easily be shown that, in the Debye-Huckel approximation,

$$\Pi_{e\ell}(h) = \frac{\epsilon^{II}}{8\pi}\,\kappa^2\psi^3\,(1-\tanh^2(\kappa h/2)) \tag{64}$$

The first term in (63) may be shown to be equal to $-\frac{d\Pi_{e\ell}}{dh}$; thus, instability of the SQ mode to long waves is determined by the sign of $-\frac{d}{dh}(\Pi_{e\ell}+\Pi)$, a classical result first surmized through free energy arguments (cf. Vrij (50) and Vrij and Overbeek (51)) and later rigorously derived by Felderhof (21) in a linear, hydrodynamic stability study.

For the stretching mode, the term $1-\bar{\alpha}\tanh(\kappa h/2)\coth(\kappa\bar{\alpha}h/2)$ in Eq. (53) is always less than zero for all wavenumbers and consequently the double layer within the film acts, with respect to this mode, to destabilize the system to all disturbance wavelengths. For very long waves, the sum $\Lambda_1^1 + \Lambda_2^1$ tends to a limiting value:

$$2\lim_{k\to 0}(\Lambda_1^1+\Lambda_2^1) = 2\sigma_\infty - \frac{\bar{A}}{4\pi h^2} - \frac{\epsilon^{II}\kappa\Psi^2(\sinh(\kappa h)-\kappa h)}{4\pi(\cosh(\kappa h)+1)} \tag{65}$$

The last term in the above equation may be shown to be equal to the base state integrated electric film tension, α_e, where

$$\alpha_e = -\frac{1}{4\pi}\int_{-\infty}^{\infty}\epsilon|E_{(o)}|^2 dz_3 = -\frac{1}{4\pi}\int_{-h/2}^{h/2}\epsilon^{II}|E^{II}_{(o)}|^2 dz_3 \tag{66}$$

Hence instability of the ST mode at long waves occurs when $2\sigma_\infty - \bar{A}/4\pi h^2 + \alpha_e < 0$.

An interesting conclusion which may be deduced from the above stability criteria concerns the influence of $\bar{K}$, the parameter which measures the differential variation in the surface charge density with respect to the interfacial potential. For the ST mode the long wave stability criteria is independent of this parameter. However, for the SQ mode, constant surface charge density conditions ($\bar{K} = 0$) are more stabilizing than constant potential conditions ($\bar{K} \to \infty$), since for these two regimes, the stability criteria are, from Eq. (63),:

$$\frac{\epsilon^{II}\Psi^2\kappa^3}{4\pi \sinh(\kappa h)} > \frac{\bar{A}}{2\pi h^2} \quad \text{(constant surface density)} \tag{67}$$

$$\frac{\epsilon^{II}\Psi^2\kappa^3\tanh^2(\kappa h/2)}{4\pi \sin(\kappa h)} > \frac{\bar{A}}{2\pi h^2} \quad \text{(constant potential)} \tag{68}$$

Obviously, the difference becomes more significant as κh becomes smaller (i.e., the Debye length becomes large in comparison to the film thickness).

The pronounced stabilization of the SQ mode at long waves by double layer repulsion within the film is of course not encountered in systems in which a dielectric film is surrounded by electrolytes. Rather, such systems (for both the SQ and ST modes) are generally destabilized by the double layers in the surrounding phases. This destabilization is usually weak, contributing to a lowering in the interfacial tension (σ_∞) in a manner similar to the destabilization of the ST mode of an electrolyte film. However when a potential difference exists across a dielectric film in the base state, a marked destabilization of the SQ mode is encountered. All of these effects are examined below.

Consider first the case of equipotential film interfaces (Fig. 2a). The electrostatic terms in the expressions for $\overset{1}{\Lambda}_1 - \overset{1}{\Lambda}_2$ and $\overset{1}{\Lambda}_1 + \overset{1}{\Lambda}_2$ are both bounded, negative, monotonically decreasing (in absolute value) functions of the wavenumber k. Consequently, the double layers extending into the surrounding phases of the film act to destabilize the system, with the maximum effect realized when the wavelengths of the disturbance are very long. Since the van der Waals contribution to the marginal stability criteria of the ST mode also monotonically decreases (in absolute value) with k, a necessary and sufficient condition for the ST mode to become unstable is

$$2\sigma_\infty - \frac{\bar{A}}{4\pi h^2} - \frac{\overset{I}{\epsilon}\kappa\psi_s^2}{4\pi} < 0 \qquad (69)$$

where the last term in the above relation represents twice the limiting value (as $k \to 0$) of the electrostatic term in Eq. (49) and may also be shown to be equal to the electric film tension (Eq. (66)) of the base state illustrated in Fig 2a. For the SQ mode, since the van der Waals term in Eq. (48) diverges (negatively) as $k \to 0$, there always exists a critical wavelength such that the film is unstable to all waves greater than this critical value; the influence of the double layers of charge in this mode is only to reduce the value of the critical wave.

More complicated behavior results when a potential difference exists across the dielectric in the base state (Figs. 2b-1 and 2b-2). Consider first the ST mode; shown in Figs. 8a and 8b are diagnostic plots of values of the group $\sigma_\infty/(\bar{A}/(8\pi h^2))$ $(=\alpha)$ which satisfy the marginal stability relation ($\overset{1}{\Lambda}_1 + \overset{1}{\Lambda}_2 = 0$, Eq. (51)), as a function of $k_c h$. Unstable points lie beneath the curves since such points correspond to $\overset{1}{\Lambda}_1 + \overset{1}{\Lambda}_2 < 0$. The limiting value of the α group as $k \to 0$ can easily be shown to be equal to $1 - \frac{1}{2}\ \alpha_e/(\bar{A}/(8\pi h^2))$ where α_e is the base state integrated electric film tension defined in general by Eq. (66) and given explicitly below for the base states of Fig. 2b-1 and 2b-2.

$$\alpha_e = -\frac{\overset{II}{\epsilon}}{4\pi}\left(\frac{\Delta\Phi}{h}\right)^2 h - \frac{\overset{I}{\epsilon}\kappa\psi_s}{4\pi} \qquad (70)$$

The inflections in the curves of Figs. 8a and 8b arise from the $\psi_s\Delta\Phi$ term in Eq. (51) which is the only term which is not monotonic in k. Fig. 8a pertains to the base state of Fig. 2b-1 where $\psi_s\Delta\Phi > 0$. The plot indicates that for intermediate values of $|\Delta\Phi|$, the condition $2\sigma_\infty - \bar{A}/(4\pi h^2) + \alpha_e < 0$ is sufficient, but not necessary, for instability because values of σ_∞ for which this inequality is not satisfied are unstable to a narrow band of wavenumbers (see the dotted line in the figure). For the case in which $\psi_s\Delta\Phi < 0$ (Fig. 8b), the stability loci are also not monotonic for intermediate values of $|\Delta\Phi|$. However, the local maxima occurring at nonzero k are less than the limiting value as $\kappa \to 0$. Consequently the condition $2\sigma_\infty - \bar{A}/(4\pi h^2) + \alpha_e < 0$ is necessary and sufficient for instability of the ST mode for the circumstance in which $\psi_s\Delta\Phi < 0$.

The marginal stability equation for the SQ mode ($\overset{1}{\Lambda}_1 - \overset{1}{\Lambda}_2 = 0$, Eq. (50)) is dominated at long wavelength disturbances by the $(\Delta\Phi/h)^2$ term in Eq. (50) which diverges negatively as $k \to 0$. (The rest of the electrostatic terms in Eq. (50) remain finite in this limit.) Consequently, even in the absence of van der Waals destabilization, the SQ mode is always unstable to sufficiently long wavelength perturbations when a potential difference exists across the film.

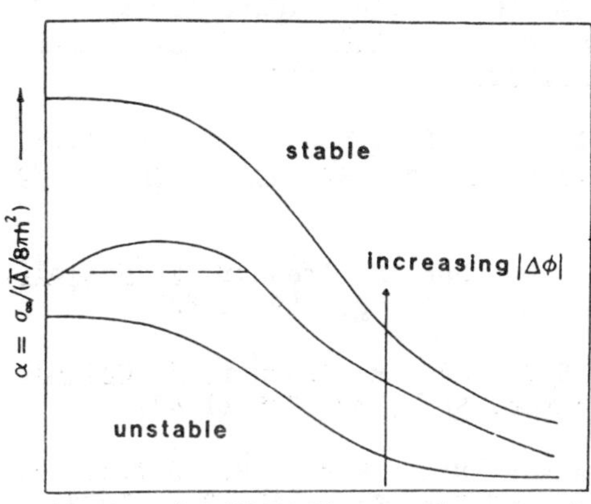

Fig. 8a

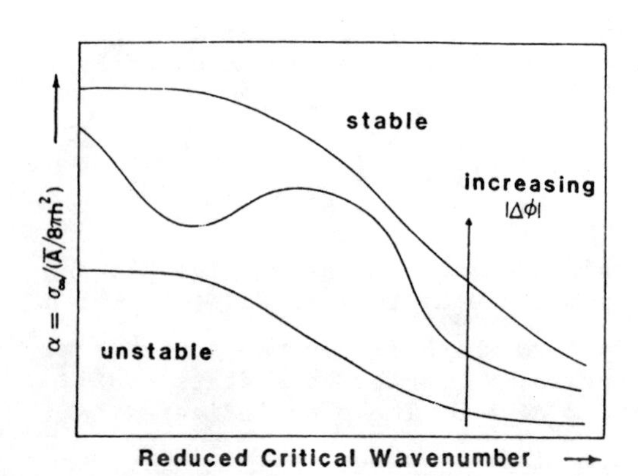

Figure 8. ST mode nonoscillatory marginal stability curves for a dielectric film surrounded by identical electrolytes; (a) $\psi_S \Delta\Phi > 0$ (see Fig. 2b-1) and (b) $\psi_S \Delta\Phi < 0$. (See Fig. 2b-2.)

ACKNOWLEDGEMENT

This work was supported in part by a grant from the National Science Foundation (Thermodynamics and Transport Phenomena Program; CBT-84200098) and a grant from the Department of Energy (DE-AC02-80ER10599)).

LITERATURE CITED

1. Scheludko, A., Adv. Colloid and Int. Sci., 1, 391 (1967).

2. Aveyard, R. and B. Vincent, Prog. in Sur. Sci., 8, 59 (1977).

3. Corbett, R.P., Dechema-Monographie, 72, 1 (1974).

4. Cross, J.A., in Surface Contamination, Mittal, K.L., (Ed.), Vol. 1, Plenum, N.Y., 1979.

5. Dunn, C.S., Doctoral Dissertation, SUNY at Buffalo, N.Y., 1978.

6. Jain, R,K., C. Maldarelli and E. Ruckenstein, AIChE Symp. Ser. Biorheology, 74, 120 (1978).

7. Wendel, H., P.M. Bisch and D. Gallez, Colloid and Polymer Sci., 260, 425 (1982).

8. Lin, S.D. and H. Brenner, J. Colloid and Polymer Sci., 85, 59 (1982).

9. Lin, S.D. and H. Brenner, J. Colloid and Polymer Sci., 89, 226 (1982).

10. Williams, M.B. and S.H. Davis, J. Colloid and Int. Sci., 90, 220 (1982).

11. Chen, J. and J.C. Slattery, AIChE J., 28, 955 (1982).

12. Chen, J., P.C. Hahn and J.C. Slattery, AIChE J., 30, 622 (1984).

13. Joosten, J.G.H., J. Chemical Physics, 82, 2427 (1985).

14. Ivanov, I.B., B.P. Radoev, E.D. Manev and A.D. Scheludko, Trans. Faraday Soc., 66, 1262 (1970).

15. Ivanov, I. and D.S. Dimitov, Colloid and Polymer Sci., 252, 983 (1974).

16. Ivanov, I., Doctoral Dissertation, University of Sophia, Bulgaria, 1977.

17. Ivanov, I., R.K. Jain, P. Somasundaran and T.T. Traykov, in Solution Chemistry of Surfactants, Mittral, K.L., Ed., Vol. 2, Plenum, N.Y., 1979.

18. Ivanov, I. and R.K. Jain, in Dynamics and Instability of Fluid Interfaces, Sorensen, T.S., Ed., Springer Verlag, N.Y., 1979.

19. Ivanov, I.B., Pure and Appl. Chem., 52, 1241 (1980).

20. Levich, V.G., Physicochemical Hydrodynamics, Prentice Hall, N.J., 1962.

21. Felderhof, B.U., J. Chem. Phys., 49, 44 (1968).

22. Maldarelli, C., R.K. Jain, I.B. Ivanov and E. Ruckenstein, J. Colloid Interface Science, 78, 118 (1980).

23. Toshev, B.V. and I.B. Ivanov, Colloid and Polymer Sci., 253, 558 (1975).

24. de Feijter, J.A., J.B. Rijnbout and A.J. Vrij, J. Colloid and Int. Sci., 64, 258 (1978).

25. Marmur, A., J. Colloid and Int. Sci., 93, 18 (1983).

26. Ivanov, I.B. and B.V. Toshev, Colloid and Polymer Sci., 253, 593 (1975).

27. McConnell, A.J., Applications of Tensor Analysis, Dover, N.Y., 1981.

28. Ono, S. and Kondo, S., in Handbuch der Physik, Flugge-Marberg, Ed., Vol. 10, Springer-Verlag, Berlin, 1960.

29. Joseph, D., Stability of Fluid Motions II, Springer-Verlag, N.Y., 1976.

30. Lucassen, J., M. van den Tempel, A. Vrij and F. Hesselink, Proc. Kon. Ned. Akad. Wet., B73, 108 (1970).

31. Vrij, A., F. Hesselink, J. Lucassen and M. van den Tempel, Proc. Kon. Ned. Akad. Wet., B73, 124 (1970).

32. Maldarelli, C. and R.K. Jain, J. Colloid and Int. Sci., 90, 233 (1982).

33. Maldarelli, C. and R.K. Jain, J. Colloid and Int. Sci., 90, 263 (1982).

34. Steinchen, A., D. Gallez, and A. Sanfeld, J. Colloid and Int. Sci., 85, (1982).

35. Sorensen, T.S., M. Hennenberg, A. Steinchen-Sanfeld and A. Sanfeld, Progr. Colloid and Polymer Sci., 61, 64 (1976).

36. Dalle Vedove, W. and A. Sanfeld, J. Colloid and Int. Sci., 84, 318 (1981).

37. Dalle Vedove, W. and A. Sanfeld, J. Colloid and Int. Sci., 84, 328 (1981).

38. Dalle Vedove, W. and A. Sanfeld, J. Colloid and Int. Sci., 95, 299 (1983).

39. Scriven, L.E., Chem. Eng. Sci., 12, 98 (1960).

40. Slattery, J.C., Chem. Engr. Sci., 12, 98 (1960).

41. Wendel, H., D. Gallez and P.M. Bisch, J. Colloid and Int. Sci., 84, 1 (1981).

42. Bisch, P.M., H. Wendel and D. Gallez, J. Colloid and Interface Sci., 14, 251 (1981).

43. Gallez, D., P.M. Bisch and H. Wendel, J. Colloid and Interface Sci., 92, 296 (1983).

44. Sche, S., J. Electrostatics, 5, 71 (1978).

45. Sche, S. and H.M. Fijnaut, Sur. Sci., 76, 186 (1978).

46. Ruckenstein, E. and D. Prieve, J. Theor. Biol., 56, 205 (1976).

47. Ruckenstein, E. and R.K. Jain, Faraday Trans. 2, 70, 32 (1974).

48. Jain, R.K. and E. Ruckenstein, J. Colloid and Int. Sci., 54, 108 (1976).

49. Jain, R.K. and C. Maldarelli, Annals of the N.Y. Academy of Sciences, 404, 89 (1983).

50. Vrij, A., Discuss. Faraday Soc., 42, 23 (1966).

51. Vrij, A. and J. Overbeek, J. Amer. Chem. Soc., 90, 3074 (1968).

52. Chandrasekhar, S., Hydrodynamic and Hydromagnetic Stability, Dover, N.Y., 1981.

FILM INSTABILITIES ARISING FROM STEEP CONCENTRATION OR TEMPERATURE GRADIENTS

Bruce A. Nerad and William B. Krantz ■ Department of Chemical Engineering, Campus Box 424, University of Colorado, Boulder, CO 80309

Steep temperature or concentration gradients normal to an interface on a length scale characteristic of the attractive intermolecular forces can generate spontaneous free convection by a newly discovered mechanism that is quite distinct from the classic Marangoni instability. Steep gradients generate intermolecular potential energy gradients and thereby can induce free convection by a mechanism somewhat analogous to that operative in thin liquid film instabilities. The effects of this newly identified instability mechanism are usually not observable since the required steep gradients quickly decay because of molecular transport. However, if these steep gradients occur in a system which rapidly solidifies, the spatial patterns resulting from this instability mechanism contribute to the observable morphology. This paper reviews the fundamentals of the steep-gradient-driven instability mechanism and establishes for the first time that temperature as well as concentration gradients can initiate this instability.

This paper discusses a new instability mechanism which arises from steep temperature or concentration gradients normal to an initially planar free surface. As such, this newly identified instability mechanism applies to processes which are undergoing unsteady-state heat or mass transfer and which possess a free surface. Such processes include the casting of polymeric membranes, the drying of paints and varnishes, and the thermal casting of polymer films, glasses, ceramics and metals, to mention but a few.

This new instability mechanism is distinct from the classical Marangoni instability mechanism (1), which also arises because of temperature or concentration gradients normal to an initially planar free surface, for several reasons. First, the driving force for this instability mechanism appears as a body force in the equations of motion which then can be transformed so that it appears in the surface normal stress boundary condition. In contrast, the driving force for the Marangoni instability appears in the surface tangential stress boundary condition. Second, the instability of interest here can generate spontaneous free convection under conditions for which a system is stable with respect to Marangoni-driven free convection. Third, the addition of surface-active solutes can be shown to enhance the growth of this newly identified instability mechanism, whereas surfactants are known to stabilize Marangoni-driven free convection. Finally, this new instability mechanism can grow significantly faster than the classical Marangoni instabilities, thereby allowing it to be manifest in systems which solidify very rapidly such as those mentioned above.

Since a temperature or concentration gradient induces a corresponding gradient in the excess intermolecular potential energy, this new steep gradient-driven instability mechanism is somewhat analogous to the thin liquid film instability (2). This analogy was recognized only recently by Ray (3) and Ray et al. (4) who first established that very steep concentration gradients can generate spontaneous free convection in systems having a constant surface tension. The fact that steep temperature gradients also can generate spontaneous free convection by this same novel mechanism is a unique contribution of this present paper.

The thrust of this present paper then is to elucidate the physicochemical nature of this newly identified instability mechanism insofar as it can be initiated by steep temperature or concentration gradients. The scope of this paper is as follows. First, the analogy between the classical thin film instability and this new instability mechanism will be explored. Then, an idealized example of this instability driven by a steep temperature gradient will be developed. This is followed by a complementary example for which the driving force is a steep concentration gradient. A discussion of the implications of this newly discovered instability mechanism then follows. This discussion focuses on the conditions under which this instability is manifest and the resulting characteristics of the observable morphology. In particular, the morphological features of asymmetric polymeric membranes explained by this new instability mechanism are reviewed. Finally, the possible occurrence of this newly identified instability mechanism in other technologies is considered.

THIN FILM INSTABILITY ANALOGUE

Pioneering studies of the classical thin liquid film instability mechanism include those of Frenkel (5), Scheludko (6), Vrij (7), Lucassen et al. (8), and Ruckenstein and Jain (9), as well as others. An excellent comprehensive review of the thin liquid film instability literature is given by Maldarelli (10).

The analogy between the thin film and steep gradient-driven instability mechanisms is seen most clearly by considering the system analyzed by Ruckenstein and Jain (9). This is shown in Figure 1 and consists of a thin (10 to 100 nm) liquid film overlying a nonwetting solid. "nonwetting" implies that the stronger intermolecular attractions are in the liquid phase rather than between the liquid and solid phases. The physicochemical nature of this thin film instability mechanism is best explained by considering the effects of perturbing or corrugating the upper free surface of the film. This corrugated interface is shown by the solid wavy line in Figure 1, whereas the unperturbed interface is shown by the dashed line.

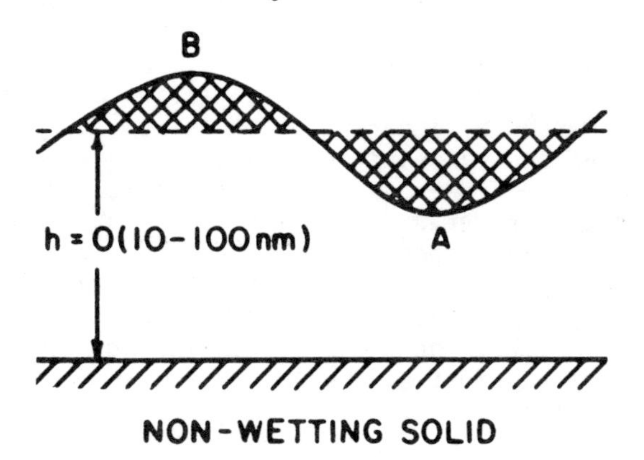

Figure 1. Schematic of a thin liquid film overlying a nonwetting solid.

The net effect of corrugating the interface is to have moved liquid from below the dashed line, thereby creating a trough in region A, in order to create a crest in region B. This packet of liquid now in region B, is located further from the nonwetting solid thereby permitting it to interact with a thicker layer of the liquid phase beneath it. Since the attraction between molecules in the liquid is stronger than between molecules in the liquid and solid, the interfacial perturbation serves to increase the magnitude of the intermolecular potential energy (a negative quantity for attractive forces), thereby lowering the free energy of the system; this implies that the system is potentially unstable in that minute spontaneous corrugations of the interface can reduce the system free energy by increasing in amplitude. Note however, that if the film were thicker than the range of the attractive intermolecular forces (10 to 100 nm), a perturbation of the interface would not lead to a perceptible decrease in the system free energy, thereby implying that a thicker film overlying a nonwetting solid will be stable to all but the longest (and perhaps unrealizable) wavelength perturbations. Hence, for a thin film instability to occur, it is essential that a perturbation cause a change on a length scale characteristic of the range of influence of the intermolecular forces, namely 10 to 100 nm. Note also that a thin liquid film on a wetting surface will be stable to all wavelength perturbations since an interfacial corrugation will move liquid further away from the solid to which it is more strongly attracted, thereby increasing the system free energy.

The driving force for the thin liquid film instability mechanism appears as a body force in the equations of motion and is expressible as the gradient of the excess intermolecular potential energy. The analogy between the thin liquid film instability and the steep gradient instability of interest here is apparent when one realizes that there are ways other than having two interfaces in close proximity to create a gradient in the excess intermolecular potential energy. In particular, a significant gradient in the excess intermolecular potential energy will exist in the neighborhood of an interface if temperature or concentration gradients are imposed which are steep on a length scale characteristic of the attractive intermolecular forces. Such steep gradients indeed are possible especially during the initial moments of unsteady-state heat or mass transfer. In particular, they frequently occur during materials fabrication processes which employ either rapid thermal quenching or diffusion-controlled solvent casting.

TEMPERATURE GRADIENT-DRIVEN INSTABILITY

Now that the physical basis for the thin film and analogous gradient-driven instability mechanisms has been explored, let us examine more closely how in particular a steep temperature gradient normal to an interface can initiate spontaneous free convection. We will begin by by examining how a steep temperature gradient can cause a gradient in the excess intermolecular potential energy. We then will analyze the consequent gradient-driven instability for a somewhat idealized unsteady-state heat transfer process.

Physical Explanation

Figure 2 shows a schematic of a liquid layer, not necessarily thin, undergoing heat transfer (as depicted by the wavy arrows) to a surrounding noninteracting medium. This heat transfer induces a temperature as well as a density gradient in the liquid, as is suggested by the more dense packing of the molecules (shown by the circles) near the interface in Figure 2. Since the effective magnitude of the intermolecular attractive forces decreases with increasing temperature (which is the reason most liquids eventually boil), there are stronger attractive intermolecular interactions near the cooler interface than in the warmer bulk liquid. Note that this is analogous to the thin liquid film instability for which the intermolecular interactions were stronger near the interface because it was further from the nonwetting solid surface. Now let us consider a minute spontaneous perturbation or corrugation of the free surface about its unperturbed location shown by the dashed line in Figure 2.

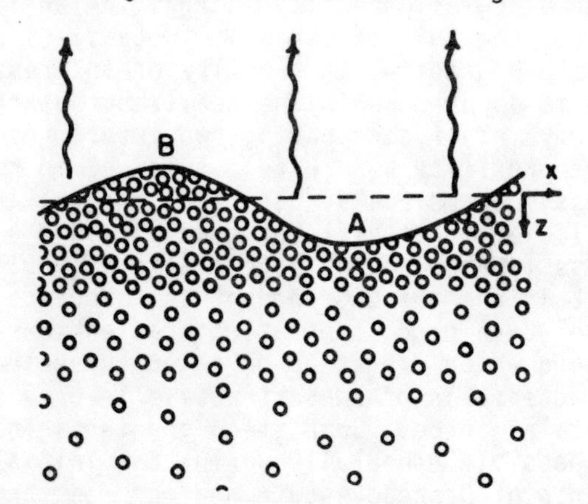

Figure 2. Schematic of a liquid sustaining a heat loss at a free surface.

Again the net effect of corrugating the interface is to have moved liquid from below the dashed line, thereby creating a trough in region A, in order to create a crest in region B. The packet of liquid now in region B, is in closer proximity to a more dense packing of the liquid phase (because of the temperature gradient). Because the attractive intermolecular forces are stronger near where the packing is denser, this spontaneous corrugation of the interface reduces the system free energy, thus implying that this system is potentially unstable. Note however, that this liquid layer would be stable to spontaneous interfacial corrugations if heat were being transferred to rather than from it.

Thus, one sees that the same physical argument can explain why both the thin film and the analogous temperature gradient-driven instability occur. In general, spontaneous free convection can be initiated in liquid layers when the attractive intermolecular forces decrease into the bulk liquid away from the free surface. Now that the physicochemical nature of the driving force for this newly discovered instability mechanism has been established, let us explore the characteristics of this gradient-driven instability by analyzing a simple system.

Mathematical Analysis

The philosophy to be pursued in the mathematical analysis developed here is to consider the simplest possible heat transfer situation involving a liquid layer, which incorporates only as much physics as is necessary to demonstrate that this newly identified instability mechanism is quite distinct from either the Rayleigh-Benard or Marangoni mechanisms. The complicating effects of these latter instabilities will be suppressed by ignoring gravitational body forces and by assuming a constant surface tension. If one can demonstrate that a steep temperature gradient normal to an interface can initiate spontaneous free convection even in the absence of any forces which could lead to Rayleigh-Benard or Marangoni instabilities, then one has convincing proof that a new instability mechanism has been identified.

Suppressing Rayleigh-Bernard and Marangoni instabilities in addition to invoking several assumptions necessary to obtain a simple closed-form analytical solution for the dispersion relation between the perturbation growth rate and the parameters of the problem, necessarily precludes applying the results of this simplified analysis to any real system. Real systems which might display free convection as a result of this new instability mechanism, typically involve nonconstant physical properties and strong coupling between the equations of motion and the energy (or conservation of species) equation. The linear stability problem for these real systems in general must be solved numerically. This obscures seeing the mathematical essence of this new instability mechanism which is readily apparent from the analytical solution to be obtained here. In a companion paper, however the authors (11) have numerically analyzed the nature of this new instability mechanism insofar as it appears to influence the morphology of cast asymmetric polymeric membranes. A comparison between this latter numerical solution, and the simple analytical results to be obtains here, suggests that the idealized problem considered here contains sufficient physics to elucidate the nature of this new instability mechanism.

The mathematical approach presented here then employs linear stability theory in order to assess the propensity of the ideal system to develop infinitesimal surface disturbances of specified wavelengths and consists of three basic steps. The first is the development of a basic-state solution which characterizes the heat transfer for the ideal, unperturbed system. Next, linearized equations describing the heat transfer and fluid dynamics associated with the corrugated interface are derived. Finally, these linearized equations are solved in order to determine the conditions under which an infinitesimal disturbance can grow.

Basic State

The model system is assumed here to be a semi-infinite liquid having constant physical properties undergoing heat loss in a zero-g environment. The assumption of a semi-infinite geometry is valid when the the thickness of the liquid is much greater than the diffusive heat transfer length scale (κ/h). This assumption simplifies the mathematical analysis by reducing the dimension of the matrix of integration constants from six to three. Assuming constant physical properties for the model system appears to be inconsistent with the contention that this new instability arises from the increase in density with a decrease in temperature. This apparent inconsistency is treated by invoking a Boussinesq-type approximation (12) whereby the variable density affects only the driving force but not the other terms in the governing equations. The assumptions of a zero-g environment and constant surface tension allow both Rayleigh-Benard and Marangoni-induced convection to be mathematically excluded and simplify the mathematical analysis by eliminating part of the coupling in the linearized disturbance equations.

In the basic-state analysis of this model system, it is further assumed that the heat transfer is one-dimensional unsteady-state conduction in a liquid having a constant temperature initially. The heat loss to the surrounding medium is assumed to be describable by Newton's Law of Cooling. Under these assumptions, the following form of the energy equation and boundary conditions describe the system:

$$T^*_{0t^*} = \alpha T^*_{0z^*z^*} \tag{1}$$

$$T^*_0 = T_i \qquad \text{at } t^* = 0 \tag{2}$$

$$T^*_0 = T_i \qquad \text{at } z^* \to \infty \tag{3}$$

$$\kappa T^*_{0z^*} = h(T^*_0 - T_a) \qquad \text{at } z^* = 0 \tag{4}$$

where T is the temperature, α is the thermal diffusivity, t is time, z is the coordinate normal to the planar interface and measured positive into the liquid, κ is the thermal conductivity, and, h is the heat transfer coefficient of the surrounding medium. The subscripts 0, i, and a, indicate a basic-state quantity, initial temperature, and, ambient temperature, respectively. The subscripts t and z stand for partial differentiation with respect to the indicated variable. Finally, the superscript * indicates a dimensional quantity.

Scales were found for this dimensional system of equations in the following manner. The reference temperature was chosen as T_a in order to make Equation (4) homogeneous. This then implied, along with Equations (2) and (3), that the temperature scale was $T_i - T_a$. By balancing the conductive and convective terms in Equation (4), the thermal length scale was found to be κ/h. Finally, the time scale of the basic state was determined to be $\kappa^2/h^2\alpha$ by balancing terms in Equation (1). Therefore, in nondimensional form, Equations (1) to (4) become:

$$T_{0t'} = T_{0zz} \tag{5}$$

$$T_0 = 1 \qquad t' = 0 \tag{6}$$

$$T_0 = 1 \qquad z \to \infty \tag{7}$$

$$T_{0z} = T_0 \qquad z = 0 \tag{8}$$

where t' is nondimensional time scaled with the basic-state time scale. The solution to the basic-state Equations (5) through (8) is found in Carslaw and Jaeger (13) and in nondimensional form is

$$T_0 = \text{erf}\left[\frac{z}{2\sqrt{t'}}\right] + e^{z+t'}\,\text{erfc}\left[\frac{z}{2\sqrt{t'}} + \sqrt{t'}\right] \tag{9}$$

where erf and erfc are the error function and complementary error function, respectively. Other basic-state relationships needed later in the analysis are the first- and second-derivatives of the temperature evaluated at the interface:

$$T_{0z}|_{z=0} = e^{t'}\,\text{erfc}(\sqrt{t'}) \tag{10}$$

$$T_{0zz}|_{z=0} = e^{t'}\,\text{erfc}\left[\sqrt{t'} - \frac{1}{\sqrt{\pi t'}}\right] \tag{11}$$

The most important property of the the basic-state solution to note is that the first derivative of the temperature, Equation (10), monotonically decreases to zero in time. Since the temperature gradient is related to the driving force for this instability, the propensity for this instability to occur will decrease in time. Thus, this instability will be observable only if the result of the heat transfer is to rapidly solidify the interfacial region of the liquid, thereby "freezing" in the interfacial corrugation.

Linearized Disturbance Equations

After the basic-state solution is obtained, the next step is to examine its stability via linear stability theory. Because the development of an instability involves the displacement of fluid packets (i.e., fluid motion), the linear stability analysis necessarily introduces the Navier-Stokes equations. To simplify the mathematical analysis, only two-dimensional normal mode disturbances are considered. This is not a serious limitation, because, in some cases, the three-dimensional results can be obtained from the two-dimensional via Squire's Transformation (14). The x- and z-components of the Navier-Stokes equations for a constant physical property fluid in a zero-g environment are given below:

$$\rho(u^*_{t^*} + u^* u^*_{x^*} + w^* u^*_{z^*}) = -P^*_{x^*} - W^*_{x^*} + \mu(u^*_{x^*x^*} + u^*_{z^*z^*}) \tag{12}$$

$$\rho(w^*_{t^*} + u^* w^*_{x^*} + w^* w^*_{z^*}) = -P^*_{z^*} - W^*_{z^*} + \mu(w^*_{x^*x} + w^*_{z^*z^*}) \tag{13}$$

where ρ is the density, u and w are the velocities in the x- and z-directions, respectively, P is the pressure, W is the excess intermolecular potential energy per unit volume, and μ is the viscosity.

The form of the Navier-Stokes equations given above is the same as that used in the analysis for a thin film instability (2) and differs from the conventional form of these equations only by the term, W_z, the gradient of the excess intermolecular potential energy. It is inserted into the equations of motion to account for the difference in molecular environment experienced by the molecules in the displaced packet of fluid. The calculation of the excess intermolecular potential energy, W, can be found in the appendix.

The proper sign for the intermolecular potential energy term perhaps is not obvious. The easiest way to determine this sign is by analogy to the pressure gradient. In Figure 3a, a higher pressure is postulated at the top of the figure; therefore, the fluid flow is downward or in the opposite direction of the pressure gradient. Thus, the pressure gradient term, P_z, comes into the equations of motion with the opposite sign of the flow or inertial terms. If stronger intermolecular attractions are postulated at the top of Figure 3b than at the bottom, fluid would flow upward and one might expect the proper sign of the intermolecular potential energy term in the equations of motion to be positive.

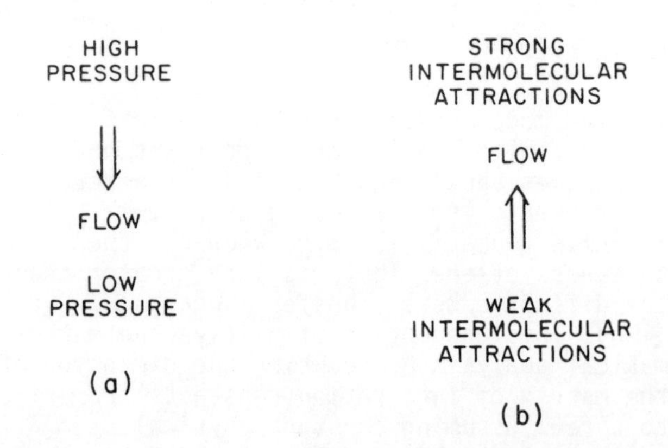

Figure 3. Explanation of the sign convention for the excess intermolecular potential energy term.

However, by convention, attractive intermolecular forces are defined to be negative;

hence, the intermolecular potential energy term enters into Equations (12) and (13) with the same sign (negative) as the pressure gradient. In the work of Ray (3) and Ray et al. (4), the sign preceding the intermolecular potential energy term is incorrectly shown as positive.

Besides the Navier-Stokes equations, other equations and boundary conditions used in this analysis are the following:

the energy equation:

$$T^*_{t^*} + u^* T^*_{x^*} + w^* T^*_{z^*} = \alpha(T^*_{x^*x^*} + T^*_{z^*z^*}) \quad (14)$$

the continuity equation:

$$u^*_{x^*} + w^*_{z^*} = 0 \quad (15)$$

the boundary conditions far into the bulk liquid:

$$w^* = u^* = 0 \quad T^* = T_i \qquad \text{at } z^* \to \infty \quad (16)$$

and at the free surface; Newton's Law of Cooling:

$$\kappa T^*_{z^*} = h(T^* - T_a) \qquad \text{at } z^* = \zeta \quad (17)$$

and the tangential and normal stress balances:

$$u^*_{z^*} + w^*_{x^*} = 0 \qquad \text{at } z^* = \zeta^* \quad (18)$$

$$P^* - P_a - 2\mu w^*_{z^*} - \sigma\zeta^*_{x^*x^*} = 0 \quad \text{at } z^* = \zeta^* \quad (19)$$

The amplitude of the surface corrugation and the velocity are related by the kinematic surface condition:

$$w^* = \zeta^*_{t^*} \qquad \text{at } z^* = \zeta^* \quad (20)$$

where σ is the surface tension and ζ is the amplitude of the surface corrugation relative to the plane of the unperturbed interface and measured positive in the z-direction. Note in Equation (18) that the driving force for a Marangoni instability, namely a lateral surface tension gradient, does not appear. By assuming that the surface tension is constant, Marangoni convection has been mathematically eliminated. Rayleigh-Benard convection also has been excluded in Equation (13) by assuming a zero-g environment. Therefore, if spontaneous free convection is predicted under these assumptions which preclude both Rayleigh-Benard and Marangoni instabilities, then it can only be a result of the temperature or intermolecular potential energy gradient; thus, a new instability mechanism will have been established.

Now let us scale, perturb and linearize Equations (12) through (20). The reference temperature and scale again are the ambient temperature, and the difference between the initial and ambient temperatures, respectively. By comparing Equations (15) and (18), the velocity scales as well as the length scales in both the x- and z-directions are found to be equivalent. As in the basic state, the length scale is found to be κ/h by balancing the convective and conductive heat transfer terms in Equation (17). The velocity scale σ/μ and the pressure scale $\sigma h/\kappa$ are determined by balancing the second and third terms, and the first and third terms in Equation (19). The scale for the intermolecular potential energy is assumed to be equivalent to the pressure scale. The time scale for the perturbed state, $\kappa\mu/h\sigma$, comes from balancing terms in the kinematic condition. If the ratio of the perturbed to basic-state time scales is much less than one, then changes in the perturbed state occur much more rapidly than for the basic state and the frozen time analysis assumed here can be used. A "frozen time analysis" refers to the treatment of the basic-state temperature profile as time-independent relative to the rate at which disturbances grow in the perturbed state. The assumption of a frozen time analysis simplifies the mathematical analysis such that the time dependence of the governing differential equations is exponential.

The perturbation step in the linear stability analysis assumes that all dependent variables are equal to their basic-state values plus a small correction due to the instability or corrugation of the interface. Mathematically, this is written as follows:

$$W = W_0 + \varepsilon W_1 \quad (21)$$

$$P = P_0 + \varepsilon P_1 \quad (22)$$

$$T = T_0 + \varepsilon T_1 \quad (23)$$

$$w = \varepsilon w_1 \quad (24)$$

$$u = \varepsilon u_1 \quad (25)$$

$$\zeta = \varepsilon \zeta_1 \quad (26)$$

The velocities in the x- and z-directions and the surface amplitude of the corrugation, ζ, have basic-state values of zero because the heat transfer in the basic state is assumed to be pure conduction in a fluid with a planar interface ($\zeta_0 = 0$). Equations (21) through (26) are substituted into the scaled forms of Equations (14) through (20) and linearized by retaining only those terms of order ε (which is assumed to be small). The result is

$$\frac{\rho\sigma\kappa}{\mu^2 h} u_{1t} = -P_{1x} - W_{1x} + u_{1xx} + u_{1zz} \tag{27}$$

$$\frac{\rho\sigma\kappa}{\mu^2 h} w_{1t} = -P_{1z} - W_{1z} + w_{1xx} + w_{1zz} \tag{28}$$

$$T_{1zz} + T_{1xx} = \frac{\sigma\kappa}{\mu\alpha h} (T_{1t} + T_{0z} w_1) \tag{29}$$

$$u_{1x} + w_{1z} = 0 \tag{30}$$

with the following boundary and kinematic surface conditions:

$$w_1 = u_1 = T_1 = 0 \qquad \text{at } z \to \infty \tag{31}$$

$$T_{1z} = T_1 \qquad \text{at } z = \zeta_1 \tag{32}$$

$$u_{1z} + w_{1x} = 0 \qquad \text{at } z = \zeta_1 \tag{33}$$

$$P_1 - 2w_{1z} - \zeta_{1xx} = 0 \qquad \text{at } z = \zeta_1 \tag{34}$$

$$w_1 = \zeta_{1t} \qquad \text{at } z = \zeta_1 \tag{35}$$

By employing a frozen time, normal mode analysis, the above coupled partial differential equations were reduced to a system of coupled ordinary differential equations in the independent variable z. A normal mode analysis refers to the fact that the equations are separable and that the lateral or x-dependence of all the perturbed (subscript 1) dependent variables is of the form exp[ikx] where k is the nondimensional wave number or 2π times the reciprocal nondimensional wavelength. Frozen time refers to the assumption that the basic-state temperature gradient, T_{0z}, does not change significantly in time on a scale for which the perturbations grow significantly. This has the consequence that the time dependency of the perturbed dependent variables is of the form exp[βt] where β is the nondimensional growth coefficient of the interfacial disturbance. A positive growth coefficient implies that the surface disturbance will grow in time, whereas a negative β means decay of the disturbance. If a positive β can be found for some k, then the system is said to be unstable.

The resulting system of ordinary differential equations can be converted into two ordinary differential equations in the perturbed normal velocity, w_1, and perturbed temperature, T_1, by first eliminating the perturbed pressure in Equations (27) and (28) by cross-differentiation, and then by eliminating the pressure between Equations (27) and (34). The perturbed lateral velocity, u_1, and the perturbed surface amplitude, ζ_1 can be eliminated by the use of Equation (32) and Equation (35), respectively. Finally, the boundary conditions at the free surface are expanded in a Taylor series about the planar interface (z=0) to yield the following:

$$[D^2 - k^2]\left[D^2 - k^2 - \frac{\rho\sigma\kappa}{\mu^2 h}\beta\right]\tilde{w} = 0 \tag{36}$$

$$\left[D^2 - k^2 - \frac{\sigma\kappa}{\mu\alpha h}\beta\right]\tilde{T} = -\frac{\sigma\kappa}{\mu\alpha h} T_{0z}\tilde{w} \tag{37}$$

with boundary conditions

$$D\tilde{w} = \tilde{w} = \tilde{T} = 0 \qquad \text{at } z \to \infty \tag{38}$$

$$\beta(D-1)\tilde{T} + (T_{0zz} - T_{0z})\tilde{w} = 0 \qquad \text{at } z = 0 \tag{39}$$

$$(D^2 + k^2)\tilde{w} = 0 \qquad \text{at } z = 0 \tag{40}$$

$$\beta\left[D^3 - 3k^2 D - \frac{\rho\sigma\kappa}{\mu^2 h}\beta D\right]\tilde{w} - k^2\beta\tilde{W} + (k^4 - k^2 W_{0z})\tilde{w} = 0 \qquad \text{at } z = 0 \tag{41}$$

where D^n is operator notation for the nth-derivative with respect to z and the tilde indicates that a normal mode frozen time analysis has been employed, i.e., the variables designated with a tilde are functions only of z. Note that the intermolecular potential energy, W, has disappeared from the resulting form of the equations of motion given by Equation (36) and has reappeared in the normal stress balance given by Equation (41). Since W, the driving force for this new instability mechanism, occurs in a boundary condition at the free surface, this new instability mechanism is an interfacial instability. Furthermore, if the interface between the fluid and the surrounding medium were not a free surface, then W would not have reappeared in the problem as currently formulated and spontaneous free convection would not be possible.

Solution of Disturbance Equations

Equations (36) through (39) represent the system of equations to be solved in order to determine when an instability will result. However, this system of equations can be solved only numerically unless further assumptions are made. Recall that an analytical solution is sought in order to illustrate, as simply as possible, the nature of this instability mechanism. The term necessitating a numerical solution is the one on the right-hand-side of Equation (37). This term contains the basic-state temperature gradient, T_{0z}, which depends on z and hence introduces nonconstant coefficients. In the present analysis, it is assumed that this term on the right-hand-side of Equation (37) is negligible. This assumption is justifiable when the coefficient, $\sigma\kappa / \mu\alpha h$, is small, i.e., for a large viscosity, thermal diffusivity, and heat transfer coefficient, and small surface tension and thermal conductivity. Neglecting this term does allow an analytical solution to be attained. However, this analytical solution is implicit in the growth coefficient, β. Thus, it is difficult to infer from this solution when a surface disturbance will grow ($\beta>0$), or decay ($\beta<0$). An explicit solution can be found provided that the terms containing β in Equations (36) and (37) as well as the third term in Equation (41) are small. This will be true not only when β is small, but even when β reaches moderate values since the coefficients which multiply these terms are small under the conditions which allow neglecting the nonconstant coupling term (right-hand-side of Equation (37)). Furthermore, for large wavenumbers, the previously identified terms containing β can become considerably larger and still be negligible relative to the terms proportional to k^2. For these assumptions, the explicit analytical solution for the dispersion relation giving the nondimensional growth coefficient as a function of the nondimensional wave number is found to be:

$$\beta = \frac{(-k^2 + W_{0z})(k+1) + \tilde{W}(T_{0zz} - T_{0z})}{2k(k+1)} \quad (42)$$

There are several important features apparent from Equation (42). First, the above dispersion relation is not valid at small wavenumbers because of neglecting some terms containing the growth coefficient. This can be seen by evaluating the above expression as the wave number goes to zero; an unrealistic infinite value of β results. Second, the denominator in Equation (42) is always positive; hence, the sign of β (whether an instability grows or decays) depends solely on the numerator. At large wave numbers, β is always negative (implying stability) because the k^2 term dominates. The k^2 term is traceable to the effect of curvature in the normal stress balance; thus, the decay of small wavelength disturbances is due to surface tension. On the other hand, an instability will result whenever the intermolecular potential energy terms, W_{0z} and $\tilde{W}$, assume positive values, or, as recognized in the physical explanation, when there are stronger intermolecular attractions near the surface than in the bulk of the fluid. Note though, if the intermolecular potential energy terms are positive, but small, that only small wave number, long wavelength disturbances can grow. In most cases, the wavelength of these weakly-driven disturbances will be so long such that they will be larger than the container dimension and hence, would not be observable. Another feature of Equation (42) to note is that the relationship for the neutrally stable wave number for a thin film overlying a nonwetting solid can be recovered. (A neutrally stable wavenumber has a zero growth coefficient.) For a thin film instability, there is no heat transfer and the temperature is not perturbed. The implication of not perturbing the temperature of a thin film overlying a nonwetting solid is that there is no perturbation of the intermolecular potential energy either, i.e., $\tilde{W}=0$. Thus, the critical wavenumber, k_c, is:

$$k_c = W_{0z}^{1/2} \quad (43)$$

exactly the result of Ruckenstein and Jain (9).

The dispersion relation given by Equation (42), is plotted in Figure 4, which shows the log of the nondimensional growth coefficient as a function of the nondimensional wave number for various nondimensional basic-state times. Since Figure 4 is a plot of the explicit dispersion relation, it is not valid at small wave numbers as evidenced by the growth coefficient becoming infinite for small wavenumbers. If the implicit form of the dispersion relation were plotted instead, k=0 would be found to be neutrally stable. Hence, the curves would exhibit a maximum growth rate at small but nonzero wave numbers.

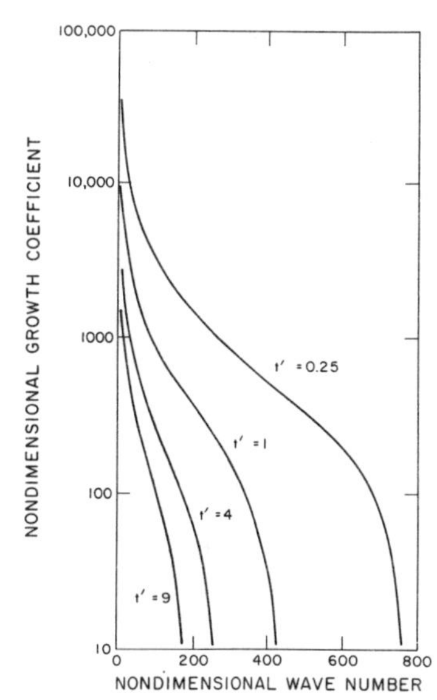

Figure 4. Nondimensional growth coefficient versus the nondimensional wave number for various nondimensional basic-state times.

Recall earlier it was stated that the basic-state temperature gradient at the surface decreased monotonically with time (see Equation (10)). Since the driving force for this new instability mechanism is related to this time-dependent temperature gradient, the spontaneous convection arising from this instability should also die out in time. This dying out of the driving force for this instability manifests itself in Figure 4 in two ways. First, the growth rate for a specific wavelength disturbance decreases in time. Second, the width of the band of unstable wave numbers decreases in time. The time shift is toward less highly amplified longer wavelength disturbances. Therefore, the effect of this instability mechanism can be seen only if the result of the heat transfer is to rapidly solidify the liquid and thereby freeze in the interfacial disturbance as surface structure. However, there is the additional effect of the spontaneous free convection increasing the heat transfer rate caused by this instability mechanism which can be important even if the interfacial corrugations are not frozen into the final structure.

At this point, the physical basis for why both thin films and liquid films subject to steep temperature gradients can be unstable should be clear. Hopefully, the simplified mathematical model developed above illustrates effectively the nature of the steep temperature gradient-induced instability mechanism. It remains then to examine the possibility that steep concentration gradients also can cause spontaneous free convection in liquid layers.

CONCENTRATION GRADIENT-DRIVEN INSTABILITY

The analysis of the possibility that spontaneous free convection can be initiated by steep concentration gradients normal to an initially planar free surface of a fluid having a constant surface tension proceeds in a manner similar to that of the previous section. First, the same physical explanation for why both thin films and liquid films subject to steep temperature gradients can be unstable with respect to surface disturbances will be applied to a liquid layer undergoing mass loss, to ascertain the conditions under which an instability will result. Then, an idealized physical system which incorporates only the essential features of the concentration gradient-driven instability mechanism will be modeled mathematically. For the assumptions of this idealized model system, the mathematical description of the steep concentration gradient-driven instability problem, when cast into nondimensional form, will be found to be identical to that of the steep temperature gradient-driven problem.

The physical explanation of how a steep concentration gradient can induce an instability is aided by Figure 5 which shows a binary system undergoing mass transfer.

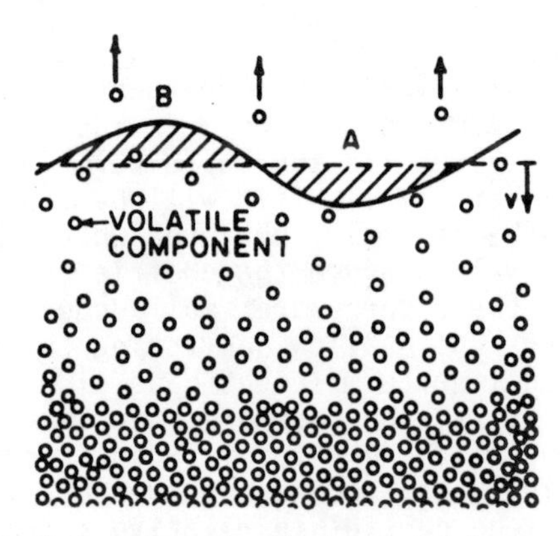

Figure 5. Schematic of a liquid sustaining evaporative mass transfer at a free surface.

Because of the mass loss, shown schematically by the circles and arrows above the interface, there exists a concentration gradient of the volatile component as suggested by the decrease in circle density near the interface. The net mass loss also implies that the interface will drop in time. This motion of the interface must be accounted for in the mathematical modeling of the system. The other important possible source of interfacial motion in this system arises because of interfacial perturbations or corrugations. However, before determining whether these corrugations will grow, it is necessary to state which of the several intermolecular attractions are the strongest. Let us first assume that the strongest interactions are between two solvent molecules. In this instance, the displacement of a packet of fluid from position A to position B moves solvent molecules further from the concentrated solvent in the bulk; hence, this situation is stable (W_{0z} negative). On the other hand, if the strongest attractions are between two nonvolatile component molecules, then spontaneous free convection can result (W_{0z} positive). There is yet another case to consider, namely when the mass transfer is into a medium which strongly interacts with the liquid layer, e.g., a precipitation bath. In this case, if the strongest interactions are between the volatile component and the precipitant, spontaneous free convection also can occur.

Now that the physicochemical basis of the steep concentration gradient instability has been established, let us explore its characteristics by again considering a model system. The basic state for this model system corresponds to binary diffusion in a semi-infinite liquid having constant physical properties. The mass loss is described by a lumped parameter or film theory boundary condition. The interfacial velocity due to the mass loss is assumed to be small. This assumption is reasonable when the total density of the solution is large or when the volatile component is dilute.

For these assumptions, the equations describing the steep concentration gradient-induced instability are identical to those governing the steep temperature gradient-driven instability mechanism. The only distinction between the two is the manner in which the intermolecular potential energy term arises. Thus, all conditions cited for when spontaneous free convection was predicted for the temperature gradient case apply also to the concentration gradient-driven instability mechanism. These conditions are summarized in the next section.

DISCUSSION OF RESULTS

In the previous two sections, simplified mathematical models were developed to illustrate how a steep gradient in either temperature or concentration normal to a free surface could cause an instability. This newly identified instability mechanism can be initiated for two general transfer situations. The first is when the heat or mass transfer is into a noninteracting medium. In this instance, an instability is possible when the attractive intermolecular forces decrease into the bulk liquid away from the free surface. On the other hand, when the transfer is into a strongly interacting medium, then the conditions for initiating this instability can be greatly enhanced. However, as the heat or mass transfer diminishes the gradients in the system, the free convection resulting from this instability becomes progressively less significant. Thus, the directly observable effects of this instability are apparent only when the heat or mass transfer causes rapid solidification of the corrugated interface.

The characteristic size of these surface structures depends on the particular physical parameters of the system. However, an estimate of the transfer conditions required to see 10 nm diameter structures is a temperature gradient of 10^5 K/cm or a concentration gradient of 10^3 g/cm^3/cm. The manner in which the physical parameters affect the structure is as follows. A decrease in the thermal length scale, κ/h, implies an increase in the magnitude of the temperature gradient, and thereby more prominent, smaller diameter surface structures. If the viscosity of the liquid layer is increased, the increased viscous dissipation will retard the generation of spontaneous convection and less prominent larger diameter surface structures will be seen. If the intermolecular attractive forces are strengthened, for example, by changing one of the solution components, then a smaller concentration or temperature gradient can yield the same excess intermolecular potential energy gradient and hence, similar surface structure. Finally, if a surfactant is added to the system, the decrease in surface tension decreases the restoring capillary force associated with the curvature and the instability can be enhanced. Note that this in contrast to the Marangoni instability where the addition of a surfactant always counteracts the surface tension gradient driving force, thereby exerting a stabilizing influence. The dependence of both the characteristic size and prominence of the surface structure on these system parameters was used by Ray (3) to establish that the surface morphology of asymmetric polymeric membranes is a result of the concentration gradient-driven, rather than the Marangoni instability mechanism.

APPLICATIONS

Asymmetric Membrane Morphology

The unique structure of asymmetric polymeric membranes shown in Figure 6 is responsible for the commercial success of the desalination process.

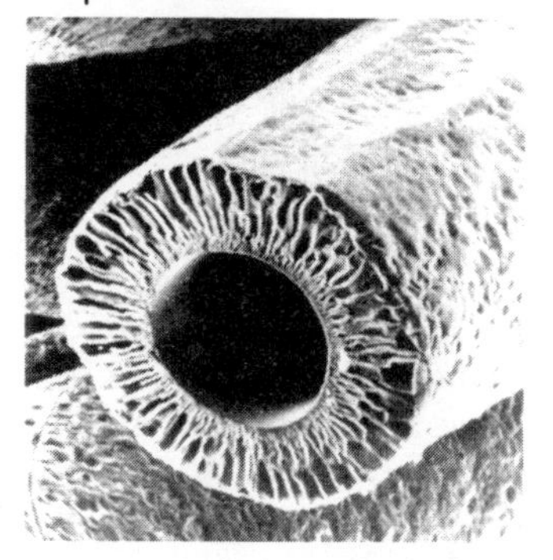

Figure 6. Scanning electron micrograph of a polysulfone hollow fiber membrane having a diameter of approximately 1000 μm (courtesy of H. K. Lonsdale, Bend Research, Inc., 1982).

This figure shows a scanning electron micrograph of a 1000 μm diameter polysulfone hollow fiber membrane. The large voids or fingers in the wall of the membrane provide sufficient structural integrity to withstand the high pressures (up to 70 bar) employed in reverse osmosis, without impeding the permeation through the membrane. The nodules or bumps on the outer surface of the hollow fiber are believed to control the selectivity of the membrane (15). This all-important structure consisting of fingers and nodules has been reasonably well established to arise from the concentration gradient-induced instability mechanism (3,4).

An asymmetric polymeric membrane of the Loeb-Sourirajan type is made by dissolving a polymer in a suitable solvent (or solvent mixture) and casting the solution as a thin film on a smooth solid surface. The cast film is sometimes subjected to an air-evaporation step prior to immersion into a precipitation bath (usually water). The film is allowed to gel or solidify and the solidified film is annealed in hot water to decrease the porosity of the finished membrane. The nodular and fingered structure of the membrane forms during the air evaporation or precipitation bath immersion step when the rapid solvent loss produces a large concentration gradient. For most common membrane casting solutions, the interactions between two solvent molecules or between the solvent and the precipitation bath are the strongest. In the case when the strongest attractions are between the solvent molecules, the physical arguments developed earlier in this paper suggest that an isothermal air evaporation step is stable. However, an instability is predicted for the precipitation step whenever the intermolecular attractions are dominant between the solvent and precipitant. These interactions are generally dominant, otherwise the solvent would not dissolve in the precipitation bath.

Ray (3,4) modeled the air evaporation step of the asymmetric membrane fabrication process as isothermal. In the basic state, Ray accounted for the variable density and diffusion coefficient as well as the finite fluid depth and the nonuniform shrinkage due to both the mass loss and excess volume of mixing effect. Ray then perturbed the equations of motion, assuming constant density but variable viscosity, and solved the resultant system of equations numerically. His results are summarized in Figure 7 which plots the nondimensional growth coefficient as a function of the nondimensional wave number for several nondimensional excess intermolecular potential energy gradients. This graph possesses a striking feature which allows for easy verification of Ray's theory, namely that the most highly amplified wave number (k=2.11) is independent of the value of the driving force. This implies that the most prominent surface structure should have a finger spacing-to-membrane thickness ratio of 4.19. Experimental corroboration of this prediction as well as other evidence in support of the gradient-driven instability mechanism for explaining membrane morphology will be presented in a companion paper in this symposium volume (11).

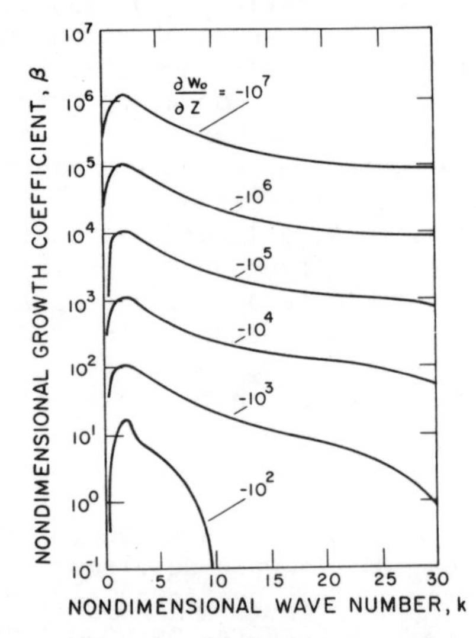

Figure 7. Nondimensional growth coefficient versus nondimensional wave number for varying W_z (from Ray et al. (*4*)).

However, there is a problem with the analysis of Ray which, except for a fortuitous coincidence, should invalidate Ray's theory. First, the analysis in the present paper, in contrast to Ray's findings, predicts that the precipitation bath immersion not the air evaporation step is unstable. This discrepancy is due to the error in Ray's analysis concerning the sign of the intermolecular potential energy gradient term in the equations of motion. The effect of correcting this sign error is merely to change the sign of W_{0z} from negative to positive in Ray's results (Figure 7). The important conclusion that the most highly amplified wave number is independent of the value of the driving force and equal to 2.11 remains unchanged, although now positive values of the intermolecular potential energy gradient are required. Positive values of W_{0z} do occur for the precipitation step and because all of Ray's data are for membranes formed via a precipitation step, the remarkable agreement between Ray's predictions and the experimental data is understandable.

Other Potential Applications

Even though the evaporative casting of membranes has been shown to be stable with respect to the concentration gradient-induced instability mechanism, membranes made in this manner, most notably those of cellulose acetate, possess a nodular skin structure. Experimental evidence has shown that the evaporation step is far from isothermal, but involves very rapid surface cooling of 15 to 20° C (16). Therefore, the nodular structure of evaporatively cast membranes could be due instead to the temperature gradient-driven instability mechanism.

The temperature gradient-driven instability mechanism may also be operative in other cases where evaporative transport occurs in applied liquid films, for example in the application of varnishes, paints, floor waxes, and nail polishes. The characteristic diameter of the nodular structures induced by the temperature gradient inherent in the drying of these materials in fact may dictate whether a glossy or flat sheen is obtained.

Another example where the temperature gradient-induced instability mechanism may arise is in the thermal casting of molten metals. The simplified mathematical model of the thermal instability developed here may apply reasonably well to the fabrication of superior spheres for ball bearings in the microgravity environment of space. If this is the case, then perfectly round featureless spheres cannot be made in space. The spheres should still possess some degree of surface microstructure which will contribute the residual friction in the ball bearing assembly.

This newly discovered gradient-induced instability mechanism may have significant implication for materials processing in the low gravity environment of space. Materials processing in space is attractive since it avoids settling problems, for example, in crystal growth, and minimizes spurious free convection effects associated with Rayleigh-Benard and Rayleigh-Taylor instabilities. However, since materials processing in space frequently involves unsteady-state processes with large temperature or concentration gradients, clearly one needs to be concerned about the possibility of spurious free convection arising from the gradient-driven instability mechanism discussed in this paper.

The examples mentioned above are meant to be suggestive of the myriad of potential applications of this new instability mechanism. Hopefully, they will stimulate further discussion and research in this area. Clearly more work is necessary.

CONCLUSIONS

- Interfacial free convection and surface structure can arise via an instability mechanism initiated by steep temperature or concentration gradients normal to an initially planar free surface in a liquid having constant surface tension.
- A gradient-driven instability can be initiated during unsteady-state heat or mass transfer to a noninteracting medium whenever the intermolecular attractive forces decrease into the bulk liquid from the free surface. Mass transfer into a strongly interacting medium can exert a significant destabilizing influence.
- The generation of surface structures having diameters of about 10 nm requires either a temperature gradient of approximately 10^5 K/cm or a concentration gradient of about 10^3 g/cm^3/cm.
- Only during the initial moments of unsteady-state processes are the gradients sufficiently large to produce submicron size structures.
- The interfacial disturbances arising from this gradient-induced instability mechanism are observable only if the effect of the heat or mass transfer is to rapidly solidify the interface.
- This instability mechanism is operative in the casting of asymmetric polymeric membranes and may be important in many other unsteady-state processes as well.

ACKNOWLEDGMENTS

The authors acknowledge the following sources of financial support of this work: the NSF Graduate Fellowship Progam, the NSF Grant No. CPE 8121841, the National Bureau of Standards Cooperative Agreement No. NB84RAH45130, and the National Center for Atmospheric Research Advanced Study Program.

APPENDIX: CALCULATION OF THE EXCESS INTERMOLECULAR POTENTIAL ENERGY

The purpose of this appendix is to illustrate how to calculate the excess intermolecular potential energy gradient, W_{0z}, and the perturbed energy, $\tilde{W}$. The manner in which this is done is to pick a molecule i located in the liquid (see Figure A1), assume an interaction potential function (such as the Lennard-Jones 6-12 Law), and sum the interaction exerted on the molecule i by all the other molecules j in the liquid and surrounding medium.

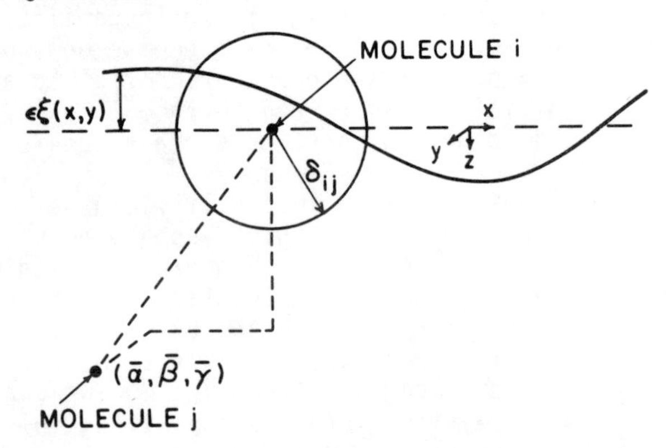

Figure A1. Diagram for the calculation of the intermolecular potential energy.

This then yields the intermolecular potential energy at a given point in the liquid. To obtain the excess intermolecular potential energy, the same procedure is applied to the reference state of constant concentration or temperature, and the difference taken. The quantities needed in the dispersion Equation (42) are related to the excess intermolecular potential energy by (all variables in the appendix are dimensional quantities, although the superscript stars have been omitted for convenience):

$$W_z\,|_{z=0} = \left[\frac{\partial}{\partial z} W\,|_{\varepsilon=0}\right]_{|z=0} \tag{A1}$$

$$W_1\,|_{z=0} = e^{ikx}\,\tilde{W}\,|_{z=0} = \left[\frac{\partial}{\partial \varepsilon} W\,|_{z=0}\right]_{|\varepsilon=0} \tag{A2}$$

Following this procedure yields:

$$\begin{aligned} W = &\int_{-\delta_{ij}}^{\delta_{ij}} \int_{-[\delta^2_{ij}-\bar\beta^2]^{1/2}}^{[\delta^2_{ij}-\bar\beta^2]^{1/2}} \int_{[\delta^2_{ij}-\bar\alpha^2-\bar\beta^2]^{1/2}}^{\infty} I_1(\bar\alpha,\bar\beta,\bar\gamma)\, d\bar\gamma\, d\bar\alpha\, d\bar\beta \\ &+ \int_{-\delta_{ij}}^{\delta_{ij}} \int_{[\delta^2_{ij}-\bar\beta^2]^{1/2}}^{\infty} \int_{\varepsilon\zeta}^{\infty} I_1(\bar\alpha,\bar\beta,\bar\gamma)\, d\bar\gamma\, d\bar\alpha\, d\bar\beta \\ &+ \int_{-\delta_{ij}}^{\delta_{ij}} \int_{-\infty}^{-[\delta^2_{ij}-\bar\beta^2]^{1/2}} \int_{\varepsilon\zeta}^{\infty} I_1(\bar\alpha,\bar\beta,\bar\gamma)\, d\bar\gamma\, d\bar\alpha\, d\bar\beta \\ &+ \int_{\delta_{ij}}^{\infty} \int_{-\infty}^{\infty} \int_{\varepsilon\zeta}^{\infty} I_1(\bar\alpha,\bar\beta,\bar\gamma)\, d\bar\gamma\, d\bar\alpha\, d\bar\beta \\ &+ \int_{-\infty}^{-\delta_{ij}} \int_{-\infty}^{\infty} \int_{\varepsilon\zeta}^{\infty} I_1(\bar\alpha,\bar\beta,\bar\gamma)\, d\bar\gamma\, d\bar\alpha\, d\bar\beta \end{aligned} \tag{A3}$$

where

$$I_1(\bar\alpha,\bar\beta,\bar\gamma) = \frac{-A_{ij}[\rho_{0i}(z)+\varepsilon\rho_{1i}(z)]\,[\rho_{0j}(\bar\gamma)+\varepsilon\rho_{ij}(\bar\gamma)]}{[\bar\alpha^2+\bar\beta^2+\bar\gamma^2]^3} - I_2(\bar\alpha,\bar\beta,\bar\gamma) \tag{A4}$$

where

$$I_2(\bar\alpha,\bar\beta,\bar\gamma) = \frac{-A_{ij}[\rho_{0i}(0)+\varepsilon\rho_{1i}(0)]\,[\rho_{0j}(0)+\varepsilon\rho_{ij}(0)]}{[\bar\alpha^2+\bar\beta^2+\bar\gamma^2]^3} \tag{A5}$$

where δ_{ij} is the distance of closest approach between two molecules i and j; $\bar\alpha$, $\bar\beta$, $\bar\gamma$ are dummy integration variables in the coordinate directions x, y, and z, respectively; and, A_{ij} is a constant (not to be confused with the Hamaker constant) proportional to the strength of the intermolecular attractions between molecules i and j. The exact intersection of the sphere of closest approach and the surface corrugation is δ_{ij} plus a correction of order higher than ε, thus, the correction does not appear in the limits of integration because only terms of order ε or less contribute to the final result. Also, the molecule i is placed at the origin. This causes no loss of generality because the reference state is such that only the z-dependence of the density profiles is important. Finally, note that the above formulation includes only the interactions in the liquid layer. If the molecular interactions with the surrounding medium are important, then additional terms of the same form of Equation (A3) are required, but with limits extending from $-\infty$ to $-[\delta^2_{ij}-\bar\alpha^2-\bar\beta^2]^{1/2}$ for the first volume integration, and from $-\infty$ to $-\varepsilon\zeta$ in the other four volume integrals.

The combination of Equations (A1), (A3), (A4) and (A5), yields after conversion to more convenient cylindrical coordinates:

$$W_{oz}|_{z=0} = 2\pi \int_{\delta_{ij}}^{\infty} \left[\frac{-A_{ij}[\rho_{oi}(0)]_z \rho_{oj}(\bar{\gamma})}{4\bar{\gamma}^4} - \frac{A_{ij}\rho_{oi}(0)[\rho_{oj}(\bar{\gamma}) - \rho_{oj}(0)]}{\bar{\gamma}^5} \right] d\bar{\gamma}$$

$$+ 2\pi \int_0^{\delta_{ij}} \frac{-A_{ij}[\rho_{oi}(0)]_z \rho_{oj}(\bar{\gamma})}{4\delta^4_{ij}} d\bar{\gamma} \qquad \text{(A6)}$$

where $[\rho_{0i}(0)]_z$ is the derivative with respect to z evaluated at z=0. For linear gradients of the form

$$\rho_{oi} = \rho_i^s - b_i z \qquad \text{(A7)}$$

in a multicomponent fluid surrounded by a noninteracting medium, the result is

$$W_{oz}|_{z=0} = 2\pi A_{ij} \left[\frac{b_i \rho_j^s}{3\delta_{ij}^3} - \frac{b_i b_j}{4\delta_{ij}^2} + \frac{\rho_i^s b_j}{3\delta_{ij}^3} \right] \qquad \text{(A8)}$$

Note for a thermal gradient, b_i is approximately equal to the thermal coefficient of expansion multiplied by the reference or surface density. It should be stated that the decrease in density with increasing temperature is probably not the only effect causing the excess intermolecular potential energy gradient. However, this effect clearly shows the analogy between the steep gradient-induced and thin film instability mechanisms and demonstrates, at least theoretically, that a temperature gradient can induce spontaneous free convection. The exact nature of the relationship between the intermolecular potential energy and temperature gradients is currently being investigated.

For the fabrication of a membrane, either the solvent-solvent or solvent-precipitant interactions are the strongest. If the solvent-solvent attractions are the strongest and the excess intermolecular energy gradient is approximated by Equation (A8), then $W_{0|z=0}$ is negative since $b_{solvent}$ is negative; that is, the system is stable. However, if the membrane casting solution is bounded by an interacting medium of constant density, then a term

$$-2\pi A_{ik} \frac{b_i \rho_k^s}{3\delta^3_{ik}} \qquad \text{(A9)}$$

where the subscript k refers to the precipitant, must be added to Equation (A8). Clearly, an instability can result if this additional term outweighs the stabilizing terms of Equation (A8). This occurs when the interactions between the solvent and precipitant are dominant.

The perturbed excess intermolecular potential energy is found by application of Equation (A2) to Equations (A3) through (A5). In normal mode form, it is

$$\tilde{W} = \int_{-\delta_{ij}}^{\delta_{ij}} \int_{-[\delta^2_{ij}-\bar{\beta}^2]^{1/2}}^{[\delta^2_{ij}-\bar{\beta}^2]^{1/2}} \int_{[\delta^2_{ij}-\bar{\alpha}^2-\bar{\beta}^2]^{1/2}}^{\infty} I_3(\bar{\alpha},\bar{\beta},\bar{\gamma})\, d\bar{\gamma} d\bar{\alpha} d\bar{\beta}$$

$$+ \int_{-\delta_{ij}}^{\delta_{ij}} \int_{[\delta^2_{ij}-\bar{\beta}^2]^{1/2}}^{\infty} \int_0^{\infty} I_3(\bar{\alpha},\bar{\beta},\bar{\gamma})\, d\bar{\gamma} d\bar{\alpha} d\bar{\beta}$$

$$+ \int_{-\delta_{ij}}^{\delta_{ij}} \int_{-\infty}^{-[\delta^2_{ij}-\bar{\beta}^2]^{1/2}} \int_0^{\infty} I_3(\bar{\alpha},\bar{\beta},\bar{\gamma})\, d\bar{\gamma} d\bar{\alpha} d\bar{\beta}$$

$$+ \int_{\delta_{ij}}^{\infty} \int_{-\infty}^{\infty} \int_0^{\infty} I_3(\bar{\alpha},\bar{\beta},\bar{\gamma})\, d\bar{\gamma} d\bar{\alpha} d\bar{\beta}$$

$$+ \int_{-\infty}^{-\delta_{ij}} \int_{-\infty}^{-\infty} \int_0^{\infty} I_3(\bar{\alpha},\bar{\beta},\bar{\gamma})\, d\bar{\gamma} d\bar{\alpha} d\bar{\beta} \qquad \text{(A10)}$$

where

$$I_3(\bar{\alpha},\bar{\beta},\bar{\gamma}) = \frac{-A_{ij}\left[\rho_{oi}\tilde{\rho}_{1j}(\bar{\gamma})e^{ik_x\bar{\alpha}+ik_y\bar{\beta}} + \rho_{oj}(\bar{\gamma})\tilde{\rho}_{1i}(0)\right]}{[\bar{\alpha}^2+\bar{\beta}^2+\bar{\gamma}^2]^3} \qquad \text{(A11)}$$

The leading behavior, or long wavelength value of $\tilde{W}$ for linear concentration profiles of the form of Equation (A7) yields:

$$\tilde{W} \leq 2\pi A_{ij} \left[\frac{b_j}{4\delta_{ij}^2} - \frac{\rho_i^s + \rho_j^s}{3\delta^3_{ij}} \right] \qquad \text{(A12)}$$

Equations (A8), (A12), (10) and (11) were used in Equation (44) to calculate the growth rate of a particular wavelength disturbance, i.e., Figure 4.

NOMENCLATURE

Roman

A	Hamaker Constant
A_{ij}	Constant proportional to the strength of the intermolecular attractions
b	Slope of linear density profile
D	Operator Notation
h	Thickness of thin film, or, heat transfer coefficient
k	Wave number
P	Pressure
T	Temperature
t	Time
u	Velocity in x-direction
W	Excess intermolecular potential energy
w	Velocity in z-direction
x	Lateral coordinate
z	Coordinate perpendicular to planar surface

Greek

α	Thermal diffusivity
$\bar{\alpha}$	Dummy integration variable
β	Growth rate of the surface disturbance
$\bar{\beta}$	Dummy integration variable
$\bar{\gamma}$	Dummy integration variable
δ_{ij}	Distance of closest approach between molecules i and j
ε	A small quantity
κ	Thermal conductivity
λ	Wavelength of the surface disturbance
μ	Viscosity
ζ	Coordinate of surface disturbance
ρ	Density
σ	Surface tension

Subscripts

a	Ambient
c	Critical
i	Initial, or, molecular index
j	Molecular index
k	Molecular index
t	Partial derivative with respect to time
x	Partial derivative with respect to x
z	Partial derivative with respect to z
0	Basic state quantity
1	Perturbed quantity

Superscripts

s	Value at the surface
*	Dimensional quantity
~	Variable in normal mode, frozen time form
'	Indicates time was nondimensionalized using the basic-state time scale

LITERATURE CITED

1. C. V. Sternling and L. E. Scriven, AIChE J., 5(4), (1959), 514.

2. C. Maldarelli, R. Jain, I. Ivanov, and E. Ruckenstein, J. Colloid Interface Sci., 78(1), (1980), 118.

3. R. J. Ray, "Interfacial Instabilities Arising from Excess Intermolecular Potential Gradients: Application to Membrane Morphology", Ph.D. Thesis, University of Colorado, Boulder, 1983.

4. R. J. Ray, W. B. Krantz, and R. L. Sani, J. Mem. Sci., 23, (1985), 155-82.

5. J. Frenkel, "Kinetic Theory of Liquids", Dover, New York, 1955.

6. A. Scheludko, Proc. Kon. Ned. Akad. Wet., B65, (1962), 87.

7. A. Vrij, Disc. Faraday Soc., 42 (1966), 23.

8. J. Lucassen, M. van den Tempel, A. Vrij, and F. Hesselink, Proc. Kon. Ned. Akad. Wet., B73, (1970), 108.

9. E. Ruckenstein and R. Jain, Faraday Trans., II, 70, (1974), 132.

10. C. Maldarelli, "The Linear Hydrodynamic Stability of Interfacially Perturbed Thin Films with Applications to the Onset of Small Scale Biological Cell Membrane Motions", Ph.D. Thesis, Columbia University, 1981.

11. B. A. Nerad, K. J. Gleason, W. B. Krantz, and R. J. Ray, Thin Liquid film Phenomena, AIChE Symp. Series (this volume), (1986).

12. S. Chandrasekhar, Hydrodynamic and Hydromagnetic Stability, Clarendon Press, Oxford, 1947.

13. H. S. Carslaw, and J. C. Carslaw, Conduction of Heat in Solids, Clarendon Press, Oxford, 1947.

14. H. B. Squire, Proc. Roy. Soc., A142, (1933), 621.

15. R. E. Kesting, Synthetic Polymer Membranes, McGraw-Hill, New York, 1971.

16. N. B. Gallagher and W. B. Krantz, Proceedings of the 61st Annual Meeting of the Southwestern and Rocky Mountain Division of AAAS, Tuscon, AR, March 19-23, 1985, p. 45.

THE ROLE OF WETTABILITY IN THE BREAKUP OF LIQUID FILMS INSIDE CONSTRICTED CAPILLARIES

P. A. Gauglitz and C. J. Radke ■ Department of Chemical Engineering, University of California
Berkeley, CA 94720

To understand the role of wettability on gas-foam generation in porous media, this work considers the effects of conjoining/disjoining pressure on the dynamics of a liquid film forming an unstable collar in both straight and constricted cylindrical capillaries. A nonlinear evolution equation is derived to describe the dynamic response of a thin viscous film under the influence of surface tension and conjoining or disjoining forces. The thin-film limit of the new evolution equation in straight capillaries is solved with a linear stability analysis. Finite element analysis solves the full, nonlinear evolution equation for the growth of large amplitude collars in constricted capillaries. Time to breakup depends on the pore shape, the strength of the conjoining/disjoining forces, the initial film thickness, and also, on the fluid viscosity, interfacial tension, and unconstricted pore radius which combine to form a characteristic scaling time.

Results show that both conjoining (intermediate wettability) and disjoining (strongly wetting) forces inhibit breakup. We propose the criterion that these forces inhibit snap-off when $Ca \leq |3A^*/\sigma R_T^{*2}|^{3/8}$, where A^* is proportional to the Hamaker constant and σ denotes the surface tension. Thus, wettability inhibits snap-off at very low bubble capillary numbers, Ca, or in very small radius capillaries, R_T^*.

Thin liquid films wetting solid surfaces arise in many physical situations. Of primary interest here are oil-bearing underground porous media. When two phases are present in a porous medium, the distribution of the wetting and nonwetting phases within the pore spaces is crucial to the motion of these fluids (1). Often, wetting liquid flows into collars at pore constrictions that may become unstable and snap-off to form liquid lenses (2). The lenses (i.e., liquid lamellae), if stabilized by surfactants, may block flow paths and increase the resistance to flow through the porous medium. Foam occurs in the porous medium when large numbers of liquid films break up to form lenses which are stabilized against rupture. The addition of surfactants, to stabilize the lenses, is known to improve mobility control and reduce gravity override in steam displacement processes for increasing heavy oil recovery (3,4).

An important mechanism for foam bubble generation is snap-off at pore necks (5). The four key steps in bubble snap-off are depicted in Figure 1. In the first two steps, the nonwetting bubble moves through a liquid-filled constriction depositing a film of wetting liquid. Due to a difference in interfacial curvature, liquid is driven into the growing collar by surface tension forces as shown in step 3. Eventually, sufficient liquid collects in the collar, and it becomes unstable and breaks up to form a liquid lens as shown in step 4. This final step is bubble snap-off to form a foam lamella.

The objective of this work is to determine how wettability influences the above snap-off mechanism in constricted cylindrical capillaries. By wettability, we refer here to the macroscopic, static contact angle, Θ, gauged through the liquid phase. Contact angles less than 90° reflect partial wetting while those greater than 90° correspond to partial nonwetting. Complete wetting and nonwetting occurs at 0° and 180° contact angles, respectively.

As shown by Ivanov and co-workers (6), and Mohanty (2), and discussed by others (7,8), macroscopic wettability is directly related to the molecular conjoining/disjoining pressure. Conjoining and disjoining forces arise in thin fluid films (e.g., less than 1 μm thick) where the close proximity to a solid wall modifies the pressure in the liquid film. Deryagin (9,10) defined the disjoining pressure as the excess pressure arising in thin films above the pressure of the same fluid as a bulk phase. A positive disjoining pressure corresponds to a lower pressure in the film than the same fluid in its bulk state. This situation leads to a spontaneous thickening of the film. With a liquid film and a positive disjoining pressure function, complete wetting (i.e., $\Theta = 0$) occurs.

Conversely, a negative disjoining pressure, called conjoining pressure, corresponds to a higher pressure in the film. Here, local thinning of the film occurs which eventually leads to dewetting of the solid surface (11,12,13). For a liquid film with a negative disjoining function, contact angles greater or equal to 90° arise. This results in partial nonwetting and a gas film adjacent to the solid surface.

Disjoining and conjoining forces also play a role in the dynamics of the snap-off phenomena. Teletzke (14) has shown that conjoining/disjoining pressure alters the film thickness deposited by a bubble. This result is important since the film thickness governs the rate of liquid flow into the growing collar shown in step 3 of Figure 1. In addition, the flow rate in the liquid film must also be influenced by conjoining/disjoining pressure. Therefore, we include conjoining/disjoining forces when describing the film flow depicted in Figure 1. This permits us to investigate the role of liquid wettability during gas bubble snap-off in constricted cylindrical capillaries.

PREVIOUS WORK

The shape and static stability of wetting collars (Figure 2a) has been investigated by Everett and Haynes (15) and Mohanty (2). Their results show that thin collars are stable, but sufficiently thick collars are unstable and break up into lenses (Figure 2b). Mohanty (2) considered straight and constricted capillaries, and included disjoining and conjoining forces in his analysis. For a fixed capillary pressure (i.e., the pressure difference across the collar interface due to curvature and conjoining/disjoining pressure), Mohanty found that pure disjoining forces stabilize an otherwise unstable collar, and conversely, a combination of disjoining and conjoining forces (partial wetting) destabilize an otherwise stable collar. These static results are useful, but to determine the size and rate of bubbles generated by snap-off, we must consider the dynamics of collar growth and breakup.

Dynamic breakup of liquid films in straight capillaries (when conjoining/disjoining forces are negligible) has also received attention. The linear stability analysis performed by Goren (16) demonstrates that infinitesimal sinusoidal disturbances grow exponentially in liquid films of uniform thickness wetting the inside wall of a capillary. This is in accordance with the experimental data of Goren (16), Goldsmith and Mason (17), and Gauglitz (18). More recently, Hammond (19) developed a nonlinear analysis, restricted to films that are thin compared to the capillary radius, to follow the fate of disturbances when their amplitude becomes large. Since the film thickness must be small compared to the capillary radius, Hammond's solution only permits films to evolve into stable collars, such as the one depicted in Figure 2a. Thus, Hammond's analysis unfortunately does not permit breakup, which is step 4 in Figure 1. The experimental data of Goren (16), Goldsmith and Mason (17) and Gauglitz (18) clearly show that liquid films indeed do break up into lenses, such as the one depicted in Figure 2b. The inadequacy of Hammond's thin-film approximation lies in the inaccurate estimate of the circumferential curvature during the latter stages of the breakup to form the liquid lens. In this paper, we employ an extended evolution equation utilizing a more correct estimate of the circumferential curvature (18,20) in addition to including conjoining and disjoining forces.

The influence of conjoining/disjoining pressure on the motion of liquid films in cylindrical geometries, the main theme of this work, has not been very thoroughly examined. Jo (21) considered the stability of an annular film in straight capillaries by employing directly the results of Ruckenstein and Jain (11). However, only liquid motion on a planar solid surface was considered by Ruckenstein and Jain. Thus, Jo neglects completely the crucial circumferential curvature in his analysis. Both Hammond (19) and Goren (16) show that the growth of disturbances is influenced by the delicate competition between the transverse and circumferential radii of curvature. Therefore, we include here both principle radii of curvature and the conjoining/disjoining pressure in our analysis.

In the theory section which follows, we derive a nonlinear evolution equation for axisymmetric liquid films in constricted capillaries. Next, to provide useful physical insight into the role of fluid wettability in the breakup process, the adjoining section presents a linear stability analysis in the thin-film limit of the evolution equation for straight capillaries. Following this, the nonlinear evolution section gives the full numerical solution to the nonlinear evolution equation for liquid films in constricted capillaries. Since conjoining/disjoining forces influence both the deposited film thickness and the evolution of the film, the final section elucidates the combined effect of these two phenomena on bubble snap-off, and summarizes the role of wettability.

THEORY

Consider a viscous film of liquid wetting the inside wall of a constricted capillary, as shown in Figure 3. Cylindrical coordinates (r,x) are used with the axial origin at the neck of the constriction. The radius of the unconstricted pore is R_T^*. The shape of the constriction is prescribed by an arbitrary function, $\lambda(x)$, whose slope is small so the radius of the capillary changes slowly with position. Variables and dimensions in Figure 3 are nondimensionalized as follows:

$$r = r^*/R_T^* \qquad x = x^*/R_T^* \ , \tag{1}$$

$$p = p^*/(\sigma/R_T^*) \qquad \Pi = \Pi^*/(\sigma/R_T^*) \ , \tag{2}$$

$$\tau = t^*/(3\mu R_T^*/\sigma) \ , \tag{3}$$

$$\kappa = r^*/R_T^* = \lambda - h \text{ at the interface.} \tag{4}$$

The superscript * represents dimensional quantities; all unsuperscripted quantities are dimensionless. Time is scaled by a characteristic time obtained from ratio of the length R_T^* over the characteristic velocity σ/μ (19). The important unknown is κ, the radial position of the film interface, or equivalently h, the local film thickness. For the results reported here, we define the pore shape with a cosine function as follows:

$$\lambda(x) = 1 - 0.4\left[1 + \cos(x\pi/10)\right]. \tag{5}$$

To follow the liquid motion in the film, a nonlinear evolution equation is derived for the dynamic film position κ (19,22). We utilize here, an extension of the small-slope evolution equation proposed by Gauglitz (18) and Gauglitz and Radke (23) to include conjoining/disjoining pressure.

According to Figure 3, liquid flows in the film due to a gradient in the mean curvature of the interface along the capillary, and also due to a gradient in the disjoining pressure arising from variations in the film thickness. If the film is thin, and the slope of the pore wall is small, the classical lubrication approximation applies for the fluid motion in the film. In addition, inertial effects are neglected (19), and rectangular coordinates are employed. We modify the equations of motion by including the conjoining/disjoining forces as a body force (12,24). Within the lubrication approximation, Teletzke (14) has justified this approach theoretically. The velocity profile, v_x^* in Figure 3, is determined easily after imposing the boundary conditions of no-slip at the capillary wall and no-stress at the gas-liquid interface, since the gas is presumed inviscid. Integrating the velocity profile over the thickness of the film gives the volumetric flow as a function of the total pressure driving force as follows:

$$\frac{Q^*}{R_T^*\sigma/\mu} = \lambda\left[-\frac{\partial p}{\partial x} + \frac{\partial \Pi}{\partial x}\right] h^3 \ , \tag{6}$$

where Π is the nondimensional conjoining/disjoining pressure. The following simple functional form for the disjoining pressure (25) proves sufficient for this investigation:

$$\Pi^*(h^*) = \begin{cases} -A^*/h^{*3} & \text{conjoining} \\ A^*/h^{*3} & \text{disjoining} \ , \end{cases} \tag{7}$$

where $\pm A^*$ is a dimensional scaling coefficient. A^* is related to the Hamaker constant and typically has values near 10^{-21} J (25). Other simple functional forms are available for Π, such as an inverse quadratic or an exponential dependence on the film thickness (2,14,21,25).

As noted above, Ivanov and co-workers (6) and Mohanty (2) have shown that the conjoining/disjoining pressure determines the contact angle. If the disjoining pressure is always greater than zero, the liquid-phase contact angle is zero. However, for a conjoining/disjoining pressure function which is both positive at small film thicknesses and negative at larger thicknesses, intermediate wettability exists with a contact angle Θ of $0 < \Theta < \pi/2$. The disjoining pressure functions given by Equation 7, correspond to a perfectly wetting liquid (= 0) for $+A^*/h^{*3}$. Following the analysis of Mohanty (2), we find that the conjoining pressure function, $-A^*/h^{*3}$, results in $\Theta \geq \pi/2$, which is a partially to completely nonwetting liquid. Although simple, the inverse cubic dependence for the conjoining/disjoining pressure given by Equation 7 represents both wetting and nonwetting liquids and will suffice here.

The pressure driving force in Equation 6 is obtained from the normal stress balance at the film interface, which is the Young-Laplace equation within our approximation. We impose the following small-slope approximation to the curvature which was proposed by Gauglitz (18) and Gauglitz and Radke (23) as a simple, yet accurate approximation:

$$- p = \frac{1}{\kappa} - \frac{\partial^2 \kappa}{\partial x^2} , \tag{8}$$

where p is the liquid pressure, and the gas pressure is set to zero since it is assumed constant.

To obtain an evolution equation, we combine the derivatives of Equations 7 and 8 with an integral form of the continuity equation and the kinematic condition (14,19), and substitute the flow relation, Equation 6, to yield:

$$\frac{\partial \kappa}{\partial \tau} = \frac{1}{\kappa} \frac{\partial}{\partial x} \left\{ \left[- \frac{1}{\kappa^2} \frac{\partial \kappa}{\partial x} - \frac{\partial^3 \kappa}{\partial x^3} + 3\tilde{A} \frac{1}{(\lambda - \kappa)^4} \left(\frac{d\lambda}{dx} - \frac{\partial \kappa}{\partial x} \right) \right] \lambda (\lambda - \kappa)^3 \right\} , \tag{9}$$

where $\tilde{A} = \pm A^*/(\sigma R_T^{*2})$ is a dimensionless measure of the conjoining/disjoining pressure. This group is sometimes referred to as the Scheludko number (aside from a factor of $1/6\pi$ (26)).

Equation 9 is the desired evolution equation for the dynamic interface position in a constricted capillary. This equation is a highly nonlinear partial differential equation that is first order in time and fourth order in axial position. We must further simplify this relation to obtain analytic information, otherwise, the only amenable solution is numerical. To give useful physical insight, we consider now a linear stability analysis of the proposed evolution equation, Equation 9.

LINEAR STABILITY ANALYSIS

Qualitative understanding of the role of conjoining/disjoining forces emerges by considering the fate of infinitesimal sinusoidal disturbances with a normal modes analysis (12,19,27). To obtain a simple base state, we consider only a straight capillary with a wetting film of uniform thickness which is initially at rest. As a further simplification, we adopt only the thin-film limit of the evolution equation (18,19) obtained by substituting for κ in Equation 9 the following:

$$\kappa = 1 - \varepsilon \bar{h} , \tag{10}$$

where ε is a small constant reflecting the initial film thickness in the straight capillary, and $\bar{h} = h^*/h_o^*$ is a scaled film thickness. Expanding in a Taylor series for $\varepsilon \to 0$ and keeping leading order terms in ε gives:

$$\frac{1}{\varepsilon^3} \frac{\partial \bar{h}}{\partial \tau} = - \frac{\partial}{\partial x} \left\{ \bar{h}^3 \left[\frac{\partial^3 \bar{h}}{\partial x^3} + \frac{\partial \bar{h}}{\partial x} + \frac{3\tilde{A}}{\varepsilon^4} \frac{1}{\bar{h}^4} \frac{\partial \bar{h}}{\partial x} \right] \right\} . \tag{11}$$

This is a direct extension of Hammond's (19) nonlinear evolution equation to include conjoining/disjoining forces. Consider now sinusoidal disturbances of the form,

$$\bar{h} = 1 + \beta \exp{(\alpha \tau + ikx)} , \tag{12}$$

where β is a small constant reflecting the amplitude of the infinitesimal disturbance. Substituting Equation 12 into Equation 11 and linearizing yields the following dispersion relation:

$$\frac{\alpha}{\varepsilon^3} = -k^4 + k^2 \left(1 - 3 \frac{\tilde{A}}{\varepsilon^4} \right) , \tag{13}$$

where α is the growth-rate factor. That is,

instabilities grow only for positive α values. The wavelength of the fastest growing disturbance, Λ_{max}, is given by

$$\Lambda_{max} \equiv 2 \frac{\pi}{k_{max}} = \frac{2^{3/2}\pi}{\left(1 - \frac{3\tilde{A}}{\varepsilon^4}\right)^{1/2}} . \qquad (14)$$

Results: Linear Stability

The linear analysis determines whether sinusoidal disturbances grow or decay. As portrayed in Figure 4, the dispersion relation (Equation 13) gives a one parameter family of curves relating the dimensionless growth rate to the wavelength of the disturbance. At each parameter value, the maximum in the curve corresponds to the fastest growing disturbance which dominates film breakup. The crucial parameter determining the role of conjoining/disjoining pressure is $3\tilde{A}/\varepsilon^4$ which shows the important influence of the initial film thickness ε and the size of the pore, R_T^*, since $\tilde{A} = A^*/(\sigma R_T^{*2})$. For $3\tilde{A}/\varepsilon^4 = 0$, we recover the results of Hammond (19) where $\Lambda_{max} = 2^{2/3}\pi$.

Conjoining/disjoining pressure affects the both the value of the growth-rate parameter and the wavelength of the fastest growing disturbance. For $0 < 3\tilde{A}/\varepsilon^4 < 1$, disjoining pressure resists local thinning of the film, and accordingly, the growth rate is less. The dominate wavelength shifts towards longer waves corresponding to a decrease in the wave number, k. When $3\tilde{A}/\varepsilon^4 > 1$, all disturbances decay, and a stable film uniformly wets the inside wall of the capillary.

Conjoining pressure, corresponding to $3\tilde{A}/\varepsilon^4 < 0$, increases the growth rate, and shifts the fastest growing disturbance towards shorter wavelengths. Analogous to Ruckenstein and Jain (11), Williams and Davis (12), and Kheshgi (13), the growth of a disturbance indicates the initial dynamics of a thin film evolving to either a dry patch (dewetting) or a liquid lens. Dewetting is detrimental to snap-off as we discuss further in the next section.

When disjoining forces are sufficiently large (i.e., $3\tilde{A}/\varepsilon^4 \geq 1$), the liquid is strongly wetting: all disturbances are damped since the growth-rate factor defined by Equation 13 is negative for all wavelengths. When $3\tilde{A}/\varepsilon^4 << -1$, Equation 13 demonstrates that conjoining pressure is the dominate driving force for breakup. That is, the factor of unity in the denominator on the right side of Equation 13, which arises from the circumferential radius of curvature, is small compared to the conjoining pressure term. Accordingly, the growth rate α is established by the competition between the stabilizing transverse curvature (i.e., $-k^4$ in Equation 13) and the driving force which is the conjoining pressure (i.e., $-k^2 3\tilde{A}/\varepsilon^4$ in Equation 13). For this limiting situation, our results simplify to those of Jo (21) and Ruckenstein and Jain (11) since the geometry is now essentially planar. The nonlinear analysis of Williams and Davis (12) applies in this limiting case as well.

The important results of the linear stability analysis are that disjoining pressure makes all disturbances stable if $3\tilde{A}/\varepsilon^4 \geq 1$, and that conjoining forces become important when $3\tilde{A}/\varepsilon^4 \leq -1$. Combining these criteria, we find that conjoining/disjoining pressure must be considered when $3|\tilde{A}|/\varepsilon^4 \geq 0$ (1).

The linear stability analysis also demonstrates that two types of breakup exist: 1) snap-off due to the circumferential curvature, and 2) spontaneous film rupture (i.e., dewetting) due to the conjoining pressure. The linear stability analysis cannot determine which type of breakup will predominate, snap-off or dewetting, since it only indicates whether infinitesimal sinusoidal disturbances grow or decay. To determine which occurs first, snap-off or dewetting, we must perform a nonlinear stability analysis which involves solving Equation 9.

NONLINEAR EVOLUTION

Consider the evolution of a film of uniform thickness coating the inside wall of a smoothly constricted capillary as prescribed by Equation 5. Since conjoining/disjoining forces depend on the thickness of the film, a range of initial thicknesses is investigated. In addition to the conjoining/disjoining forces becoming more important in thinner films, the time required for liquid to flow into the growing collar depends strongly on the thickness of the liquid film. With thicker films, liquid flows more quickly since viscous resistance is relatively less, and collars snap off faster. Conversely, with thinner films, snap-off occurs more slowly since the viscous resistance to flow is larger in the thinner films. This is evident in Equation 6 where the flow rate in the film has a cubic dependence of the film thickness h.

Numerical Solution

The evolution equation given by Equation 9 demands numerical solution. It is parabolic in nature, requiring an initial condition and four boundary conditions. The initial profile is a film of constant uniform thickness $h_0 = h_0^*/R_T^*$ which results in the following initial condition:

$$\kappa(x,0) = \lambda(x) - h_0 \; . \qquad (15)$$

Note that h_0 in the constricted capillary is equivalent to ε in the straight capillary. The boundary conditions of symmetry at the pore neck (x = 0) indicate that the first and third derivatives are zero:

$$\frac{\partial\kappa(x,\tau)}{\partial x} = \frac{\partial^3\kappa(x,\tau)}{\partial x^3} = 0 \quad \text{at } x = 0 \; . \qquad (16)$$

A long distance from the pore, we require that the film have a constant thickness and zero slope:

$$\kappa(x,\tau) = 1 - h_0, \; \frac{\partial\kappa(x,\tau)}{\partial x} = 0 \quad \text{at } x = +16 \; . \qquad (17)$$

At a position of x = +16, the boundary conditions given by Equation 17 have no influence on the collar evolution. Applying the boundary conditions of Equation 17 farther away from the constriction causes no change in the results, indicating that indeed x = +16 is a long distance from the constriction. In addition, we have investigated applying symmetry conditions at x = +16 where the first and third derivatives are zero as in Equation 16. There is no difference in the results, so our analysis applies to periodically constricted capillaries as well.

To obtain a numerical solution, we employ the Galerkin finite element method (28,29) to discretize the axial derivatives in Equation 9, and the Crank-Nicholson method (30) to discretize the time derivatives. The resulting system of nonlinear algebraic equations is solved at each time step with Newton-Raphson iteration. Extensive details of the numerical procedures are given in Gauglitz (18). We report next, the results for a range of initial thicknesses which show the influence of conjoining/disjoining forces.

Results: Constricted Capillary

We focus on the effect of the initial film thickness because of its strong influence on the time to breakup. Time to breakup is defined as the instant when the collar crosses the centerline of the pore. Collar growth accelerates rapidly at the end of the breakup process, and the calculation is stopped before the neck radius of the collar becomes identically zero. Interrupting the calculation when the collar neck radius becomes less than $0.05R_T^*$ gives an error of less than 1% in the calculated time to breakup. Rapid growth of the collar once it reaches a critical thickness has been elucidated in a previous study (18,23) and is observed experimentally (17). Results without thin-film forces will be discussed first, then the effects of conjoining/disjoining forces are included.

The initial condition of a uniform initial film along the pore wall is shown in Figure 5. A sequence of film profiles is shown in Figure 6 for a uniform initial film of thickness h_0 = 0.0124 with $\tilde{A}$ set to zero. To observe the collar shape more carefully, we plot the solution only near the constriction neck. The first profile at τ = 1500 shows the initial film growing thicker at the pore neck as liquid collects in a growing collar. In the second and third profiles, local thinning at the sides is noticeable as the collar becomes large. This arises from the collar drawing liquid in rapidly from the sides as the collar accelerates its growth. Indeed, the collar grows at an ever increasing rate as it becomes larger, with breakup occurring at τ_b = 5496 which is soon after the third profile in Figure 6 at τ = 5300.

The effect of the initial film thickness on the time to breakup, without conjoining/disjoining forces, is summarized in Figure 7 by the line labeled by $\tilde{A} = 0$. Time to breakup increases rapidly for thinner initial films due to the cubic dependence of the flow rate on the film thickness as given in Equation 6.

When $\tilde{A}$ is positive, disjoining pressure resists local film thinning and slows collar growth. If sufficiently strong, disjoining pressure completely inhibits snap off. The dynamic evolution results for $\tilde{A} = 3.33(10^{-6})$ are shown in Figure 8 for an initial thickness of $h_o = 0.01$. Although the collar forms slower under the influence of disjoining forces, it eventually does snap-off. The initial profile at $\tau = 0$ shows the uniform initial thickness. At $\tau = 3.0(10^4)$, liquid fills in at the constriction. The collar continues to grow as liquid flows in from the sides of the capillary, as seen in the profiles. However, in comparison to the profiles without disjoining pressure, more liquid fills in before the short-wavelength collar emerges in the film. This can be understood from the linear stability analysis and Figure 4. When disjoining forces become important, the maximum in each curve of Figure 4 decreases (and shifts to smaller wave numbers). This maximum corresponds to the fastest growing wavelength which now grows slower compared to when $\tilde{A} = 0$. Hence, the collar in Figure 8 emerges slower from the thickening film. Further, the local thinning at the sides of the collar is inhibited by the disjoining pressure which resists thinning. Thus, the small dips apparent at the sides of the collar in Figure 6 (where $\tilde{A} = 0$) do not occur at the sides of the collar in Figure 8.

For comparison, results for different $\tilde{A}$ values are summarized in Figure 7. For positive values of $\tilde{A}$, disjoining pressure increases the time to breakup when the initial film thickness decreases below a critical value. As expected, for larger $\tilde{A}$ values (such as a smaller radius capillary), the film thickness when disjoining pressure becomes important is larger than when $\tilde{A}$ is smaller. We see this in Figure 7 by the curve for $\tilde{A} = 1.33(10^{-5})$ deviating from the $\tilde{A} = 0$ curve at a larger initial film thickness than the curve for $\tilde{A} = 3.33(10^{-6})$. When disjoining forces dominate, they completely inhibit snap-off. The liquid film is stable since a collar will not grow, and accordingly, the breakup time becomes infinite. This feature is also apparent from the linear stability analysis (Equation 13). When $\tilde{A}$ is positive (disjoining pressure) and sufficiently large (i.e., $3\tilde{A}/\varepsilon^4 \geq 1$), the growth rate is always negative, dictating a stable film. This corresponds to the breakup time becoming infinite in Figure 7. We can further compare the nonlinear results in Figure 7 with the linear stability analysis by noting the equivalence of h_o and ε. We find from Figure 7 that disjoining pressure becomes important when $3\tilde{A}/h_o^4 \approx 10$. This is in agreement with the linear stability analysis which gives a criterion of $3\tilde{A}/\varepsilon^4 \approx 1$.

When conjoining forces (i.e., $\tilde{A} < 0$) are incorporated into the film evolution, we find that collars grow more rapidly in the initial stages of evolution. This also agrees with the linear stability analysis which demonstrates that conjoining pressure increases the growth rate of a disturbance. Although the linear stability analysis indicates an increased growth rate, it cannot determine which type of breakup will occur: snap-off or dewetting. The numerical solution of the nonlinear evolution equation for constricted capillaries (Equation 9) demonstrates that below a critical initial film thickness, dewetting breakup occurs before snap-off. When conjoining forces are unimportant (smaller $\tilde{A}$ values or thicker initial films), snap-off is completely unaffected by the conjoining pressure.

Dewetting begins at the sides of the collar where local thinning occurs, as displayed in Figure 6. We further depict this thinning and film collapse in Figure 9. Once local thinning begins, conjoining forces drive fluid away from the thin area causing film rupture. This catastrophic event corresponds to dewetting or the formation of a dry patch. As mentioned in the linear stability analysis section, the conjoining pressure function employed (Equation 7) reflects a contact angle of $\Theta \geq \pi/2$. Once conjoining collapse occurs, numerical integration of the evolution equation is stopped since we believe the collar will no longer grow because liquid cannot flow into the collar from the sides of the capillary. Thus, dewetting inhibits snap-off.

The results including conjoining forces for a series of different negative $\tilde{A}$ values are shown in Figure 7. We indicate with an open circle the critical initial thickness below which dewetting occurs for each value of $\tilde{A}$. As expected, at each negative value of $\tilde{A}$, dewetting occurs for the same initial film thickness at which disjoining forces slow the collar evolution. In accordance with disjoining effects, for larger values of $|\tilde{A}|$, dewetting occurs in thicker initial films. It is also apparent in Figure 7 that the time to breakup (i.e., snap-off) is unaffected by the conjoining pressure until dewetting occurs since the curves for all negative $\tilde{A}$ values lie essentially on the curve labeled $\tilde{A} = 0$.

The two previous sections have considered the affects of conjoining/disjoining forces (wettability) on the dynamic fluid motion in a film that uniformly coats the inside wall of a constricted capillary. The results show that both conjoining and disjoining pressure inhibit snap-off. To complete our study of fluid wettability and snap-off, we must consider the role of conjoining/disjoining pressure as it affects how a liquid film is initially deposited in the capillary according to step 2 in Figure 1. Teletzke (14) has shown that when a bubble (or drop) displaces a viscous fluid from a capillary, the thickness of the film deposited by the bubble depends on the conjoining/disjoining pressure. In the next section, we combine the results of the two preceding sections with the results of Teletzke (14) to develop criteria for when, and in what manner, fluid wettability affects snap-off.

RAMIFICATIONS OF WETTABILITY ON SNAP-OFF

If a bubble or drop displaces a perfectly wetting liquid from a capillary, a film of liquid is deposited on the inside wall of the capillary (31,32). The film thickness depends on the capillary number which is defined as follows:

$$Ca \equiv \frac{\mu U^*}{\sigma} , \qquad (18)$$

where U^* is the velocity of the bubble front. When the bubble (or drop) displaces a fluid of intermediate wettability with $0 < \theta < \pi$ (due to a conjoining pressure contribution), a film of liquid may or may not be deposited. If a liquid film is not deposited, snap-off can never occur.

As indicated in the introduction, the sign and magnitude of the conjoining/disjoining forces determine the fluid wettability (i.e., the contact angle) for static fluid interfaces (2,6), as well as the dynamic contact angle when a gas-liquid interface moves along a solid surface (14). Pure conjoining forces cause a liquid to nonwet partially a solid ($\pi/2 \leq \theta \leq \pi$). Conversely, pure disjoining forces indicate complete wetting ($\theta = 0$). With conjoining forces, Teletzke (14) notes that for bubbles moving through capillaries, there exists a critical capillary number below which no film is deposited, thus preempting the possibility of snap-off. However, for any fluid regardless of its wettability, a liquid film will always be deposited for sufficiently high capillary numbers and snap-off may occur.

Teletzke has calculated the deposited film thickness for the conjoining/disjoining pressure functions given by Equation 7. We can obtain an expression for the capillary number (i.e., the film thickness) when conjoining/disjoining pressure becomes important by ascertaining the magnitude of the terms involving the conjoining/disjoining forces in Teletzke's (14) film-profile equation. Teletzke's equation is given in Appendix A. There, we show that conjoining/disjoining pressure becomes important during film deposition if,

$$3|\tilde{A}|Ca^{-2} \geq O(1). \qquad (19)$$

We can compare this to the role of wettability in the film evolution. The dispersion relation of the linear stability analysis, Equation 13, and the nonlinear evolution calculations in Figure 7, demonstrate that conjoining/disjoining forces become important when $3|\tilde{A}|/\varepsilon^4 \geq O(1)$. The initial film thickness, ε, is related to the capillary number by $\varepsilon = O(Ca^{2/3})$ (32,33). Combining these relations, we discover that conjoining/disjoining forces become important during film evolution when,

$$3|\tilde{A}|Ca^{-8/3} \geq O(1). \qquad (20)$$

Notice that Equation 20 has a different capillary-number dependence than Equation 19. Comparison shows that for $Ca \ll 1$, conjoining/disjoining forces affect film evolution at a higher capillary number where film deposition remains unaffected.

With Equations 19 and 20, capillary-number criteria can now be developed to predict when snap-off occurs for fluids with any wetting characteristics. We summarize these criteria in Table 1. For conjoining forces, we first consider at what values of the capillary number liquid films are deposited. Equation 19 shows that for $Ca < |3\tilde{A}|^{1/2}$ no film is deposited. We signify this by a "NO" in the "Film Deposition" row. In the second row called "Snap-off", we indicate by a "NO" that snap-off can never occur without a liquid film. When $Ca > |3\tilde{A}|^{1/2}$, a liquid film is deposited by a bubble (indicated by a "YES" in Table 1), and a bubble can snap off. For these higher values of Ca, we must consider the evolution of the film and the role of wettability as given by Equation 20.

For $Ca < |3\tilde{A}|^{3/8}$, a film of liquid is deposited, but it dewets during evolution and does not snap off. Only when $Ca > |3\tilde{A}|^{3/8}$ can a bubble snap off as shown by the "Yes". Snap-off can occur for these values of the capillary number because a finite film of liquid is deposited and during film evolution dewetting does not occur.

For a disjoining force or complete wetting, Table 1 presents Teletzke's (14) result that a film is deposited at all capillary numbers. However, in the evolution of these films, disjoining forces inhibit snap-off. When $Ca < |3\tilde{A}|^{3/8}$, snap-off does not occur since disturbances will not grow. Conversely, when $Ca > |3\tilde{A}|^{3/8}$, snap-off can occur as indicated by the "YES".

Since the specific conjoining/disjoining pressure function given by Equation 7 considers both perfectly wetting and partially to completely nonwetting liquids, Table 1 considers all possible combinations of wettability affecting film deposition and evolution. For both disjoining and conjoining forces, snap-off only occurs for $Ca > |3\tilde{A}|^{3/8}$, which implies that a sufficiently thick initial film must be deposited for snap-off to occur.

Experimental results for the time to breakup of gas bubbles moving through constricted capillaries approximately 0.1 cm in diameter are reported elsewhere (18). For parameter values of $A^* = 10^{-21}$ J, σ = 30 mN/m, and $R_T^* = 5(10^{-4})$ m, the criterion developed above gives that snap-off can occur if $Ca > 2(10^{-5})$. Thus, wettability inhibits snap-off only at very low capillary numbers. All experiments were performed at Ca values greater than $2(10^{-5})$. The data agree quantitatively with the theory for $\tilde{A} = 0$ indicating that wettability is not important in capillaries of this size. In addition, the shape of the growing collar in Figure 6 accurately represents the experimentally observed shape (18).

For capillary numbers greater than about 10^{-3}, the discovered snap-off criterion dictates that snap-off can occur in pores larger than 1 µm. Therefore, wettability inhibits snap-off in cylindrical capillaries only in very small pores. Thus, the calculations presented in this work, and the experimental results in Gauglitz (18) demonstrate that wettability does not affect snap-off in typical oil-bearing porous media ($R_T^* \sim 10^{-5}$ m) for capillary numbers greater than 10^{-3}.

Interestingly, other visual observations by Gauglitz (18) of snap-off in constricted square capillaries indicate that dewetting does occur and inhibits snap-off in 0.1 cm I.D. square capillaries for capillary numbers in the range of 10^{-5} to 10^{-3}. Thus, square capillaries do not follow the criterion developed in the previous section. Further study is required to obtain a complete understanding of wettability effects in constricted square capillaries.

SUMMARY AND CONCLUSIONS

The role of wettability, as determined by the conjoining/disjoining pressure, on the evolution of liquid films is investigated in both straight and constricted capillaries. Solution of a new nonlinear evolution equation demonstrates that both conjoining and disjoining forces inhibit snap-off. We propose the criterion that gas-bubble snap-off can occur in constricted cylindrical capillaries, unaffected by liquid wettability, if $Ca > |3\tilde{A}|^{3/8}$.

ACKNOWLEDGEMENTS

This research was supported by the U.S. Department of Energy under contract DE-AC03-76SF00098 to the Lawrence Berkeley Laboratory of the University of California. PAG gratefully acknowledges financial support from the Shell Development Company.

NOTATION

A^* conjoining/disjoining pressure scaling coefficient, ~ 10^{-21} J

$\tilde{A}$ $= A^*/\sigma R_T^{*2}$, dimensionless conjoining/disjoining pressure parameter

Ca capillary number (defined in Equation 18)

h^* film thickness, m

h_o^* initial film thickness, m

h $= h^*/R_T^*$, dimensionless film thickness

h_o $= h_o^*/R_T^*$, dimensionless initial film thickness

$\bar{h}$ scaled film thickness (defind in Equation 10)

$\tilde{h}$ scaled film thickness (defined in Appendix A)

i $\sqrt{-1}$

k wave number of sinusoidal disturbance

k_{max} wave number of the fastest growing disturbance

p^* liquid pressure, Pa

p $= p^*/(\sigma/R_T^*)$, dimensionless liquid pressure

r^* radial position, m

r $= r^*/R_T^*$, dimensionless radial position

R_T^* radius of unconstricted capillary, m

t^* time, s

U^* velocity of the bubble front, m/s

v_x^* velocity in fluid film, m/s

x^* axial position, m

x $= x^*/R_T^*$, dimensionless axial position

$\tilde{x}$ scaled axial position (defined in Appendix A)

y^* position from tube wall (see Figure 3), m

Greek Letters

α growth-rate factor (defined in Equation 12)

β amplitude of infinitesimal disturbance (defined in Equation 12)

ε $= h_o^*/R_T^*$, dimensionless initial film thickness for straight capillary (defined in Equation 10)

Θ contact angle through the liquid phase

κ dimensionless radial position of the film interface

λ dimensionless radial position of capillary wall

Λ dimensionless wavelength of disturbance

Λ_{max} dimensionless wavelength of the fastest growing disturbance

μ liquid film viscosity, mPa·s

π 3.14159

Π^* conjoining/disjoining pressure (defined in Equation 7), Pa

Π $= \Pi^*/(\sigma/R_T^*)$, dimensionless conjoining/disjoining pressure

σ surface tension, mN/m

τ $= t^*/(3\mu R_T^*/\sigma)$, dimensionless time

τ_b dimensionless breakup time

LITERATURE CITED

1. Scheidegger, A. E., The Physics of Flow Through Porous Media, 3rd ed., University of Toronto Press, Toronto (1974).

2. Mohanty, K. K., "Fluids in Porous Media: Two-phase Distribution and Flow," Ph.D. thesis, University of Minnesota (1981).

3. Dilgren, R. E. and A. R. Deemer, "The Laboratory Development and Field Testing of Steam/Noncondensible Gas Foams for Mobility Control in Heavy Oil Recovery," SPE 10774, Presented at the Soc. Pet. Eng. California Regional Meeting, San Francisco (March 24-26, 1982).

4. Ploeg, J. F. and J . H. Duerksen, "Two Successful Stream/Foam Field Tests, Sections 15A and 26C Midway-Sunset Field," SPE 13609, Presented at the Soc. Pet. Eng. California Regional Meeting, Bakersfield (March 27-29, 1985).

5. Mast, R. F., " Microscopic Behavior of Foam in Porous Media," SPE 3997, presented at the Soc. Pet. Eng. 47th Annual Fall Meeting, San Antonio, Texas (1972).

6. Ivanov, I. B., B. V. Toshev, and B. P. Radoev, "On the Thermodynamics of Contact Angles, Line Tension and Wetting Phenomena," in Wetting, Spreading and Adhesion, Padday, J. F. (Ed.), Academic Press, London (1978).

7. Martynov, G. A., I. B. Ivanov, and B. V. Toshev, "On The Mechanical Equilibrium of a Free Liquid Film with the Meniscus," Kolloidnyi Zhurnal, 38, 474-479 (1976).

8. Deryagin, B. V., V. M. Starov, and N. V. Churaev, "Profile of the Transition Zone Between a Wetting Film and the Meniscus of a Bulk Liquid," Kolloidnyi Zhurnal, 38, 875-879 (1976).

9. Deryagin, B. V. and M. Kussakov, "Anomalous Properties of Thin Polymolecular Films. V.," ACTA Physicochimica U.R.S.S., 10 (1), 25-44 (1939).

10. Clunie, J. S., J. F. Goodman, and B. T. Ingram, "Thin Liquid Films," in Surface and Coloid Science, Matijevic (Ed.), 3, pp. 167-240, Wiley Interscience (1971).

11. Ruckenstein, E. and R. K. Jain, "Spontaneous Rupture of Thin Liquid Films," J. Chem. Soc. Faraday Trans. II, 70, 132-147 (1974).

12. Williams, M. B. and S. H. Davis. "Nonlinear Theory of Film Rupture," J. Coll. Int. Sci., 90 (2), 220-228 (1982).

13. Kheshgi, H. S., "The Motion of Viscous Liquid Films," Ph.D. thesis, University of Minnesota (1984).

14. Teletzke, G. F., "Thin Liquid Films: Molecular Theory and Hydrodynamic Implications," Ph.D. thesis, University of Minnesota (1983).

15. Everett, D. H. and J. M. Haynes, "Model Studies of Capillary Condensation I. Cylindrical Pore Model with Zero Contact Angle," J. Coll. Int. Sci. 38 (1), 125-137 (1972).

16. Goren, S. L., "The Instability of an Annular Thread of Fluid," J. Fluid Mech., 12, 309-319 (1962).

17. Goldsmith, H. L. and S. G. Mason, "The Flow of Suspensions through Tubes II. Single Large Bubbles," J. Coll. Sci., 18, 237-261 (1963).

18. Gauglitz, P. A., "Instability of Liquid Films in Constricted Capillaries: A Pore Level Description of Foam Generation in Porous Media," Ph.D. thesis, University of California, Berkeley (1986).

19. Hammond, P. S., "Nonlinear Adjustment of a Thin Annular Film of Viscous Fluid Surrounding a Thread of Another within a Circular Cylindrical Pipe," J. Fluid Mech., 137, 363-384 (1983).

20. Gauglitz, P. A. and C. J. Radke, "An Extended Evolution Equation for Liquid Film Breakup in Cylindrical Capillaries," in preparation, 1986.

21. Jo, E. J., "Displacement of Bubbles and Drops in Fine Capillaries," Ph.D. thesis, University of Houston (1984).

22. Atherton, R. W. and G. M. Homsy, "On the Derivation of Evolution Equations for Interfacial Waves," Chem. Eng. Commun., 2 (2) 57-77 (1976).

23. Gauglitz, P. A. and C. J. Radke, "The Dynamics of Liquid Film Breakup in Constricted Capillaries," in preparation, 1986.

24. Chen, J. D., P. S. Hahn, and J. C. Slattery, "Coalescence Time for a Small Drop or Bubble at a Fluid-Fluid Interface," AICHE J., 30 (4) 622-630 (1984).

25. Kruyt, H . R., Colloid Science, Elsevier, New York (1952).

26. Gumerman, R. J . and G. M. Homsy, "The Stability of Radially Bounded Thin Films," Chem. Eng. Commun., 2 (1) 27-36 (1975).

27. Drazin, P. G. and W. H. Reid, Hydrodynamic Stability, Cambridge University Press, New York (1981).

28. Becker, E. B., G. F. Carey, and J. T. Oden, Finite Elements: an Introduction, Volume I, Prentice-Hall, New Jersey (1981).

29. Finlayson, B. A., Nonlinear Analysis in Chemical Engineering, McGraw-Hill, New York (1980).

30. Lapidus, L., Digital Computation for Chemical Engineers, McGraw-Hill, New York (1962).

31. Taylor, G. I., "Deposition of a Viscous Fluid on the Wall of a Tube," J. Fluid Mech., 10, 161-165 (1961).

32. Bretherton, F. P., "The Motion of Long Bubbles in Tubes," J. Fluid Mech., 10, 166-188 (1961).

33. Park, C. W. and G. M. Homsy, "Two-phase Displacement in Hele Shaw Cells: Theory," J. Fluid Mech., 139, 291-308 (1984).

APPENDIX A: CONJOINING/DISJOINING PRESSURE IN FILM DEPOSITION

The film-profile equation predicting the deposited film thickness as a bubble displaces a perfectly wetting fluid was first enunciated by Bretherton (32). Bretherton, and more recently Teletzke (14), considered the role of conjoining/disjoining pressure on the deposited film thickness. The film-profile equation is an approximation valid for Ca $\rightarrow$ 0. Park and Homsy (33) have shown formally, neglecting conjoining/disjoining forces, that Bretherton's solution is the leading order term in an asymptotic expansion in $Ca^{1/3}$. Park and Homsy clarified the characteristic length scales in this problem. To obtain the scaled film-profile equation, we follow Park and Homsy by substituting into Bretherton's profile equation for the dimensional position x^* the scaled length $\tilde{x} = (x^*/R_T^*)/Ca^{1/3}$, and for the dimensional film thickness h^* the scaled thickness $\tilde{h} = (h^*/R_T^*)/Ca^{2/3}$ to yield:

$$\frac{d^3\tilde{h}}{d\tilde{x}^3} - 3\tilde{A}\,\frac{1}{Ca^2}\,\frac{1}{\tilde{h}^4}\,\frac{d\tilde{h}}{d\tilde{x}} = 3\left[\frac{\tilde{h} - \tilde{h}_o}{\tilde{h}^3}\right], \qquad (A.1)$$

where $\tilde{h}_o$ is the scaled deposited film thickness. Conjoining/disjoining forces are important when the coefficient multiplying the second term of Equation A.1 is of order one or greater:

$$3|\tilde{A}|Ca^{-2} \geq O(1). \qquad (19)$$

This criterion predicts well the onset of conjoining/disjoining forces in Teletzke's (14) calculation of the deposited film thickness using the conjoining/disjoining pressure functions given by Equation 7.

Table 1

Role of Wettability on Thin-Film Breakup

($\tilde{A} = A^*/\sigma R_T^{*2}$)

Partial Nonwetting, $\pi \geq \theta \geq \pi/2$ (Conjoining Pressure $\tilde{A} < 0$)			
	$Ca < \|3\tilde{A}\|^{1/2}$	$Ca > \|3\tilde{A}\|^{1/2}$	
Film Deposition (Laydown)	NO	YES	
		$Ca < \|3\tilde{A}\|^{3/8}$	$Ca > \|3\tilde{A}\|^{3/8}$
Snap-off (Evolution)	NO	NO	YES
Perfect Wetting, $\theta = 0$ (Disjoining Pressure $\tilde{A} > 0$)			
	$Ca < (3\tilde{A})^{1/2}$	$Ca > (3\tilde{A})^{1/2}$	
Film Deposition (Laydown)	YES	YES	
		$Ca < (3\tilde{A})^{3/8}$	$Ca > (3\tilde{A})^{3/8}$
Snap-off (Evolution)	NO	NO	YES

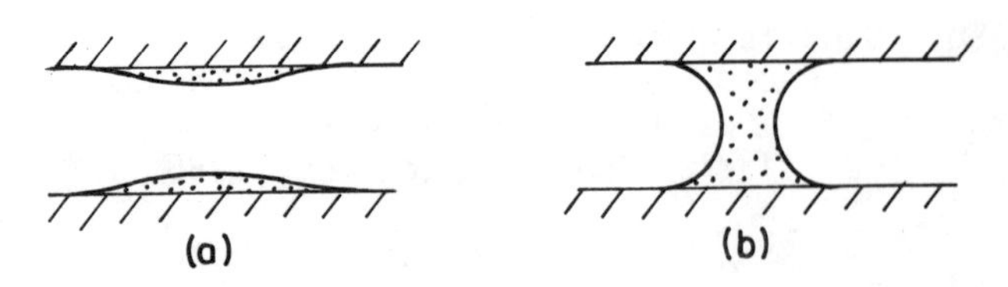

Figure 2. Schematic of a stable collar (a) and a liquid lens (b).

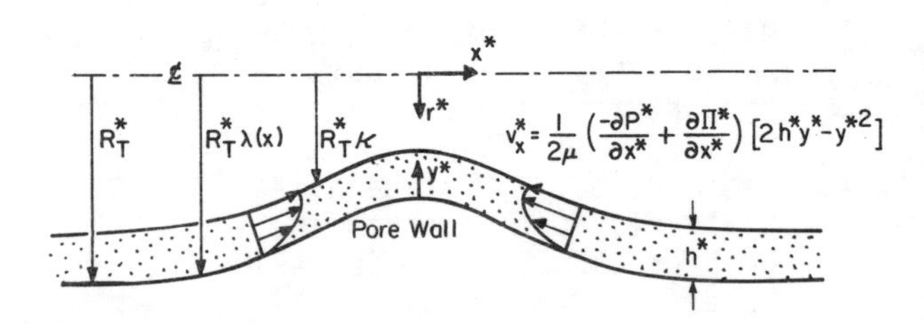

Figure 3. Liquid film wetting the inside wall of a constricted capillary.

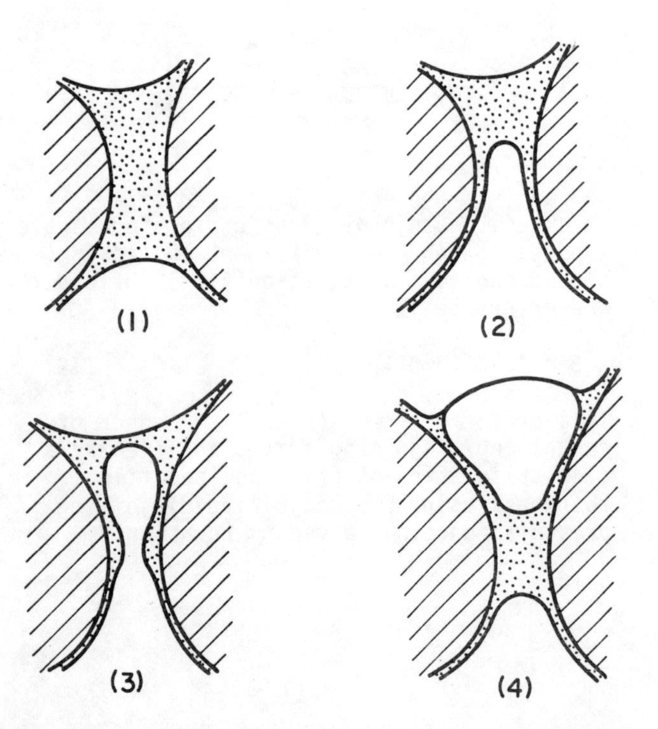

Figure 1. Pore level view of bubble generation by snap-off.

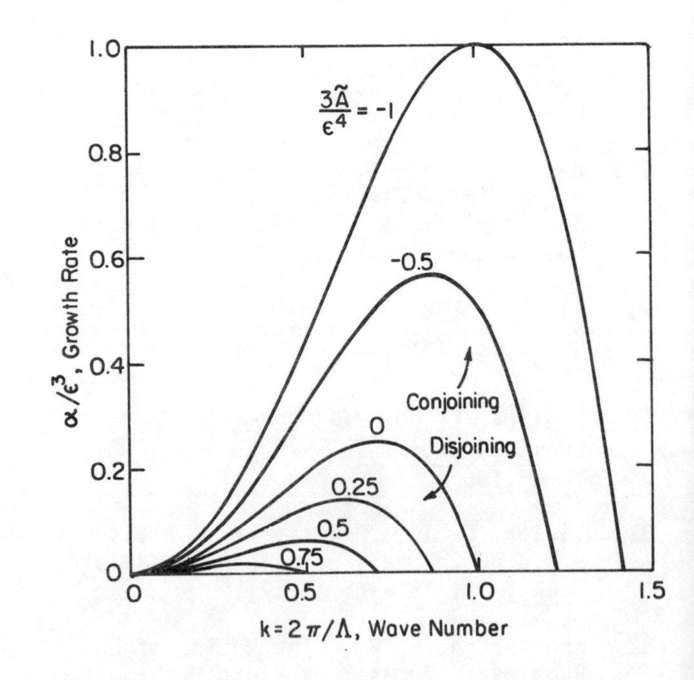

Figure 4. Thin-film growth rate for an initial sinusoidal disturbance in a straight cylindrical capillary.

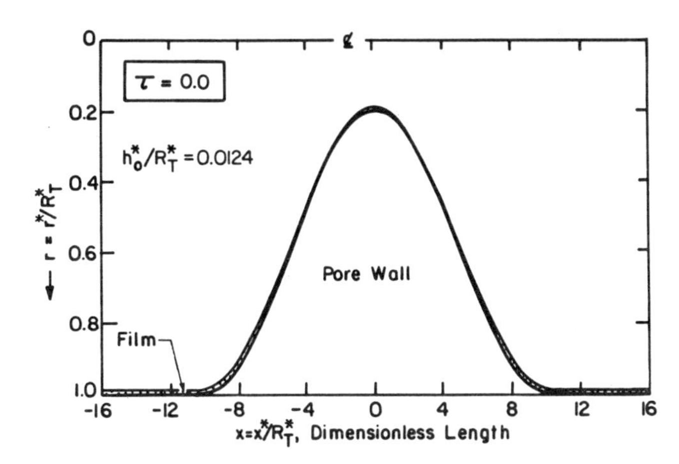

Figure 5. Initial condition of a uniform film thickness.

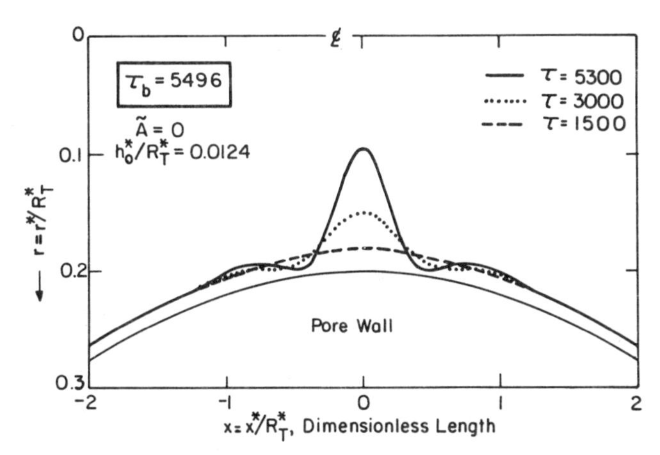

Figure 6. Profiles for liquid-film evolution in a constricted capillary.

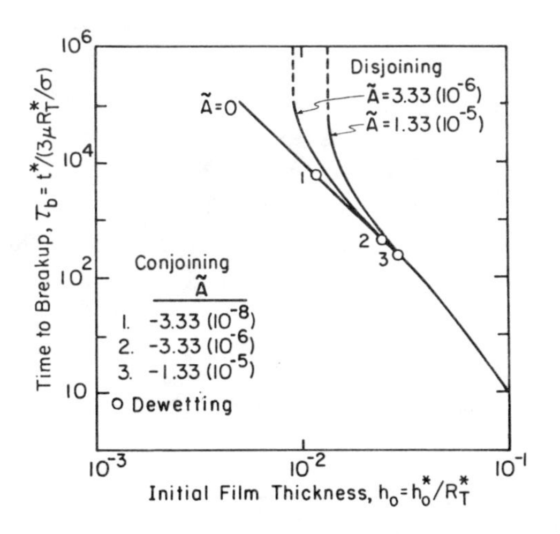

Figure 7. Effect of the initial film thickness and $\tilde{A}$ on the time to breakup; $\tilde{A} > 0$ gives a perfectly wetting liquid, while $\tilde{A} < 0$ indicates a nonwetting liquid. Open circles indicate the onset of dry patch formation.

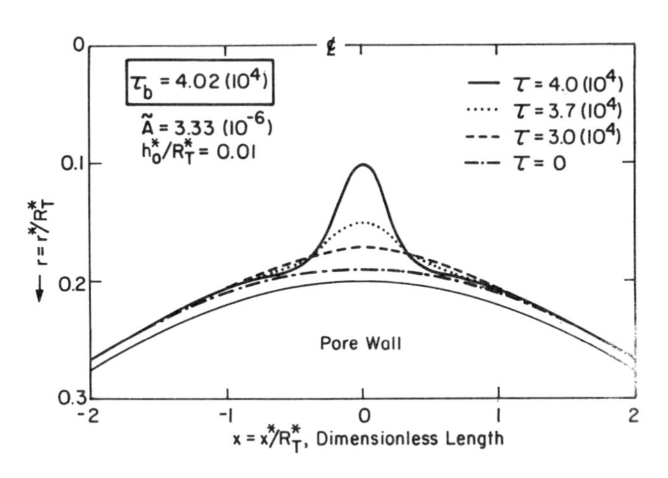

Figure 8. Profiles for liquid-film evolution with disjoining pressure in a constricted capillary.

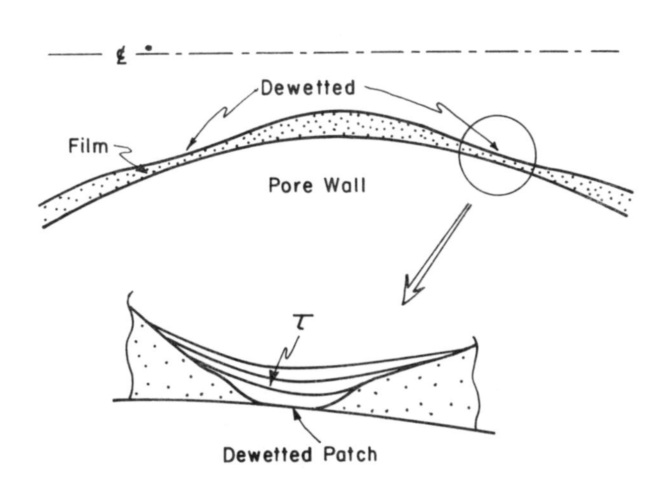

Figure 9. Dewetting film due to conjoining pressure.

INTERFACIAL PROPERTIES OF THIN LIQUID FILMS IN RELATION TO BOUNDARY AND HYDRODYNAMIC LUBRICATION IN THE EYE

S. Kalachandra and D. O. Shah ■ The Center for Surface Science and Engineering and
Departments of Chemical Engineering and Anesthesiology
University of Florida, Gainesville, FL 32611

Various surface phenomena such as spreading of oil at the air/tear interface, kinetics of thinning of a film, rate of evaporation of water, wettability of cornea and lubrication of cornea and eyelid are pertinent to normal blinking process. Using slit-lamp fluorophotometer (SLFP), we investigated the flow dynamics of the tear film between the blinks in the presence or absence of tear substitutes. It was observed that instillation of tear substitutes containing surface active polymers causes upward movement of tear fluid immediately following the opening of the eye due to Marangoni effect. The upward spreading of the film of surface active polymer and meibomian lipids causes an aqueous boundary layer to move upward. This mechanism was confirmed by the experiments conducted in vitro. It was shown that the vertical spreading of an adsorbed polymer film from a petridish on a vertical glass slide can drag a boundary layer of substantial thickness. The SLFP measurements showed that a 10 to 30 μm thick aqueous layer is dragged by a moving polymer film. The thickness of the dragged boundary layer was 13 μm for a moving monolayer of polyvinyl alcohol.

A specially designed friction testing instrument was employed to evaluate the coefficient of friction of several commercial tear substitutes on PMMA/PMMA surfaces. It was shown that these ophthalmic solutions exhibit a broad range of coefficient of friction. It is proposed that the degree of comfort in the eye is related to the coefficient of friction between the cornea and the eyelid. The smaller values of coefficient of friction should correspond to a greater degree of comfort in the eye.

Surface phenomena such as the spreading of meibomian oil at the air/tear interface, the kinetics of thinning of the tear film, the rate of evaporation of water from the tear film, wettability of the cornea with the tear fluid and the lubrication of cornea and eyelid are pertinent to the normal blinking process. Most of these processes occur every time we blink. Figure 1 schematically illustrates various surface phenomena occurring in the precorneal tear film. It is important to recognize that phenomena occurring at the cornea/tear and air/tear interfaces influence the behavior of the tear film. Surface chemistry of tear film has been reviewed in a monograph and in several pioneering papers (1) to (6).

Millions of people world wide suffer with a condition called ''dry eye syndrome''. In this situation, the thickness of the tear film decreases considerably between blinks, resulting in the formation of dry spots on cornea. Figure 2(a) shows the view of the tear film through an ophthalmologist's slit lamp. Before the examination of the eye, a paper strip dipped into fluorescein, a water soluble dye, is brought in contact with the eye surface. The presence of this dye makes the tear film appear greenish through the slit lamp photometer. However, a person suffering with the dry eye syndrome will exhibit within a few seconds, dry spots which will appear black through the slit lamp. The water recedes from these dry spots which appear black due to the lack of water and the water soluble dye. (Figures 2(a) and 2(b)). If one blinks under these conditions, there is considerable friction between the eyelid and the dry spots on the cornea. This may lead to the damage of the corneal surface. Several tear substitute solutions containing polymers are available to stabilize a thick layer of water on cornea. However, relatively very little scientific data are available for the effectiveness of these tear substitutes.

The kinetics of drainage of the tear film is very important for the normal blinking process. It is desirable to slow down the drainage of the tear film in order to retain a greater thickness of the tear film on cornea during blinking. The slit lamp fluorophotometer (SLFP) can be used to investigate the kinetics of drainage in the tear film (Figure 3). It consists of a light source with filter emitting only the excitation wavelength of 490 μm, and another filter allowing only the emission wavelength of 520 μm relevant to fluorescein. This light is then transmitted by a fiber optics bundle to a photomultiplier which amplifies the signal. The signal is then passed to a photometer and printed on a chart recorder. With the concentration of fluorescein used in this study (5 x 10^{-4} gm/ml), we were able to obtain a linear plot of the fluorescence intensity versus the thickness of the aqueous layer containing fluorescein (7). Thus, it was established that the fluorescein intensity is a linear function of the thickness of the aqueous layer for a given concentration of the fluorescein.

The subject of lubrication is concerned with the art of reducing frictional resistance occurring between two sliding solid surfaces. Any substance inserted between the two sliding surfaces for the purpose of reducing the friction is called a lubricant. There are two types of operating conditions in lubrication, namely, boundary lubrication and hydrodynamic lubrication (8). In the former case, the lubricant film can not support the load and contact occurs between the two surfaces. In this case, the coefficient of friction decreases with the viscosity and speed and increases with load (8). In the case of hydrodynamic lubrication, the two sliding surfaces are separated by a thin film of lubricant. The frictional drag is entirely due to the rheological properties of the lubricant film. The coefficient of friction in this region increases with viscosity and speed and decreases with load. The friction between the cornea and the eyelid has been assumed to be of the boundary type (9). The surface in this case, are likely to contact each other. The purpose of this paper is to report the surface chemical phenomena occurring in the precorneal tear film and lubrication properties of some commercial ophthalmic solutions.

MATERIALS AND METHODS

All commercial ophthalmic solutions were obtained from the respective manufacturers. Most pure polymers were obtained form Polysciences, Inc. All dry polymers were dissolved in a 0.9% NaCl solution made with - - distilled water from an all-glass still. Fluorescein was added to the polymer solutions to a final concentration of 5 x 10^{-4} gm/ml. The viscosity of oil solutions was measured with a Brookfield viscometer. A Rame-Hart contact angle goniometer was used to measure the advancing contact angle of solutions at equilibrium on PMMA surface and surface tension of solutions was measured by Wilhemy plate method connected to a strain transducer.

The coefficient of friction of commercial tear substitutes was measured by a specially designed friction tester. The usefulness and capability of this instrument has already been successfully demonstrated (10) by its sensitivity to low loading forces (1 to 10 g). Friction measurements for these solutions at equilibrium were made at approximately 3 to 5 minutes after the solution was delivered to the test surface. Vertical forces were applied to a Nylon or a PMMA spherical stylus, 0.25 inch in diameter, which runs in a track of lubricating solution applied to a rotating plate of PMMA. The plate rotates at constant speed with respect to the stylus. During the course of the experiment, the polymer solution is replenished every 1.5 minutes to maintain the same concentration of the solution. The values of the coefficient of friction were the average of a least six measurements of the same experiment. The standard deviation in the value of the coefficient of friction is not more than + 0.008. This paper presents a brief review of our recent studies on the surface properties and lubrication characteristics of ophthalmic solutions (11, 12).

RESULTS AND DISCUSSION

Tear Film Thickness Between Blinks

When a continuous recording of fluorescence intensity was made after instillation of 1 μl of 1.25% fluorescein in the eye, in some individuals, the Precorneal Tear Film (PTF) consistently decreased in fluorescence between blinks (Figure 4). However, under the same conditions the PTF in other individuals either maintained its thickness or decreased its thickness depending upon the day of observation. However, most of the polymer solutions tested as well as all commercial solutions allowed the PTF to increase in thickness between blinks. After a complete blink, as the upper lid opens, a film appears to move upward over fluorescent tear film. The duration of this phenomenon after instillation has not been studied extensively. However, when specifically looked for, it has been observed for as long as 20 minutes after instillation of a polymer. The magnitude of change in thickness is difficult to measure since it begins instantaneously after a blink. We observed that most of the surface active polymers tend to increase the thickness of the precorneal tear film between the blinks (Figure 5). In order to explain the increase in thickness between the blinks we have proposed an explanation which is schematically shown in Figure 6. The instillation of an eye drop containing a surface active polymer is expected to produce an adsorbed film at the air/tear interface. As one blinks, the adsorbed polymer film will be compressed. Upon opening the eyelid, the compressed polymer film is expected to spread upward due to surface tension gradient (Marangoni effect). The vertical spreading of the polymer film will also carry with it a boundary layer of the aqueous fluid. This will cause an upward flow of the aqueous layer along with the spreading of the polymer monolayer. In order to prove that indeed this mechanism is responsible for the upward dragging of the tear fluid, we performed a model experiment using a wet glass slide as shown in Figure 7.

The Water Dragging Capacity of Moving Polymer Films

Since all polymers used in our studies exhibit surface activity, it is expected that they will adsorb air/tear interface in the same manner as the meibomian oil. Such an absorbed film will be compressed as a result of a blink. If different polymers drag different amounts of water as boundary layer, as they spread, then perhaps, this could also explain the differences in the preocular fluorescence intensity observed with polymer solutions of the same viscosity (7). The following experiments were performed to simulate the spreading of a polymer film at the air/tear interface. A clean glass slide was dipped in a beaker filled with saline (containing fluorescein at a concentration of 5×10^{-4} gm/ml). The polymer solution in question was prepared at the same fluorescein concentration and placed in a petridish with a surface area of 64 cm^2. When the wet glass slide was touched to the surface of the polymer solution a film was observed to move upward on the vertical slide. If simultaneously, the change in fluorescence intensity of the slide was measured, the thickness of the boundary layer can be calculated using the calibration curve. Table 1 shows the thickness of the aqueous boundary layers dragged by the spreading of polymer films. It is obvious that spreading polymer films can drag the aqueous sublayer of thickness 9 to 30 microns. In an attempt to see if the thickness of the boundary layer can be accounted for by a moving monolayer, the following experiment was carried out. A monomolecular film of PVA was spread on the surface of a large trough by touching a glass rod, previously dipped in PVA solution, to the surface of the trough containing saline and fluorescein (5×10^{-5}

gm/ml). Polymer molecules were allowed to spread on the saline surface until no more molecules could be spread on the surface of the trough. A clean wet glass slide was touched to the surface of the trough allowing the monolayer to spread on the glass surface. The thickness of the water layer dragged by a monomolecular film of PVA was 13 microns (Table 1), similar to that produced by the bulk solution of PVA. This suggest that the thickness of the water layer dragged on a glass surface by PVA can be accounted for by a monomolecular film of the same polymer, indicating that the polymer molecules at the surface are mainly responsible for the dragged water layer and that the PVA molecules in the bulk solution do not significantly contribute to the thickness of the dragged water layer. However, it should be recognized that factors such as entanglement and aggregation of polymer molecules could considerably influence this process with other polymer types and under different spreading conditions.

Lubrication Properties of Commercial Tear Substitutes

Table 2 shows the data of coefficient of friction, viscosity, surface tension and contact angle of the ten ophthalmic solutions, namely, (1) Adsorbo Tear; (2) Tears Plus; (3) Lens Mate; (4) Adapettes: (5) Tear Gard; (6) Tears Naturale; (7) Muro Tears; (8) Neo Tears; (9) Hypo Tears and (10) Liquifilm Tears. In order to explore the possibility of a direct relationship between the coefficient of friction and viscosity, surface tension or contact angle, we plotted the coefficient of friction data these data parameters as shown in Figures 8, 9, and 10. It is obvious that there is no direct correlation between the coefficient of friction and viscosity, surface tension or contact angle of these solutions on PMMA surface. In contrast to our observations with regard to the coefficient of friction of the ophthalmic solutions and their non-dependence on the corresponding values of viscosity, surface tension and contact angle, Benedetto and co-workers demonstrated that the dynamic film thickness of the commercial ophthalmic solutions was strictly viscosity-dependent and did not depend upon the contact angle or the surface upon which the solution was deposited. However, it was found that the thickness of the aqueous layer increase as the surface tension of the solution decreased. It was concluded that the dynamic film thickness increases with the viscosity of the solutions having the same surface tension (11).

Table 2 shows clearly that the coefficient of friction of commercial tears substitutes can vary over a broad range from 0.11 to 0.213. The next phase of our research is to correlate the coefficient of friction with the degree of comfort in the eye. We propose that the degree of comfort provided by various tear substitutes in the eye is related to the coefficient of friction between eyelid and cornea in the presence of these tear substitutes. Of course, it is recognized that the ingredients of the tear substitutes are drained away continuously from the tear fluid, and therefore, various properties of the tear film including lubricity will change with time. However, we believe that the initial feeling of comfort should correlate with the coefficient of friction of these solutions as measured on the model low energy surfaces.

In summary, the results presented in this paper indicate that various surface phenomena and lubrication properties of polymer solutions are important for the normal blinking process.

Acknowledgements

The authors wish to express their sincere thanks to Dr. A. Benedetto, Dr G. Brauninger, Mr. M. O'Rear, Mr. K. Shelat and Mrs. K. Kumar for their valuable assistance in the research presented here. The financial support of Alcon Laboratories for this research is gratefully acknowledged.

LITERATURE CITED

1. Holly, F. J. and M. A. Lemp, The Preocular Tear Film and Dry Eye Syndromes, Little, Brown and Company, Boston, MA, 1973.

2. Brauninger, G. E., D. O. Shah, and H. E. Kaufman, Am. J. Ophthalmol., 73, 132-134, (1972).

3. Lemp, M. A. and F. J. Holly, Ann. Ophthalmol., 4, 15-20, (1972).

4. Holly, F. J., and M. A. Lemp, J. Contact Lens Soc. Am., 5, 12-19, (1971).

5. Holly, F. J., and M. A. Lemp, Am. Ophthalmol., 11, 239-251, (1971).

6. Lemp, M. A., C. H. Dolman, and F. J. Holly, Am. Ophthalmol., 2, 258-261, (1970).

7. Benedetto, D. A., D. O. Shah, and H. E. Kaufman, Investigative Ophthalmology, 14, 887-902, (1975).

8. Appeldoorn, J. K., and G. Barnett, Proc. Toilet Goods Association, No. 40, 28-35, 1963.

9. Ehlers, N., Acta Ophthalmol., 81 (Suppl.), 111-113, (1965).

10. Kalachandra, D. A., and D. O. Shah, Rev. Sci. Instrum., 55, 998-1000, (1984).

11. Benedetto, D. A., D. O. Shah, and H. E. Kaufman, Ann. of Ophthalmology, 10, 1-4, (1978).

12. Kalachandra, S. and D. O. Shah, Annals of Ophthalmology, 17, 708-13, (1985).

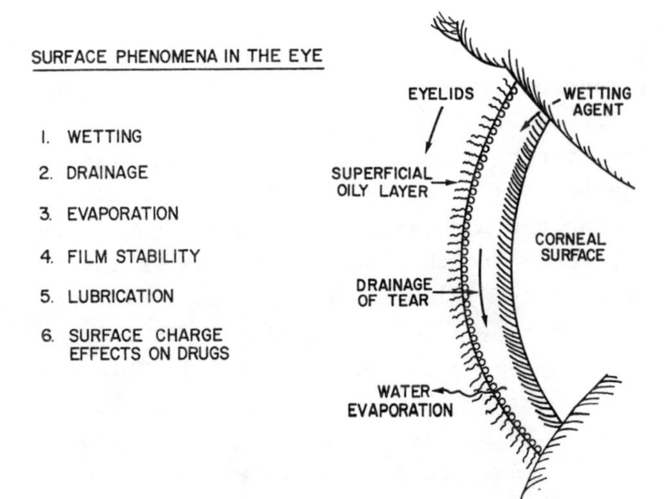

Figure 1. A schematic presentation of various surface phenomena occurring in the eye.

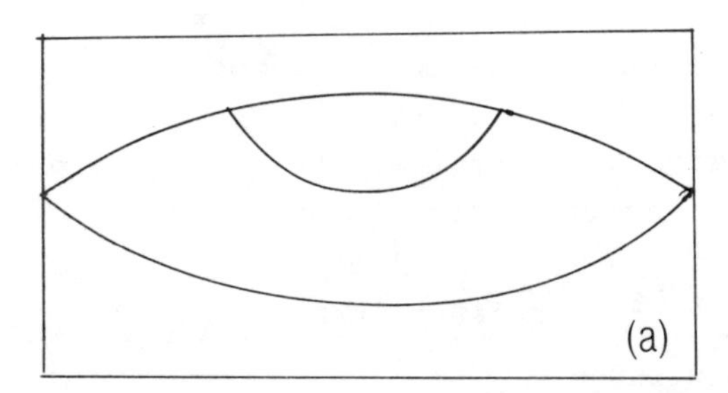

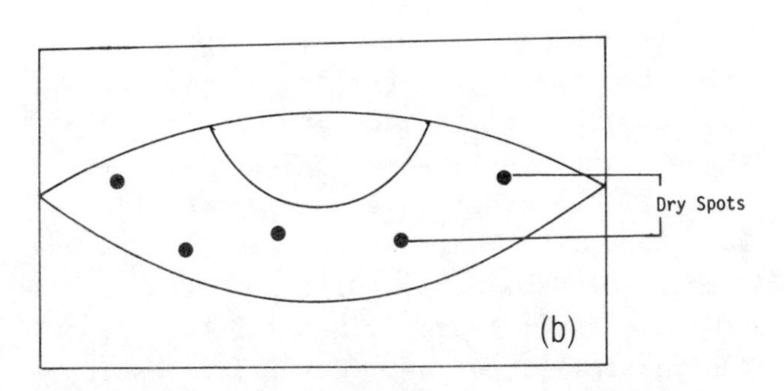

Figure 2. Formation of dry spots on eye (a) and (b).

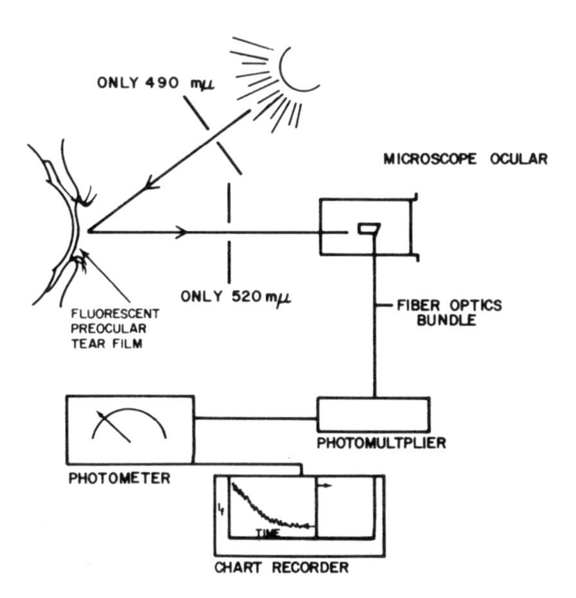

Figure 3. Instrumentation (Slit Lamp Fluorophotometer) to study flow dynamics and tear film thickness in human eye using fluorescence method.

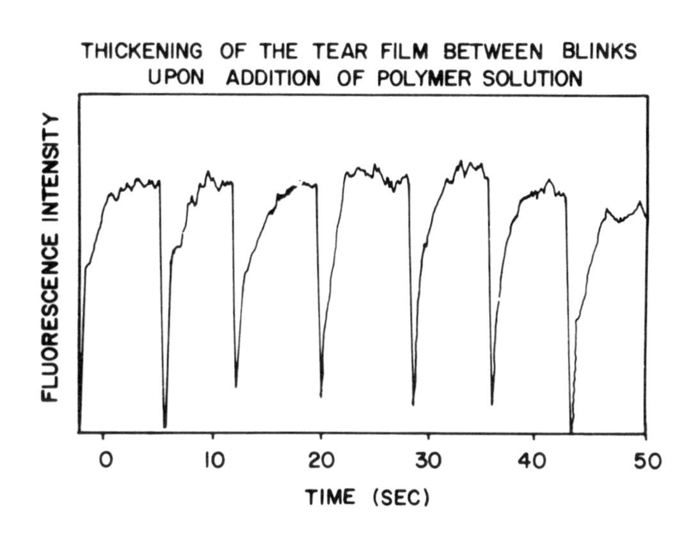

Figure 5. Thickening of the tear film between blinks upon addition of a surface-active polymer.

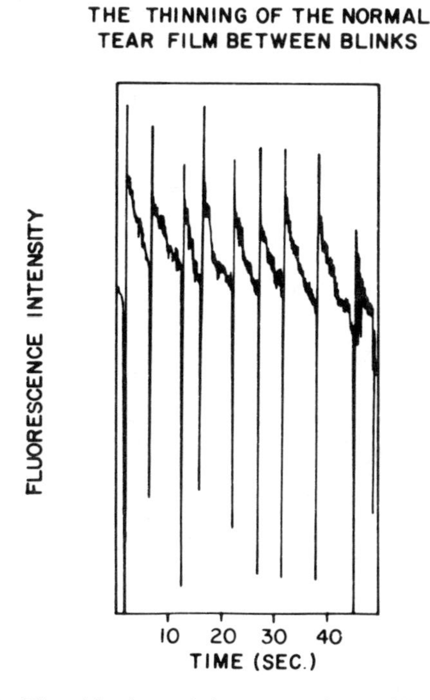

Figure 4. The thinning of the normal tear film between blinks.

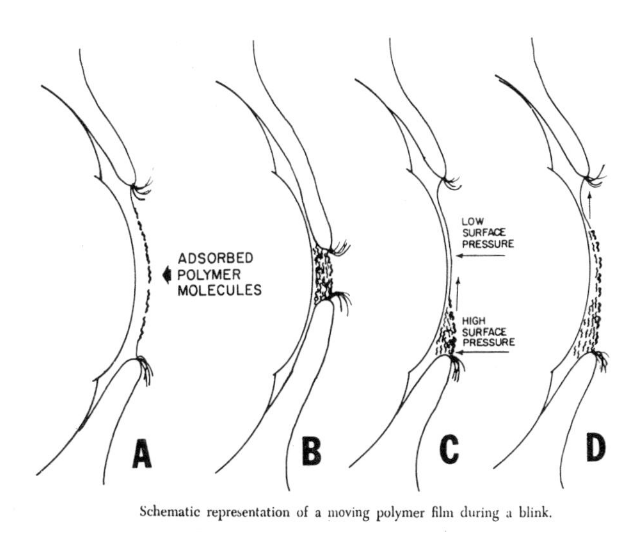

Figure 6. Schematic representation of a moving polymer film during a blink.

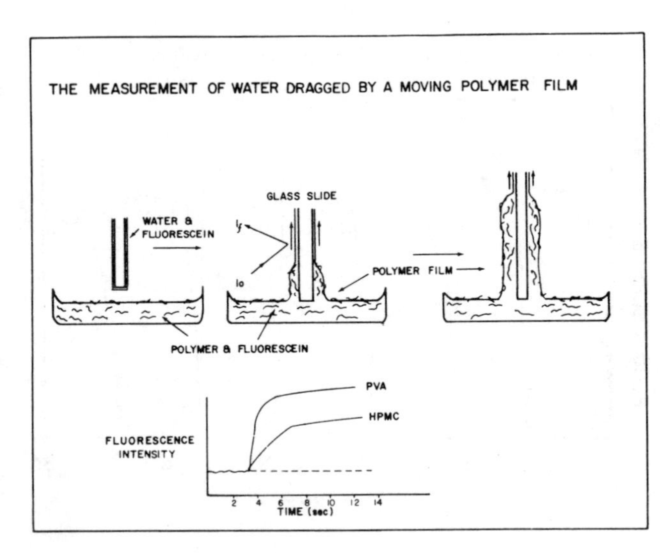

Figure 7. The slide technique used to measure the thickness of water layer dragged with a moving polymer film.

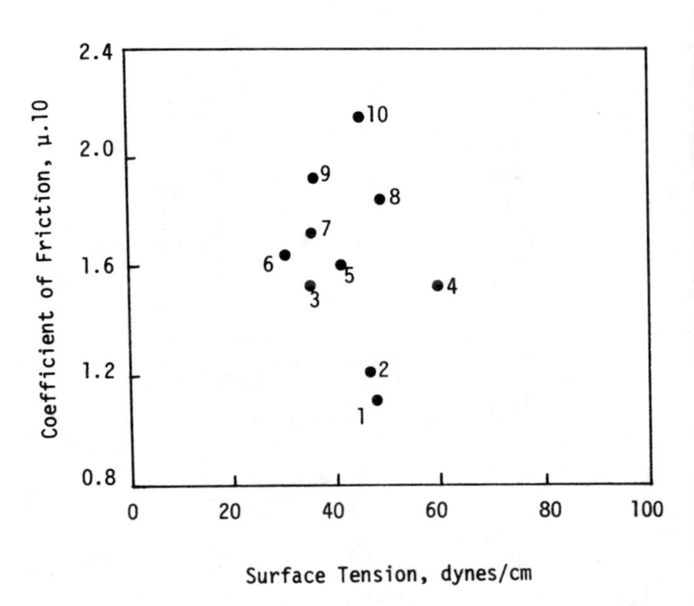

Figure 9. Coefficient of friction as a function of surface tension for various ophthalmic solutions.

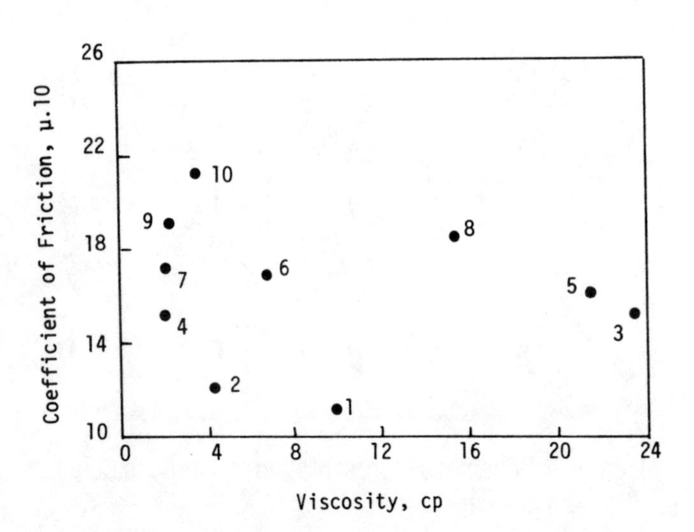

Figure 8. Coefficient of friction as a function of viscosity for various ophthalmic solutions on PMMA/PMMA surfaces.

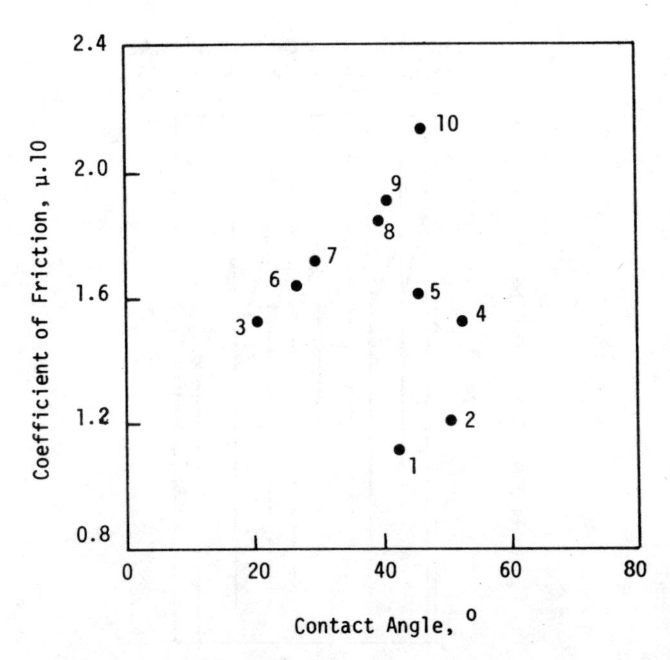

Figure 10. Coefficient of friction as a function of contact angle for various ophthalmic solutions.

Table 1. Thickness of aqueous boundary layers dragged by moving polymer films.

Polymer	Viscosity (centipoise)	Thickness (pm) of water layer dragged by polymers
Barnes-Hind wetting solution	58	11
Adapt	70	16
Presert	18	--
Lacril	28	16
Visculose	130	30
PVA	120	19
PVA	20	19
HPMC	120	11
HPMC	20	9
Monomolecular film of PVA		13

*Surface area of trough = 64 cm^2

Table 2. Surface chemical and lubrication properties of various ophthalmic solutions.

Commercial Tear Substitute	Coefficient of Friction*	Viscosity (cp)	Surface Tension (dynes/cm)	Contact Angle of the solution on Clean PMMA *θ
Adsorbotear (Alcon)	0.112	10.03	47.9	42^{0}
Tears Plus (Allergan)	0.121	4.52	46.4	50^{0}
Lens Mate (Alcon)	0.152	23.61	34.8	20^{0}
Adapettes (Alcon)	0.153	2.13	60.0	52^{0}
Tear Gard (Bio Products Ophthalmics)	0.161	21.53	41.7	45^{0}
Tears Naturale (Alcon)	0.164	7.03	30.8	26^{0}
Muro Tears (Muro Pharmaceuticals)	0.172	2.34	35.5	29^{0}
Neo-Tears (Barnes-Hind)	0.184	15.51	48.8	39^{0}
Hypotears (Coopervision)	0.191	2.43	38.6	40^{0}
Liquifilm Tears (Allergan)	0.213	3.91	45.0	46^{0}

*Coefficient of friction was measured between PMMA platae and PMMA ball of 0.25 inch in radius at a stylus velocity of 400 mm/s (80.4 revs/min) with a vertical load of 5 g.

PMMA=polymethyl methacrylate.

HYDRODYNAMIC PROPERTIES OF THIN FILMS STUDIED BY LASER LIGHT SCATTERING

J. G. H. Joosten ■ Van 't Hoff Laboratory and Department of Biophysics, University of Utrecht, The Netherlands

The dynamics of thermally excited surface waves on thin liquid films is studied by means of (laser) light scattering. The free soap films were stabilized by the cationic surfactant dodecyltrimethylammoniumbromide. Surface scattering from thin films is described as a combination of squeezing mode and bending mode scattering. Because of the large difference in characteristic time scales the dynamics of squeezing and bending modes can be studied separately. Experimental results on the bending mode were reported in (*11*) where it was shown that the surface tension of the film equals the bulk value of this quantity. In this paper the relaxation of the squeezing mode is studied as a function of film thickness and wave number. We find that the shear viscosity of the liquid comprising the film does not differ from the bulk value. Furthermore it is concluded that a layer of 0.7 nm at each interface of the film is hydrodynamically immobilized.

The dominant forces between colloids in ionic solutions are the electrical double layer repulsion and the London-van der Waals attraction (1). These forces are the constituants of the so-called DLVO theory (Derjaguin, Landau, Verwey and Overbeek). In 1954 Derjaguin and Titievskaya (2) suggested that the above mentioned forces are also operative in thin liquid films like soap films stabilized by an ionic surface active substance.

In this paper we report about soap films suspended in air having an area of some square cm. The two liquid-air interfaces are covered with a monolayer of ionic surfactant molecules. The principal forces that govern the thickness of a soap film are the above mentioned colloidal interaction forces: A repulsive force caused by the overlap of diffuse electrical double layers, built up near the interfaces by the ions in the solution. This repulsion is counteracted by the London-van der Waals attraction forces between all molecules in the film. For films in a gravity field also the hydrostatic pressure tends to make the film thinner. The equilibrium thickness, that results from the interplay of the interaction forces and the hydrostatic pressure head, can be varied to some extent by varying the ionic strength of the solution from which the films are made. Typical equilibrium thicknesses range from 5 nm (a Newton black film or a membrane structure) up to about 100 nm so they constitute systems with one dimension in the colloidal domain. Therefore, soap films are, geometrically simple, model systems for the investigation of colloidal interaction forces.

Since the time soap films were accepted as model systems for the study of these forces a considerable amount of papers appeared in literature. Most of these papers deal with studies on equilibrium thicknesses, contact angles, velocity of thinning, elasticity, and rupture of films. For reviews see (3,4,5,6).

Light has proven to be a powerful means for studying properties of thin liquid films since it can be applied in a non-destructive and hardly disturbitive way for the study of the, generally, very sensitive films. The determination of the thickness of films e.g. is almost always carried out by measuring the optical reflection coefficient. Also the so-called film contact angle (7,8) can be measured by analyzing the Fresnel pattern of visible light that emerges from the discontinuity in the transition region between the plane parallel part of the film and the Plateau border.

Interaction forces in liquid films can be studied by means of light scattering. Light is scattered by the presence of thermally excited ripples or corrugations on the two interfaces of a film. Surface ripples having a wavelength comparable with the wavelength of light give a measurable scattering (diffuse reflection) if they are illuminated by visible light. For the surface of a bulk liquid the scattered intensity is inversely proportional to the surface tension. In a thin liquid film, like a soap film, the surface ripples are not only influenced by the surface tension but also by the long-range interaction forces which reach across the film (9,10). This makes it possible to study these forces from the intensity of the scattered light.

Surface fluctuations are not static but evolute in time and this time evolution of the ripples can be studied by analyzing the fluctuations in the scattered light intensity around its mean value. The dynamics of the surface ripples on a film are not only determined by the surface tension and the interaction forces but also by the hydrodynamic flow pattern caused by the motion of the interfaces.

Based on these observations, the light scattering experiments from soap films can be divided into two types of measurements. First, the mean scattered intensity can be measured (the integral or static light scattering method). From this technique one obtains information on the interfacial tension and the interaction forces. Second, by measuring the power spectrum (or equivalently the time autocorrelation function) of the scattered light, dynamic features of the ripples can be studied. From the latter type of experiments one obtains not only information on the interaction forces and the surface tension but also on the shear viscosity of the liquid inside the film. The films studied in this paper are stabilized by the cationic surfactant dodecyltrimethylammoniumbromide (DTAB). Results concerning interaction forces in these films were reported in (11). Here emphasis will be put on the results with respect to dynamic aspects of the surface ripples.

THEORY

For a proper understanding of the equations that follow we will elucidate the basic soap film-model briefly. This model comprises a triple layer structure consisting of two layers of paraffinic chains with thickness h_1 and an aqueous layer with thickness h_2 in between. The sum $h_2 + 2h_1$ denotes the total thickness of the film (also called material thickness). Another important quantity is the so-called equivalent water thickness h_w which is an apparent thickness. The equivalent water thickness is obtained by measuring the optical reflection coefficient of the film thereby assuming that the film is a homogeneous layer of thickness h_w and with a refractive index equal to that of the bulk solution.

In the formulas for the interaction forces one needs h_2, and this thickness can be calculated from h_w by a procedure to be described in the Section Thickness Measurements.

Next to the definitions given above for the various thicknesses we introduce a hydrodynamic thickness h_s and an effective mass thickness h. These two quantities are related to the dynamics of the surface fluctuations; h_s is introduced in Equation (9) whereas h is needed in the analysis of the bending mode results (see Equation (11)).

Light Scattering

The surfaces of a liquid film are not perfectly flat, but are slightly corrugated due to the thermal motion of the molecules. The instantaneous shape of a liquid-air interface can be described by a two-dimensional spatial Fourier analysis. In a light scattering experiment, various Fourier components of the surface motion can be studied since the scattering of a Fourier component is restricted to a well defined direction. The relation between the wavevector $\underline{q}$ of a Fourier component and the wave vectors of incident and scattered light is found from the conservation of momentum in the film plane. In our experimental geometry the scattered light is detected in the plane of incidence and one has (9)

$$q = |\underline{q}| = \frac{2\pi}{\lambda_0} |\sin\theta_0 - \sin\theta|, \qquad (1)$$

where λ_0 is the free space wavelength of the incident light, θ_0 the angle of incidence, and θ the angle of observation which is defined with respect to the normal on the film.

The scattering from a liquid film can be described by a linear superposition of two eigen modes namely the so-called squeezing mode and bending mode (9). The time dependent amplitude of the Fourier component q of the scattered light, $E_q(t)$, is for small amplitudes of the surface waves given by

$$E_q(t) = \{Ls_{z,q}(t)+Kb_{z,q}(t)\}\exp[-i\omega_0 t], \qquad (2)$$

where $s_{z,q}$ and $b_{z,q}$ are the amplitudes of the squeezing and bending modes, respectively. The quantities K and L in Equation (2) are optical functions depending on λ_0, θ_0, θ, the film thickness, the index of refraction n, the state of polarisation and field strength of the incident beam. Furthermore ω_0 is the frequency of the incident (laser) beam. For the complex expressions of K and L we refer to (9,10). In a so-called time-averaged light scattering experiment the mean scattered intensity is measured. From Equation (2) it follows that the scattered intensity depends on $<|s_{z,q}(t)^2>$ and $<|b_{z,q}(t)^2>$. The latter quantities can be found from statistical mechanical considerations (9). It turns out that for the squeezing mode

$$<|s_{z,q}(t)|^2> = \frac{2k_BT}{2V'' + \gamma q^2} \qquad (3)$$

where

$$V'' = \frac{\partial^2 V_W}{\partial h_2^2} + \frac{\partial^2 V_{DL}}{\partial h_2^2}$$

with V the free energy of film formation. For films stabilized by an ionic surfactant $V = V_{DL} + V_W$ where V_{DL} is the contribution arising from the overlapping of the electrical double layers built up near the interfaces by the monolayers of ionized surfactant molecules. The double layer repulsion tends to separate the surfaces of the film whereas the London-van der Waals contribution V_W acts in the opposite direction. For the mean squared amplitude of the bending mode fluctuations one finds

$$<|b_{z,q}(t)|^2> = \frac{2k_BT}{\gamma q^2} \qquad (4)$$

From Equations (2,3,4) it follows that by measuring the scattered intensity, one can extract information on the surface tension and interaction forces. It was shown in (11) that the experimental data for DTAB films can be interpreted by the classical DLVO theory using the Lifshitz theory for the London-van der Waals part of V". We found an exponential thickness dependence of the electrical double layer contribution to V". The experimental decay length of the latter contribution agrees very well with the theoretical value of the Debye length. For the surface potential a value of 95 mV was found.

Since in this paper mainly dynamic aspects of the surface fluctuations will be discussed we will confine ourselves to dynamic light scattering and leave time-averaged light scattering out of consideration.

Because of relaxation of surface elevations, the scattered light has a broadened spectral distribution compared with the incident light. The broadening is too small to be analyzed by conventional interferometry so the technique of optical mixing (12) must be used. This technique is in literature often referred to as intensity fluctuation spectroscopy (IFS) or photon correlation spectroscopy (PCS).

Below we summarize the theory that deals with the dynamics of surface waves on thin liquid films. We will only quote the relationships relevant to the measurements and their interpretation. For a complete formulation of the theory we refer to (13,14).

The relavant quantities that are obtained by PCS experiments from thin films are the correlation functions $\phi_s(q,t)$ for the squeezing mode and $\phi_b(q,t)$ for the bending mode or the corresponding power spectra that are related to the correlation functions by a Fourier transform with respect to time (Wiener-Khintchine theorem). The correlation functions are for stationary processes given by

$$\phi_s(q,\tau) = <s^*_{z,q}(0)\ s_{z,q}(\tau)>, \qquad (5)$$

and

$$\phi_b(q,\tau) = <b^*_{z,q}(0)\ b_{z,q}(\tau)> \qquad (6)$$

where the asterisk means the complex conjugate quantity, and the brackets denote a thermal equilibrium ensemble average.

The correlation functions can be calculated by using macroscopic electrohydrodynamic equations of motion since in light scattering only Fourier components of wavelength Λ (= $2\pi/q$) large compared to molecular dimensions can be probed (Λ varies roughly between 1 and 100 μm).

An important quantity for the calculation of the correlation functions is the elasticity coefficient ε_s of the soap monolayers that is defined as

$$\varepsilon_s = -\Gamma_o \frac{\partial\gamma}{\partial\Gamma} \quad , \qquad (7)$$

where Γ is the actual surfactant concentration in the interfaces and Γ_o its mean value.

The parameter ε_s controls the balance between tangential forces resulting from inhomogeneities in the distribution of the surfactant in the interfaces and the hydrodynamic drag at the interfaces. In the limit $\varepsilon_s \to \infty$ one has the so-called no slip situation. In many practical circumstances the no slip condition is reached for $|\varepsilon_s| \geqslant \gamma$ (13,14). For $|\varepsilon_s|$ in the range 30 mN/m to 1000 mN/m it was found that in the so-called long-wavelength limit i.e. $qh \ll 1$ $\phi_s(q,\tau)$ can be approximated by

$$\phi_s(q,\tau) = \phi_s(q,o)\exp[-\beta_s|\tau|] \ , \qquad (8)$$

where

$$\beta_s = \frac{q^2 h_s^3}{24\eta}[2V'' + \gamma q^2] \qquad (9)$$

The correlation function ϕ_b for the bending mode can be approximated by

$$\phi_b(q,\tau) = \phi_b(q,o)\exp[-\beta_b|\tau|]\cos\omega_b\tau \qquad (10)$$

The damping constant β_b and the oscillating frequency ω_b for the bending mode cannot be expressed as explicit functions of the system parameters and they must be evaluated numerically. From the analysis of the dynamics of the bending mode it follows namely that the loading effect, of the medium bounding the film, on the dynamics cannot be neglected even if this is a low density gas phase like air (see (14)). The physical reason for this is that the penetration depth of the velocity field, induced by the surface waves, into the adjacent phases is of the order of magnitude of the wavelength of the ripples. Since the wavelength is in the μm range it follows that a considerable layer of the adjacent phase participates in the motion. We can elucidate this by facing an approximate expression for ω_b (14)

$$\omega_b \approx q\left[\frac{2\gamma}{\rho h + 2\rho'/q}\right]^{\frac{1}{2}} \qquad (11)$$

where ρ' is the mass density of the film bounding phase, and ρh is the mass density per unit film area. From Equation (11) one observes that the effective inertial mass per unit area equals ρh, plus $2\rho'/q$ which is the mass per unit area of the adjacent phase that moves along with the film. Since h is in the nm range and q^{-1} in the μm range the contribution $2\rho'/q$ cannot be neglected even if ρ and ρ' differ a factor of 1000. As far as the damping constant β_b is concerned we have found (14) that this quantity depends mainly on the properties of the gas phase namely the density ρ' and the viscosity η'. The correlation functions for the amplitude fluctuations, Equations (8) and (10), can be related to the autocorrelation function of the scattered field. This function $g^{(1)}(q,\tau)$ is defined as

$$g^{(1)}(q,\tau) = \langle E_q^*(0)E_q(\tau)\rangle \qquad (12)$$

Combining Equation (12) with Equations (2), (8) and (10), and given the independence of the modes, one derives for the function $g^{(1)}$

$$g^{(1)}(q,\tau) = \langle I_s\rangle\exp[-\beta_s|\tau|] + \langle I_b\rangle\exp[-\beta_b|\tau|]\cos\omega_b \ , \qquad (13)$$

where $\langle I_s\rangle$ and $\langle I_b\rangle$ are the mean scattered intensities of the squeezing and bending mode respectively. The function $g^{(1)}(q,\tau)$ is measured in the so-called heterodyne scheme of PCS. When homodyne detection scheme is used the autocorrelation of the scattered intensity $g^{(2)}(q,\tau)$ is measured. Assuming the scattering process obeys Gaussian statistics one has the following relation (12)

$$g^{(2)}(q,\tau) = (\langle I_s\rangle + \langle I_b\rangle)^2 + \sigma|g^{(1)}(q,\tau)|^2 \qquad (14)$$

where $\sigma(0 < \sigma \leqslant 1)$ depends on the optical detection configuration. The mean scattered intensities of the squeezing and bending modes are of the same order of magnitude in the reflection region. This implies that also in the homodyne scheme the frequency ω_b can be measured. That is, the squeezing mode acts as a local oscillator for the bending mode in this case.

EXPERIMENTS

Experimental Setup

The apparatus used for the light scattering experiments is described in detail in (11) and (15). The main part of this setup is a double walled stainless steel vessel, vapor tight, with a temperature control system resulting in a long-term temperature stability inside the apparatus of 27.000 ± 0.002 ^oC. The film is formed in a (grounded) glass frame, with a hole of about 2 cm in diameter. During all experiments the film in the hole is in contact with the bulk solution.

The film is illuminated by an Ar-ion laser (Spectra Physics, model 165). The laser operates at 514.5 nm, at a power of about 200 mW, and the light is polarized perpendicular to the plane of incidence. The angle of incidence was chosen as $\theta_o = 60^o$. We are able to measure the light scattering from the film at angles of about ± 3° from the specularly reflected beam and about ± 5° from the transmitted beam. The specularly reflected light intensity is recorded continuously during the light scattering experiments, and is used afterwards to calculate the film thickness at each stage. The scattered light is detected by a photomultiplier (EMI 9658 RA), the output pulses of which were either fed into a correlator (Malvern, K7025, 64 channels) or into a spectrum analyzer (Hewlet Packard, 3585 A, 20 Hz-40 MHz). The output of the correlator is proportional to the autocorrelation function of the scattered field $g^{(1)}$ or scattered intensity $g^{(2)}$ depending on the detection scheme (12). The output of the spectrum analyzer is proportional to the power-spectrum i.e. the Fourier transform of either $g^{(1)}$ or $g^{(2)}$. Although the power spectrum of the fluctuations contains the same information as the correlation function, each has its own merits, making it necessary to use either of both techniques depending on the spectral range.

The measured light scattering data together with the reflection data are collected in a microcomputer (ITT, 2020 Apple), that also controls the correlator and the spectrum analyzer. The output of the microcomputer is then transmitted to a minicomputer (DGC, NOVA 3) where the final data analysis is carried out.

To conclude this section, we remark that the whole light scattering apparatus, including the laser, is placed on a vibration isolation system Newport Research Corporation (NRC) to eliminate exterior vibrations.

Materials and Methods

Films were drawn from solutions containing 10 mmole/dm^3 of the cationic soap dodecyltrimethylammonium bromide (DTAB). This soap concentration is well below the critical micellar concentration which is ≈ 16 mmole/dm^3. All solutions were prepared from twice-distilled water and contained 1.02 mole/dm^3 glycerol ("Baker Analyzed Reagent"-grade, J.T. Baker). The solutions were filtered twice through a Millipore 220 nm pore size filter in order to remove dust particles. The surface tension, γ_o, of the solutions was measured with the Wilhelmy plate method and turned out to be 45.4 mN/m at 25.0 ^oC. The shear viscosity η_o was measured by an Ubelohde viscosimeter resulting in η_o = 1.09 mPa.s at 27.0 ^oC. The refractive index n of the solution appeared to be 1.344 at λ_o = 514.5 nm.

Thickness Measurements

The thickness of the films was calculated from the intensity of the reflected light using Airy's formula (17) for a homogeneous film with a refractive index equal to that of the surfactant solution. The intensity of the reflected beam at the last maximum is used to find the thickness at other stages in the drainage process or at equilibrium. The thickness calcualted according to this procedure is the so-called equivalent water thickness h_w.

The data concerning interaction forces should be interpreted using the thickness of the aqueous core h_2. The relationship between h_w and h_2 is given by (18)

$$h_2 = h_w - \frac{2(n_{hc}^2+2)}{n^2-1} R_{hc}\Gamma_o \qquad (15)$$

where n and n_{hc} are the refractive index of the soap solution and the surfactant layers, respectively. R_{hc} is the molar refraction of the hydrocarbon chains of the surfactant, and

Γ_o is the surface concentration of the surfactant. Since for the DTAB monolayers no data for Γ_o were available we have used the value $\Gamma_o = 3.26\times10^{-6}$ mole/m^2 which Donners et al. (19) found for hexadecyltrimethylammonium bromide monolayers. For the calculation of R_{hc} we used the procedure of Rijnbout (18) resulting in 8.74 10^{-5} m^3/mole. Taking for $n_{hc} = 1.422$ (the value for bulk dodecane) one finds from Equation (15)
$h_2 = h_w - h^{opt}_{cor}$ with $h^{opt}_{cor} = 2.8$ nm

EXPERIMENTAL RESULTS

Once a film is made, the drainage process (16) starts and continues until the equilibrium thickness is reached. During this process the reflected intensity is recorded, to find the film thickness. Below a thickness of about 250 nm the drainage process goes slow enough to perform light scattering experiments. This type of measurement is called drainage-experiment and is carried out at constant angles θ_o and θ. Once the film has reached a constant thickness, the scattering angle θ can be varied. This type of measurement is denoted as equilibrium-experiment.

A calculation of the relaxation times β_s^{-1}, β_b^{-1} and the frequency ω_b (see Equations (9) and (10)), using typical values for the different parameters, shows that β_s^{-1} and β_b^{-1} are in the millisecond region, whereas ω_b^{-1} is in the microsecond region and furthermore one has $\beta_s^{-1} \gg \beta_b^{-1}$. This large separation on the time scale makes it possible to study the dynamics of the squeezing and bending modes separately as follows from Equations (13) and (14). In view of this observation we will treat the experimental results for the squeezing and bending modes separately.

Squeezing Mode

The squeezing mode experiments were carried out using the homodyne scheme with scattering angles in the range 105° to 116°, and 125° to 155°. (All angles are defined with respect to the normal on the film at the reflection side). The small region around the transmitted beam (which is at 120°) is excluded because an unwanted mixing between this beam or spurious stray light and the scattered light should be avoided. The observed (64-point) correlation functions were fitted to a single exponential plus background by minimizing the function

$$\sum_{i=1}^{64} \left[A\exp(-\frac{\tau_i}{\tau_R}) + B - \phi_{\Delta I}(\tau_i)\right]^2 \qquad (16)$$

where $\tau_i = i\tau$, τ is the sample interval time, and $\phi_{\Delta I}(\tau_i)$ is the measured correlation function at τ_i. As follows from Equation (14) τ_R must be

$$(2\tau_R)^{-1} = \beta_s \qquad (17)$$

At first, the thinning process of the film proceeds very fast, but below a thickness of about 250 nm it becomes slow enough to permit measuring correlation functions under "constant thickness" conditions.

The microcomputer which controls the correlator can be programmed to perform a runf of several measurements. The use of the microcomputer makes it possible to minimize the time between the measurement of two correlation functions because the setting of the correlator and the read-out of the data is done automatically. Therefore also a reasonable number of functions can be obtained in the thickness range where the thinning goes relatively fast. It takes about 25 seconds to obtain a function that ensures a reliable value for τ_R. The thickness corresponding to this τ_R is determined halfway the time used for the measurement of one correlation function. Typically, the relative thickness changes were 10% at $h_w \cong 250$ nm and 1% near the equilibrium thickness.

In the thickness region where $\gamma q^2 \gg 2V''$, the quantity $(2\tau_R)^{-1/3}$ should, according to Equation (9) behave as a linear function of h_s at constant q. In Figure 1 results are depicted for $(2\tau_R)^{-1/3}$ versus the equivalent water thickness h_w for five different angle scattering angles θ. Above a thickness of $\cong$ 50 nm the experimental points, $(2\ _R)^{-1/3}$ vs h_w, can be well represented by a least squares linear fit. According to Equation (9) the slope of these lines equals $[\gamma q^4(24\eta)]^{1/3}$. The ratio γ/η can be calculated from the slopes of the curves because q is known (see Equation (1)). The resulting value for γ/η is given in Table 1. One notices that a good agreement is found with the corresponding value of this ratio for the bulk solution.

One clearly observes from the examples given in Figure 1 that the extrapolations of

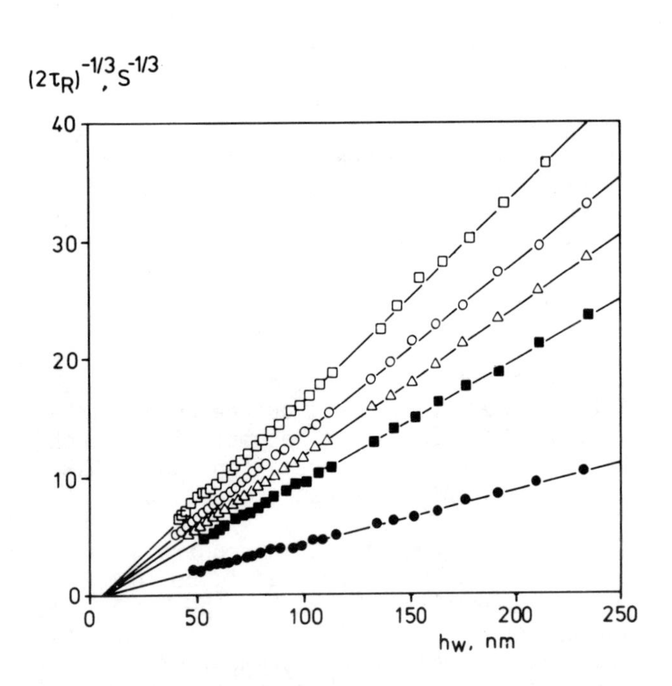

Figure 1. The cube root of the reciprocal relaxation time of the squeezing mode $(2\tau_R)^{-1/3}$ versus equivalent water thickness h_w for five different wavelengths Λ ($=2\pi/q$): □, 4.78 μm; 0, Λ = 5.44 μm; Δ, Λ = 6.09 μm; ■, Λ = 7.04 μm; •, Λ = 12.97 μm. Solid lines are linear least squares fits through data points.

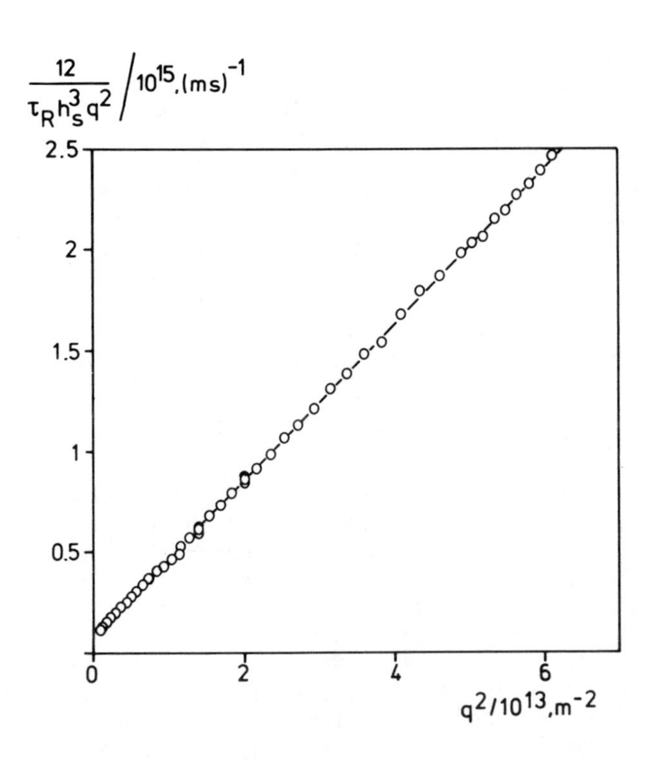

Figure 2. Values of $12/(\tau_R h_s^3 q^2)$ versus q^2 for the squeezing mode at constant thickness h_w = 29.2 nm. The thickness h_s = h_w - h_{cor} = 25.0 nm (h_{cor} is found from drainage experiments, see Figure 1). Solid line is a linear least squares fit through data points resulting in a slope γ/η = 41.0 m/s and an intercept with the $q^2 = 0$ axis $2V''/\eta = 8.45 \times 10^{13}$ $(m.s)^{-1}$.

the lines have a finite intercept with the h_w-axis. This intercept is denoted by h_{cor}. Since the curves for the different scattering angles originate from the same point h_{cor} is q independent. This implies that here exists a constant difference between the equivalent water thickness h_w and the hydrodynamic thickness h_s. We found for the DTAB films $h_{cor} = h_w - h_s = 4.2 \pm 0.2$ nm.

After the film had reached a constant thickness the scattering angle was also varied. From Equations (9) and (17) it follows that plotting $12/(\tau_R h_s^3 q^2)$ against q^2 should yield a straight line if h_s is constant. In Figure 2 we have depicted an example of such an experiment at constant thickness h_w = 29.2 nm. For the thickness $h_s = h_w - h_{cor}$ is taken, where h_{cor} is found for the drainage experiments. Clearly, the experimental data can be represented very well by a straight line. From a linear least squares fit (solid line in Figure 2) we find a slope of 41.0 m/s which, according to Equation (9), equals γ/η. The results for γ/η obtained from equilibrium experiments on the squeezing mode are given in Table 1. One again notices a good agreement with the corresponding bulk value.

Table 1. Bulk solution properties and results of dynamic light scattering experiments.

γ_o	η_o	$\frac{\gamma_o}{\eta_o}$	Squeezing mode		Bending mode	
(mN/m)	(mPa.s)	(m/s)	$\frac{\gamma}{\eta}$		γ	
			Drainage	Equilibrium	Drainage	Equilibrium
45.4	1.09	41.7	39.8 ± 0.8[a]	41.1 ± 0.4[b]	45.6 ± 0.3[c]	45.3[c]

[a] Average value of 10 different films

[b] Average value of 3 experiments

[c] Taken from (11)

Before we discuss the results in more detail let us briefly recall the experimental results related to the bending mode. For a more complete discussion of these results see (11).

Bending mode

A few years ago we succeeded in detecting the bending mode by means of PCS (20,21). Also Yound and Clark (22) published experimental data on the bending mode (which they call undulation mode).

We have studied the properties of the bending mode by measuring the power spectrum of the PMT voltage fluctuations instead of the autocorrelation function.

For reasons of poor signal-to-noise ratios the scattering angles are confined from 2° to 6° at both sides of the specularly reflected beam. This means that we gradually move over from the heterodyne to the homodyne scheme. In both schemes, however, the frequency shift ω_b can be determined from the spectrum unambiguously as follows from (the Fourier transforms of) Equations (13) and (14). From the bending mode-experiments we are able to extract a value for the surface tension of the film. To be precise, one actually measures the film tension (14) as introduced by de Feijter et al. (23). Numerically, however, the film tension of the films described here, is only slightly different from two times the surface tension of the bulk solution). We found a value of $\gamma = 45.6 \pm 0.3$ mN/m from bending mode-experiments on draining films (thickness-dependence of ω_b (see Equation (11)). Experiments on films with a constant thickness (q-dependence of ω_b) yielded a value of $\gamma = 45.3$ mN/m. Comparing these results for γ with the bulk value γ_0 (= 45.4 mN/m) for the surface tension one observes an excellent agreement.

DISCUSSION AND CONCLUSIONS

Let us first discuss the bending mode results briefly. From the experimental results on the bending mode it followed (11) that even up to frequencies of ≈ 4 MHz the surface tension of the film equals within the experimental error, the surface tension of the bulk solution. It was also shown that drainage and equilibrium experiments resulted in consistent values for γ. The availability of the surface tension of a film is of essential importance for the interpretation of the squeezing mode results as we will see below. Measuring the surface tension of the film as a function of the lifetime of the film enables us to check whether contamination effects possibly obscured our results. Particularly a surfactant solution with a surfactant concentration well below the c.m.c. will be sensitive for the presence of unwanted, highly surface active, material. Although some of our films were 24 h old, when carrying out the experiments, no effects of contamination were found.

We turn now over to a discussion of the squeezing mode results. The experimental data for the relaxation time τ_R are interpreted by means of Equations (9) and (17). Equation (9) is based on the assumption that effectively a no-slip situation exists at the surfactant-water interfaces. One can, of course, incorporate surface viscoelastic properties in the full expressions for the damping constant β_s. However, it was shown by Prins et al. (24) that for similar soap films, as we discuss here, the surface elasticity ε_s, (Equation (7)), is well above 100 N/m. Using this value we found (14) that both the surface elasticity and the surface viscosity do not affect the damping constant for the squeezing mode when compared to expression (9). Therefore, we have confidence in using Equation (9) for the interpretation of the squeezing mode data.

As we have seen the experimental result for the ratio γ/η, as obtained from the dynamic light scattering experiments on draining and equilibrium films, agrees very well with the bulk value for this ratio (Table 1). However, the value for γ/η, extracted from equilibrium experiments depends on the value that is used for h_s. This in contrast with the drainage experiments which yield γ/η without further additional assumptions to the model that leads to Equation (9). Our equilibrium results for the squeezing mode were always interpreted by using $h_s = h_w - h_{cor}$ where h_w is calculated from the (measured) reflection coefficient and h_{cor} (= 4.2 nm) is found from drainage experiments (Figure 1). In doing so one obtains consistent values for the quantity γ/η (Table 1). Since we know that γ equals the bulk value for the surface tension we infer from the experimental results that the shear viscosity, η, of the liquid comprising the film, does not differ from the bulk value. Our analysis implies, however, that we have modelled the viscosity profile inside the film by a step function i.e. η equals the bulk value up to two shear planes which are a distance h_s apart. The regions in between the shear planes and the monolayers are assumed to have an infinite viscosity. Let us now digress about these hydrodynamically immobilized layers. The aqueous core thickness of the film, h_2, can be found from the equivalent water thickness h_w (an experimental quantity) by means of Equation (15). For DTAB films we thus have $h_2 = h_w - 2.8$ nm and $h_s = h_w - 4.2$ nm. Apparently, the shear planes do not coincide with the planes in which the polar heads of the surfactant are localized. From our analysis we conclude that there exists a rigid layer at each interface of thickness = (4.2 - 2.8)/2 = 0.7 nm. We have found that depends on the chain length of the surfactant and on the ionic strength of the solution (25). As indicated above our conclusions rely on the assumption of a rectangular profile for the variation of the viscosity across the film. We could imagine another profile for the viscosity leading to the same overall effect but a model, predicting such a profile, requires a special thickness and q dependence in order to explain our experimental findings.

To conclude this section we point out that the equilibrium experiments on the squeezing mode yield a vlue for V"(h) because the intercept of the line in Figure 2 with the q^2 = 0-axis equals $2V''/\eta$ (see Equation (9)). Since we know , the quantity V" can be determined unambiguously. Comparison of the value for V" as obtained from dynamic light scattering experiments with the V" as obtained by time-averaged experiments (11), on the same film, reveals a systematic difference. The physical reason for this difference is not fully understood yet and will not be discussed here.

ACKNOWLEDGEMENTS

This work is part of a research program of the Foundation for Fundamental research of Matter (FOM) with financial support from the Netherlands Organization for Pure Research (ZWO).
The author wishes to thank Marina Uit de Bulten and Ellen Palm for assistance in typing and preparing the final manuscript.

NOTATION

$b_{z,q} = \zeta_q^I + \zeta_q^{II}$

h_2 = thickness of aqueous core

h_s = hydrodynamic thickness

h_w = equivalent water thickness

k_B = Boltzmann's constant

T = absolute temperature

$S_{z,q} = \zeta_q^I - \zeta_q^{II}$

Greek Letters

γ = surface tension

ζ_q = amplitude of surface wave perpendicular to the undisturbed interface

η = shear viscosity

LITERATURE CITED

1. Verwey, E.J.W., and J.Th.G. Overbeek, Theory of Stability of Lyophobic Colloids, Elsevier, Amsterdam (1948).

2. Derjaguin, B.V., and A.S. Titievskaya, Discuss. Faraday Soc., 18, 27 (1954).

3. Scheludko, A., Adv. Colloid Interface Sci., 1, 391 (1967).

4. Clunie, J.S., J.F. Goodman, and B.T. Ingram, "Thin Liquid Films", in Surface and Colloid Science, Vol. 3, E. Matijević(Ed.), Wiley-Interscience, New York (1971).

5. Buscall, R., and R.H. Ottewill, "Thin Films", in Specialist Periodical Reports, Colloid Science, Vol. 2, D.H. Everett (Ed.), Chemical Society, London (1975).

6. Mann, Jr., J.A., and K.C. Porzio, in International Review of Science, Physical Chemistry, Ser. Two, Vol. 7, M. Kerker (Ed.), Butterworths, London (1975).

7. Princen, H.M., J.Phys.Chem., 72, 3342 (1968).

8. de Feijter, J.A., and A. Vrij, J. Colloid Interface Sci., J. Colloid Interface Sci. 64, 269 (1978).

9. Vrij, A., J. Colloid Sci., 19, 1 (1964).

10. Vrij, A., Adv. Colloid Interface Sci. 2, 39 (1968).

11. Joosten, J.G.H., Ber. Bunsengesel. Phys. Chem., 88, 1153 (1984).

12. Berne, B.J., and R. Pecora, Dynamic Light Scattering, Wiley, New York (1976).

13. Joosten, J.G.H., J. Chem. Phys., 80, 2363 (1984).

14. Joosten, J.G.H., J. Chem. Phys., 80, 2383 (1984).

15. Vrij, A., J.G.H. Joosten, and H.M.Fijnaut, "Light Scattering from Thin Liquid Films", in Adv. Chem. Phys., Vol. 48, I.Prigogine and S.A. Rice (Eds.), Wiley-Interscience, New York (1981).

16. Mysels, K.J., K. Shinoda,and S. Frankel, Soap Films, Studies of their Thinning, Pergamon Press, New York (1959).

17. Born, M., and E. Wolf, Principles of Optics, Pergamon, London (1959).

18. Rijnbout, J.B., J. Phys. Chem. 74, 2001 (1970).

19. Donners, W.A.B., J.B. Rijnbout, and A. Vrij, J. Colloid Interface Sci., 61, 249 (1977).

20. Joosten, J.G.H., and H.M. Fijnaut, Chem. Phys. Lett., 60, 483 (1979).

21. Joosten, J.G.H., and H.M. Fijnaut, in Proceedings of Workshop on Light Scattering in Liquids and Macromolecular Solutions, V. Degiorgio, M. Corti and M. Giglio, (Eds.), Plenum Press, New York (1980).

22. Young, C.Y., and N.A. Clark, J. Chem. Phys., 74, 4171 (1981).

23. de Feijter, J.A., J.B. Rijnbout, and A. Vrij, J. Colloid Interface Sci., 64, 258 (1978).

24. Prins, A., C. Arcuri, and M. van den Tempel, J. Colloid Interface Sci., 24, 84 (1967).

25. Joosten, J.G.H., H.M. Fijnaut, and A. Vrij, to be published.

DIFFERENTIAL-INTERFEROMETRIC INVESTIGATION OF CURVED LIQUID FILMS

A. D. Nikolov, P. A. Kralchevsky,
I. B. Ivanov and A. S. Dimitrov ■ Laboratory of Thermodynamics and Physico-chemical Hydrodynamics
Faculty of Chemistry, University of Sofia, 1126 Sofia, Bulgaria

A differential-interferometric method is used for the determination of the film curvature at the top of small air bubbles, attached to a liquid surface. In addition the radius of the contact line and the equatorial bubble radius are measured by direct visual observations. From these data and from the conditions for mechanical equilibrium the film, line and transversal tensions are determined. The measured film and line tensions for bubbles formed in solutions of dodecyl sodium sulfate exhibit a strong dependence on the film curvature and unexpectedly large values of the (negative) line tension. It occurs that the transversal tension effect in the mechanical equilibrium of an attached bubble is of the same order as the disjoining pressure effect.

INTRODUCTION

Differential interferometry is a widespread, powerful and precise method for studying solid surfaces and biological subjects (1-4). The interest in this optical method for investigating fluid surfaces and thin liquid films has grown during the last decade. Zorin (5) has studied a biconcave meniscus in transmitted light. Del Cerro and Jameson (6) and Minqius and Nikolov (7) have applied differential interferometry in reflected light to floating lenses. Recently (8) this method was used for measuring the curvature of the film at the top of small bubbles attached to a liquid surface.

This application of differential interferometry has been stimulated to a large extent by the growing interest in line tension (see e.g. the references in (9). This quantity can affect the occurrence of a number of processes of practical importance, e.g. heterogeneous nucleagion (10,11), flotation of ores (12), droplet coalescence in emulsions (13) and microbial adhestion (14).

Line tension can, in principle, be measured by studying any phenomenon that is affected by it, e.g. from the rate of heterogeneous nucleation (11). This approach requires, however, knowledge (and/or fitting) of several parameters. Therefore, it is better to determine the line tension directly from the conditions for mechanical equilibrium of small particles at another interface.

FILM, LINE and TRANSVERSAL TENSIONS

Let us consider the equilibrium of a bubble (or drop) at a liquid surface - Fig. 1a. All the interfaces are represented in the figure by the corresponding surfaces of tension (see e.g.(15) which satisfy Laplace's equation of capillarity (16) with appropriate boundary conditions. One can describe macroscopically the equilibrium of the film at the top of the bubble with the surrounding phases by considering the film as a single membrane of tension γ which intersects the other two interfaces onto a line called contact line. In our case this line is a circumference of radius r_c - see Fig. 1a. In this approach, called membrane approach, the mechanical equilibrium at each point of the contact line is determined by the balance of four forces (17,9):

$$\underline{\gamma} + \underline{\sigma}_1^{\ell} + \underline{\sigma}_2^{\ell} + \underline{\sigma}_{\kappa} = 0 \qquad (1)$$

Here the vectors $\underline{\gamma}$, $\underline{\sigma}_1^{\ell}$ and $\underline{\sigma}_2^{\ell}$ act tangentially to the film and the two liquid (meniscus) surfaces, and are equal to the film tension γ_{ℓ} and the respective surface tensions σ_1^{ℓ} and σ_2^{ℓ}. The force $\underline{\sigma}_{\kappa}$ is directed toward the center of curvature (18) and is determined by the line tension κ and the radius of curvature, r_c, of the contact line: $|\underline{\sigma}_{\kappa}| = \kappa/r_c$.

The alternative of the membrane approach is to consider the film as a layer of finite thickness, bounded by two surfaces of tensions σ_1^f and σ_2^f - see Fig. 1b. This approach is called detailed (19, 20, 21). There are two contact lines in the detailed approach with line tensions $\tilde{\kappa}_1$ and $\tilde{\kappa}_2$. Each of the lines must be in equilibrium which is determined by the condition

$$\sigma_i^f + \sigma_i^\ell + \underline{\sigma}_i^k + \underline{\tau}_i = 0, \; i = 1,2, \qquad (2)$$

where $|\underline{\sigma}_i^\kappa| = \tilde{\kappa}_i/\gamma_{ci}$ and the vector $\underline{\tau}_i$, called transversal tension, acts on the contact line perpendicularily to the film surfaces inwards (21, 22). In fact the transversal tension accounts for the energy of interaction U^L of the two contact lines, more precisely for flat films $\tau = L^{-1}(\partial U^L/\partial h)$, where h is the film thickness and L is the length of the contact line (21). In this respect it is analogous to the disjoining pressure Π which takes into account the interaction of the two film surfaces (19, 23).

If one takes the horizontal and the vertical projection of the vectorial balance (1) and then solves the two resulting equations with respect to γ and κ, one gets

$$\gamma/\sigma = (\sin \phi_c + \sin \Psi_c)/\sin \theta, \qquad (3)$$

$$\kappa/\sigma = r_c\{\cos\phi_c + \cos\Psi_c - (\sin\phi_c + \sin\Psi_c)\cot\theta\}, \qquad (4)$$

where (for the case of an air bubble) we assumed $\sigma_1^\ell = \sigma_2^\ell \equiv \sigma$, and θ, ϕ_c and Ψ_c are the angles at which the film, bubble and the external meniscus surfaces meet the plane of the contact line - see Fig. 1a. By eliminating θ between (3) and (4) one can obtain a simple (approximate) relation between γ and κ:

$$\kappa/r_c = 2\sigma - \gamma/\cos d, \; \alpha = (\phi_c - \Psi_c)12. \qquad (5)$$

The contact radius r_c and the equatorial bubble radius R are easily measured by observing the bubble from above. Then the angles ϕ_c and Ψ_c can be calculated from the data for the radii R and r_c using some perturbational formulae, derived in (24), describing the shape of an axisymmetric sessile interface for small values of the capillary number

$$\beta = \rho g b^2/\sigma.$$

Here ρ is the density of the liquid, g is the gravity acceleration and b is the radius of curvature at the bottom of the bubble. The quantities b and β can be calculated from the measured value of R (24):

$$1/b = [1 - \beta/6 + \beta^2(\ell n2 - 1/6)]/R \qquad (6)$$

using interactions with zeroth approximation $b^{(o)} = R$. With these results for b and β one can calculate ϕ_c as (24):

$$\sin \phi_c = \frac{r_c}{b} - \beta\left(\frac{1}{3}\cot\frac{\phi_c}{2} - \frac{1}{6}\sin 2\phi_c - \frac{1}{2}\sin\phi_c\right)$$
$$-\beta^2\left[\left(\frac{3}{4} + \frac{1}{2}\cos\phi_c - \frac{2}{9}\sin^2\phi_c - \frac{1}{3}\ell n \sin\frac{\phi_c}{2}\right)\sin\phi_c - \frac{1}{2}\left(1 + \frac{1}{g}\cot^2\frac{\phi_c}{2}\right)\cot\frac{\phi_c}{2}\right] \qquad (7)$$

again using interations with zeroth approximation $\phi_c^{(o)} = \arcsin(r_c/b)$ then Ψ_c can be determined from the equation:

$$\sin \Psi_c = \frac{r_c}{b}\left\{1 + \frac{\beta}{2b}(z_c - h_c)\right\} - \sin\phi_c, \qquad (8)$$

where z_c and h_c are given by (24):

$$z_c = b\{1 + \cos\phi_c + \beta[\tfrac{1}{3}\sin^2\phi_c + \tfrac{2}{3}\ell_n \sin\frac{\phi_c}{2} - \tfrac{1}{2}(1+\cos\phi_c)]\};$$

$$h_c = r_c \sin\Psi_c \ell n\{4/[\gamma_e(\rho g/\sigma)^{1/2} r_c(1+\cos\Psi_c)]\};$$

$\gamma_e = 1.78\,072\,418...$ is Euler's number. The zeroth approximation to be used in the right-hand side of (8) when calculating Ψ_c by iterations is $h_c = 0$ i.e. $\Psi_c^{(o)} = 0$.

With the calculated values of the angles ϕ_c and Ψ_c one can determine the film tension γ and the line tension κ from Eqs. (3) - (4)

provided that the angle θ is known. Differential interferometry allows the measurement of the radius of curvature of the film R_f (see the next section) and hence, the calculation of θ from

$$\theta = \arc \sin(r_c/R_f) \tag{9}$$

of Fig. 1a.

In addition to the film tension γ and the line tension κ it is interesting to determine experimentally the transversal tension. This can be done with sufficient accuracy from the equation

$$F_b = \pi r_c^2 \Pi - 2\pi r_c \tau_1 \cos \theta, \tag{10}$$

where for small bubbles Π $P_c/2$ ($P_c = 2\gamma/R_f$ is the capillary pressure). Eq. (10) is in fact a force balance: The buoyancy force F_b is counterbalanced by the force $\pi r_c^2 \Pi$ due to the disjoining pressure Π acting over the lower film surface and the force due to the vertical component of τ_1 (see Fig. 1b). The buoyancy force F_b can be calculated by using the value of the angle Ψ_c determined from (8):

$$F_b = 2\pi r_c \sigma \sin \Psi_c. \tag{11}$$

<u>Determination of the film curvature from the interference pattern</u>

For our measurements of R_f we used the differential - interferometric method of "shearing" with a Epival Interphako Microscope manufactured by Carl Zeiss-Jena (for the construction of the microscope - see Refs. (4) and (25)). The basic principle of the shearing method consists in splitting the original image into two images. The light beams coming from the two images interfere, thus creating a rather complicated interference pattern. An example is shown in Fig. 2, where the shearing distance d = 12.08μm. The upper part of Fig. 3 is a sketch of the cross section (in the plane xOz) of the two images of the reflecting surfaces splitted at a distance d along the x-axis. The plane xOy coincides with the contact line (of radius r_c). The lower part of the figure is a sketch of the resulting interference pattern (cf. Fig.2). One clearly discerns three regions corresponding to the interference of light reflected by the two images of the respective surfaces: (I) meniscus-meniscus (moustaches), (II) meniscus-film (rings) and (III) film-film (streaks). In fact, all fringes are loci of points for which the distance between the reflecting surfaces satisfies the requirement

$$|z_\ell(x,y) - z_r(x,y)| = \ell_n \equiv n\frac{\lambda}{4};\ n=0,1,2,\ldots, \tag{12}$$

where subscripts "ℓ" and "r" denote left and right hand side images, λ is the light wavelength, and n is the order of interference. The equation of the film shape is

$$z_{\ell,r} = \left(R_f^2 - (x \pm d/2)^2 - y^2\right)^{1/2} - z_0, \tag{13}$$

where the upper sign refers to the left hand side image, and the lower sign to the right hand side: $z_0 = (R_f^2 - r_c^2)^{1/2}$ is the coordinate of the center. Thus, for region III, Equations (12) and (13) lead to

$$x^2/a_n^2 + y^2/b_n^2 = 1 \tag{14}$$

with

$$a_n = b_n/\varepsilon_n,\ b_n = (R_f^2 - \varepsilon_n^2 \ell_n^2/4)^{1/2},\ \varepsilon_n = (1 + d^2/\ell_n^2)^{1/2}. \tag{15}$$

Therefore, the streaks are parts of ellipses and the reason why they look like straight lines is the high eccentricity $(1-1/\varepsilon_n^2)^{1/2}$ 1. Equation (15) allows the calculation of the hat curvature $R_f = \varepsilon_n (a_n^2 + \ell_n^2/4)^{1/2}$ (ℓ_n, a_n and d are known from the experiment).

The visual determination of R_f was performed with complete splitting i.e. with $r_c < d/2$ (d was increased to 24.16μm in this case in order to avoid a gap between photographic and visual measurements). The essence of the method is to record r_c at the moment when the top ring shrinks to a point. From Fig. 4 it follows that at this moment $D_0 + Q(d) = \ell_n$, where $Q(d) \equiv Q(x=d)$ is calculated from (24):

$$Q(x) = r_c\left\{\arc\cosh\left(\frac{x}{r_c \sin\Psi_c}\right) - \arc\cosh\left(\frac{1}{\sin\Psi_c}\right)\right\} \cdot \sin \Psi_c$$

Then $R_f = (r_c^2 + D_0^2)/2D_0$.

In order to check the interferometric method we performed some measurements with objects of known curvature (26). We used small sessile water drops on a hydrophobic surface (Teflon) and a mercury drop on glass. The curvature at the top of such a drop can be determined both from the visually measured equatorial drop radius and interferometrically. The interferometrical measurement yields some value b_i for the radius of curvature at the top of such a drop and the independent visual measurement yields another value, b_r, for the same radius. Of course, if there is no systematic error in the interferometric measurements, one must obtain $b_i = b_r$. The data for b_i vs b_r from (26) is shown in Fig.5. One sees that $b_i = b_r$ in the framework of the random error of the measurements but there is not any systematic error in the curvature range studied.

EXPERIMENTS AND DISCUSSION

The experimental method we used to obtain bubbles of different radii is essentially the "shrinking bubble method" of Princen and Mason (27). A relatively large bubble at the liquid surface is allowed to decrease gradually its volume due to the escaping gas through the thin film, and its geometrical parameters are recorded optically as a function of the time.

The experiments were carried out with 0.05% (1.73×10^{-3} kmol/m^3) solutions of dodecyl sodium sulfate (Fisher Scientific, for high performance liquid chromatography) and two concentrations of NaCl (Merck, analytical grade) - 0.25 and 0.32 kmol/m^3. All experiments were carried out in a thermostated room at 22$\pm$ 0.5^{o}C. The surface tensions of the two solutions used, with 0.25 and o.32 kmol/m^3 NaCl, were 32.4 and 31.7 mN/m(dyn/cm) respectively. The essential part of the measurement cell (the one containing the solution) consists of a glass cylinder of diameter 1 cm and height 1.4 cm whose bottom is an optically plane-parallel glass. The bottom was fixed to the cylinder with glass powder heated at 500°C without using chemical seals. The air bubbles were blown out of a Hamilton syringe. The optical measurements were carried out with a microscope Epival Interphako, Carl Zeiss, Jena. The use of the shearing method was described in the previous section. The values of r_c and R were recorded visually every time when the diameter of the respective circumference became equal to an integar number of scale divisions. At suitably chosen time intervals (100-200 s) the image was splitted and measurements of R_f wer performed either by taking photographs or visually, by counting the number of interference rings.

A major experimental problem is that we need for the calculations the set of values r_c, R and R_f at a given moment, t, whereas some time elapses after the registration of each of these quantities. One possible way to find the required values is by least squares interpolation of the data for $r_c(t)$, R(t) and $R_f(t)$. We interpolated R(t) by the equation

$$R(t) = a_1(t_o-t)^q + a_2(t_o-t)t, \qquad (16)$$

where t_o, q, a_1 and a_2 are constants to be determined by the minimization of the dispersion

$$\phi_1(a_1, a_2, t_o, q) = \sum_i [R(t_i) - R_i]^2. \qquad (17)$$

Here R_i is the measured value of R at the moment t_i and $R(t_i)$ is calculated from Equation (16). Similar interpolation formula were used for $r_c(t)$ and $R_f(t)$.

We have processed in full details only 4 experiments - two for the solution with 0.25 kmol/m^3 NaCl (runs 1a and 1b) and two for 0.32 kmol/m^3 NaCl (runs 2a and 2b). The results are presented in Figs. 6 and 7.

As explained above, when calculating γ and κ we used the data for r_c, R and R_f (at the same moment t), obtained from the interpolation curves $r_c(t)$, R(t) and $R_f(t)$ - see e.g. Equation (16). The error bars in Figs. 6 and 7 denote the standard deviations of $\gamma/2\sigma$ and κ calculated by using the standard deviations of the measured radii r_c, R and R_f with respect to the interpolation curves. The points in the Figs. (calculated from the smooth curves) correspond to the moments, at which R_f was measured, and represent the most probable values of $\gamma/2\sigma$ and κ at those moments.

The most striking features in the behavior of $\gamma/2\sigma$ vs P_c are the large variations of γ (the respective values $\gamma_\infty/2$ for planar films, i.e. for $R_f \to \infty$, taken from Ref. (28), are shown on the ordinate axis of Fig. 6 by arrows) and the fact that at some capillary pressures γ is larger than 2σ (this has not been observed with planar films).

Quite unexpected are the data for κ (Fig. 7) - besides the large values of κ and the variation of κ with r_c, we must point out the change of sign of κ for both solutions of NaCl and to the smaller absolute values of κ for smaller bubbles i.e. for larger r_c^{-1}. There is a tendency of κ to level off for large bubbles ($r_c^{-1} \to 0$), which is more pronounced on the plot κ/σ vs R_f - Fig. 8.

All these findings reveal that (unlike the surface tension and similarly the disjoining pressure) γ and κ are strong functions of the geometrical parameters of the system. This is a new and unexpected result for γ. For the line tension this was predicted on theoretical gounds by many authors (29-33) but has not been observed experimentally.

There have been only a few attempts for experimental determination of the line tension for fluid systems with configuration similar to ours. The authors of Refs. (34) and (35) have studied the same system as ours (bubbles formed from solutions of sodium dodecylsulfate) but their values for κ are different from ours; for example they obtained κ = 0.85 nN for 0.32 kmol/m^3. The reason for the discrepancy between our and their results lies probably in the fact that they used incomplete experimental information (they did not measure the angle θ) and to make up for this deficiency they erroneously assumed that γ and κ remained constant for all bubble radii.

Navascues and Mederos (11) have determined κ from the nucleation rate of water drops on mercury. They found κ varying from - 0.290 to -0.393 nN for critical radii changing from 20.7 to 25.2nm. These low values of κ should not be surprising in view of the small size of the nuclei.

The only measurements with particle size close to ours were carried out by Torza and Mason (13), who determined κ from the equilibrium configurations of five doublets of emulsion droplets. They obtained five different values for κ (of the order of 10 nN) and attributed these differences to scattering caused by impurities in their system. In fact, a closer inspection of their data reveals that the variation in κ may well be due to geometrical factors. Indeed, their radius of curvature r_{13} of the interface between two droplets corresponds to R_f in our experiments, and if one plots their data for κ vs R_f one obtains, as with our data, a quite good linear dependence - see inset in Fig. 8.

We cannot for the time being completely rule out the possibility that the observed facts are due, at least in part, to non-equilibrium phenomena. Such an effect may be connected with the relatively slow surfactant desorption during the shrinking of the bubble, which can lead to $\sigma_1^\ell \neq \sigma_2^\ell \neq \sigma$ (more precisely $\sigma_1^\ell < \sigma_2^\ell$ - cf. Fig. 1a). But this effect is expected to be more efficient for small bubbles and in this way one hardly could explain the rise of $|\kappa|$ for large bubbles. Besides, such a lowering of σ is unlikely to exist for concentrations above c.m.c. (36). It is more probable that the observed effects are due to some slow irreversible processes, which change the film tension γ(e.g. either condensation of water on the film or gradual change in the film thickness). Because of the connection between γ and κ/r_c (cf. Eq. (5)) this change in γ causes a rise in κ, which counterbalances the alteration in the film tencion (cosd is practially a constant). From a microscopic point of view such a rise in κ may be due to deviations of the transition region (between the film and meniscus) from its equilibrium shape or to local changes of the surface tension in this region - cf. (23). Both effects can have hydrodynamic origin. The definitive explanation of the observed experimental facts requires however additional theoretical and experimental studies which are now under way.

Yet, whatever the origin of these effects might be, we believe to have firmly established that the attachment of small bubbles to a liquid surface gives rise to unexpectedly large line tensions accompanied with corresponding variations of the film tension. Both effects are pronounced functions of the bubble and film radii.

The data for the transversal tension provides additional new information for the equilibrium of an attached bubble. Table I shows how the terms in Eq. (10) vary during the shrinking of a bubble (Run 2a).

It is interesting to note that the buoyancy force F_b is much smaller than the other two terms in (10). In the limiting case of zero gravity (space laboratory) only the disjoining pressure and transversal tension terms will remain in this force balance. The values of the transversal tension τ_1 (last column in Table I) are close to the value 4.77 nN/m, which we calculated from the data for flat films in Ref. (28) using Eq. (17) from Ref. (21).

In conclusion, these experimental results indicate that the linear effects like the line and transversal tensions could play a much more significant role in the attachment of a fluid particle to another interface than it was believed until now.

REFERENCES

1. Françon, "Progress in Microscopy", Pergamon Press, London, 1961.

2. Françon and S. Mallick, "Polarization Interferometers", Wiley-Interscience, New York, 1971.

3. R. Hoffman and L. Gross, J. Microsc., 91, (1970), 149.

4. H. Beyer, "Theorie and Praxis der Interferenzmikroskopie", Akademische Verlageselschaft, Leipzig, 1974.

5. Z. M. Zorin, Kolloidn. Zh., 39 (1977), 1158.

6. M. C. G. del Cerro and J. Jameson, in "Wetting, Spreading and Adhesion" (J. F. Padday, Ed.), Academic Press, London, 1978.

7. J. Mingins and A. D. Nikolov, Ann. Univ. Sofia (Fac. Chimie), 75 (1981), 3.

8. A. D. Nikolov, P. A. Kralchevsky and I. B. Ivanov, J. Colloid Interface Sci. - to be published.

9. I. B. Ivanov, P. A. Kralchevsky and A. D. Nikolov, J. Colloid Interface Sci. - to be published.

10. R. D. Gretz, Surface Sci., 5 (1966), 239.

11. G. Navascues and L. Mederos, Surface Technol., 17 (1982), 79.

12. A. Scheludko, B. V. Toshev and D. T. Boyadjiev, J. Chem. Soc. Faraday I, 72 (1976), 2815.

13. S. Torza and S. G. Mason, Kolloid Z. u. Z. Polym., 246 (1971), 593.

14. B. A. Pethica, in "Microbial Adhesion to Surfaces", R. C. W. Verkeley, Editor, p. 19, Ellis Horwood, Chichester, U.K.,1980.

15. S. Ono and S. Kondo, in "Handbuch der Physik", Vol. 10, Springer-Verlag, Berlin, 1958.

16. H. M. Princen, in "Surface and Colloid Science" (E. Matijevic and F. R. Eirich, Eds,), Vol. 2, p.1, Wiley, New York, 1969.

17. P. R. Pujado and L. E. Seriven, J. Colloid Interface Sci. 40(1972), 82.

18. V. S. Veselovsky and V. N. Pertzov, Zh. Fiz. Khim., 8 (1936).

19. A. I. Rusanov, "Phase Equilibria and Surface Phenomena", Khimia, Leningrad, 1967 (in Russian); "Phasengleichgewichte und Grenzflächenerscheinungen", Akademie-Verlag, Berlin, 1978.

20. P. A. Kralchevsky and I. B. Ivanov, in "Surfactants in Solutions", K. L. Mittal, Ed., Plenum Press, New York, to be published.

21. P. A. Kralchevsky and I. B. Ivanov, Chem. Phys. Letters, 121 (1985), 111.

22. P. A. Kralchevsky and I. B. Ivanov and A. D. Nikolov, in "Proceedings of the VIth International Conference on Surface Active Substances, Akademie-Verlag, Berlin, to be published.

23. P. A. Kralchevsky and I. B. Ivanov, Chem. Phys. Letters, 121 (1985), 116.

24. P. A. Kralchevsky, I. B. Ivanov and A. D. Nikolov, J. Colloid Interface Sci., to be published.

25. H. Beyer, Jenaer Rdsch., 16 (1971), 82.

26. A. D. Nikolov, A. S. Dimitrov and P. A. Kralchevsky, Optica Acta - submitted for publication.

27. H. M. Princen and S. G. Mason, J. Colloid Sci., 20 (1965), 353.

28. J. A. de Feijter, Thesis, Univ. Utrecht, 1973; see also J. A. de Feijter and A. Vrij, J. Colloid and Interface Sci., 64 (1978), 269.

29. A. I. Rusanov, Kolloidn. Zh., 39 (1977), 704.

30. I. B. Ivanov, B. V. Toshev and B. P. Radoev, in "Wetting, Spreading and Adhesion", J. F. Padday, Editor, p. 37, Academic Press, London, 1978.

31. V. M. Starov and N. V. Churaev, Kolloidn. Zh., 42, 703 (1980).

32. N. V. Churaev, V. M. Starov and B. V. Derjaguin, J. Colloid Interface Sci., 89, 16 (1982).

33. G. Navascues and P. Tarazona, Chem. Phys. Letters, 82, 586 (1981).

34. D. Platikanov, M. Nedyalkov and V. Nasteva, J. Colloid Interface Sci., 75, 620 (1980).

35. A. Scheludko, B. V. Toshev and D. Platikanov, in "The Modern Theory of Capillarity", F. C. Goodrich and A. I. Rusanov, Editors, Akademie Verlag, Berlin, 1981.

36. K. Lunkenheimer-private communication.

Table 1. Bubble parameters during the bubble compression.

TABLE I

R (μm)	r_c (μm)	F_b (μN)	$\pi r_c^2 \Pi$ (μN)	$2\pi r_c \tau_1 \cos\theta$ (μN)	τ_1 (mN/m)
152.7	50.4	0.145	1.66	1.52	4.86
135.9	43.8	0.102	1.41	1.31	4.83
116.2	36.5	0.064	1.14	1.08	4.76
100.3	30.8	0.041	0.945	0.904	4.72
84.5	25.4	0.025	0.764	0.739	4.67
48.5	14.0	0.005	0.402	0.397	4.57
31.5	8.9	0.001	0.252	0.251	4.52

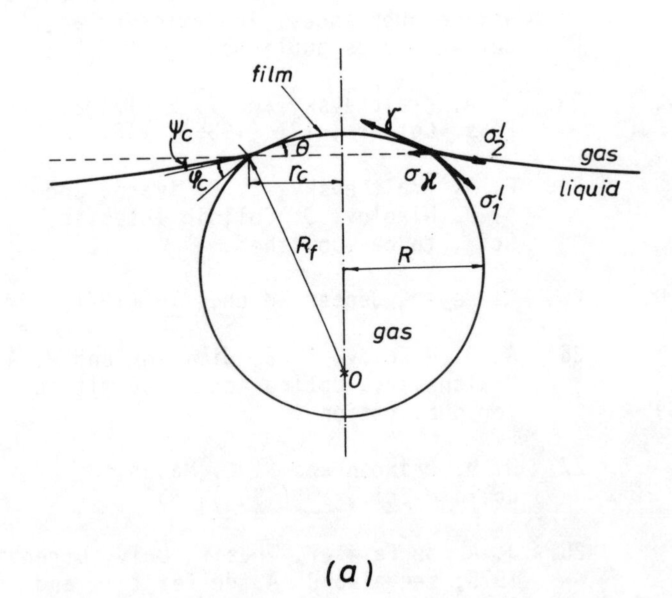

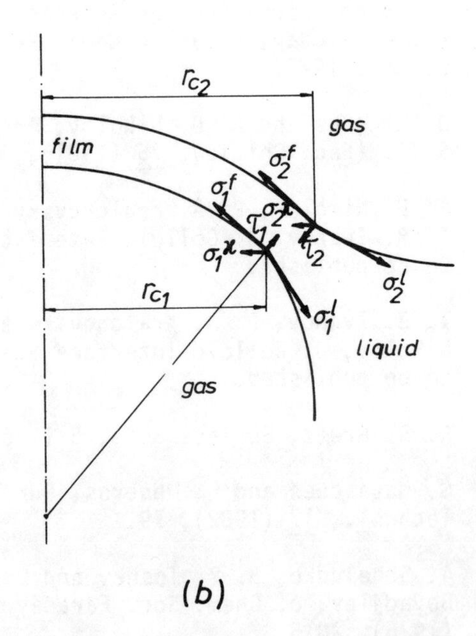

Figure 1. The force balance at each point of the periphery of a spherical thin film in the membrane approach (a) and in the detailed approach (b).

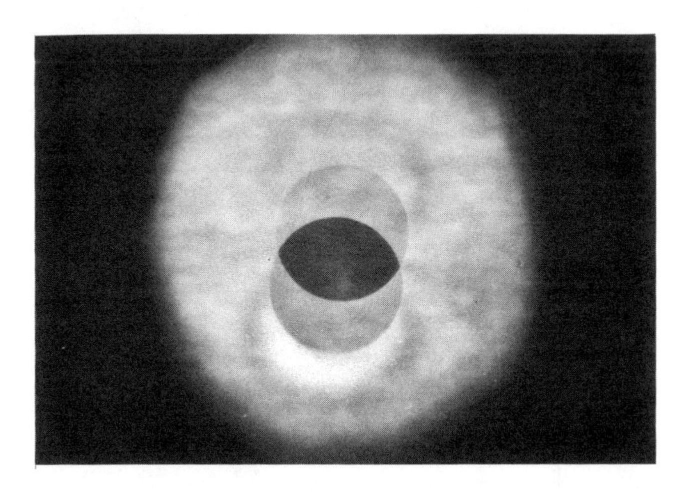

Figure 2. Differential interference pattern in light reflected from a bubble, attached to a deformed air/liquid surface (r_c = 45.4 μm, objective 25x).

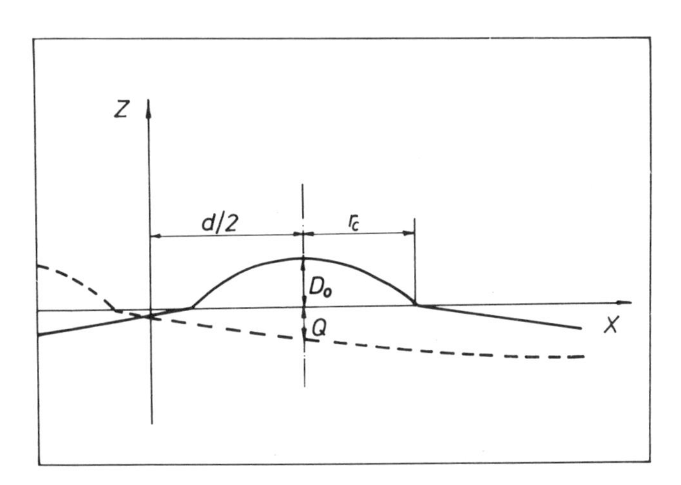

Figure 4. A scheme of the reflecting surfaces in the case of complete splitting (shearing distance $d > 2r_c$).

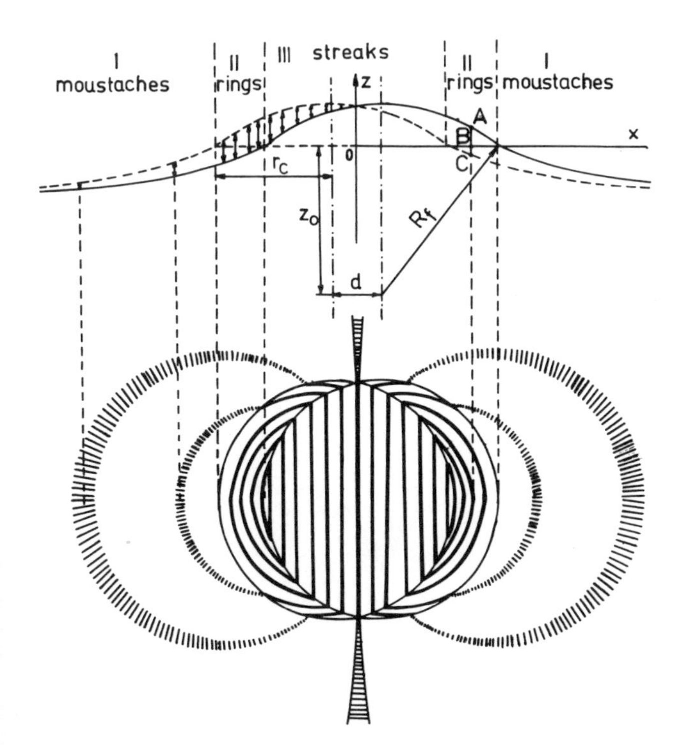

Figure 3. A sketch of the cross section of the reflecting surfaces, shifted at a distance d (upper part) and of the resulting interference pattern (lower part).

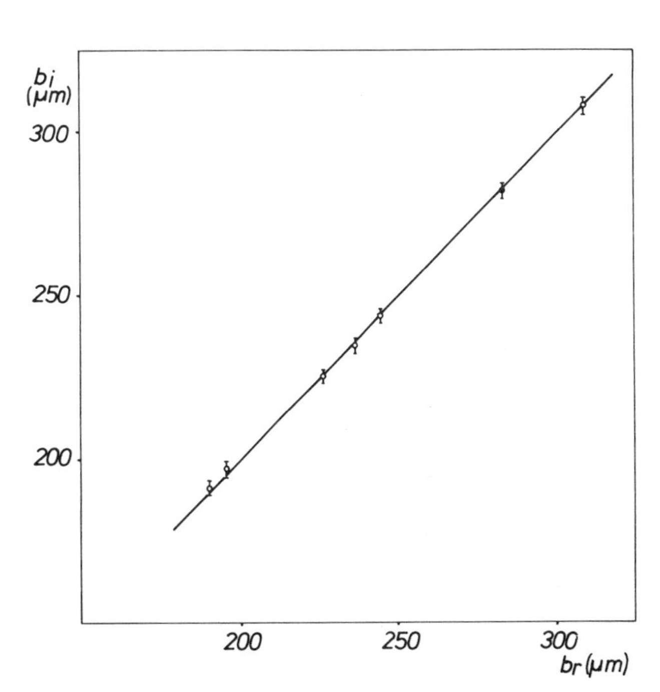

Figure 5. Interferometrically determined radius of curvature b_i vs visually determined value b_r of the same radius for 6 aqueous drops (o) and one mercury drop (●).

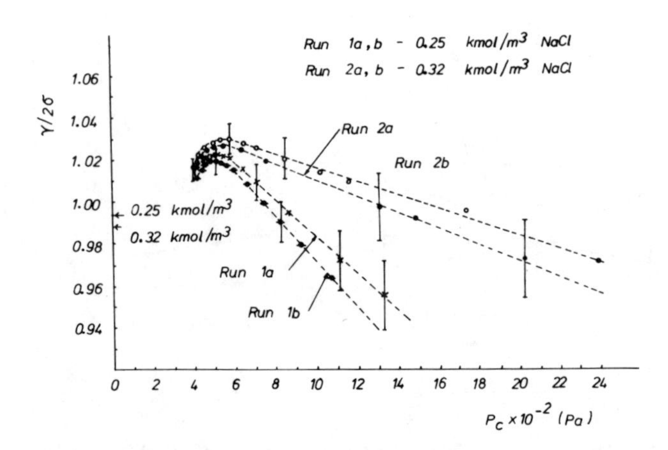

Figure 6. Dimensionless film tension $\gamma/2\sigma$ vs capillary pressure $P_c = 2\gamma/R_f$.

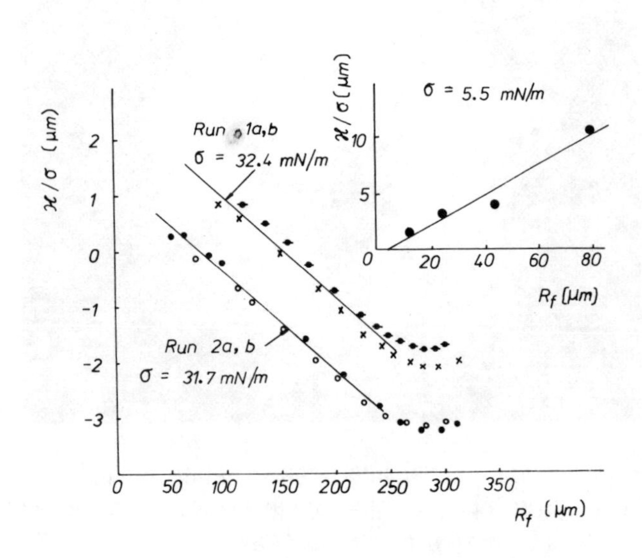

Figure 8. Dependence of κ/σ on R_f (the points correspond to runs: 1a (x), 1b (-●-), 2a (o) and 2b (●). The inset shows the same plot for the data of Torza and Mason (*13*) for doublets of emulsion droplets.

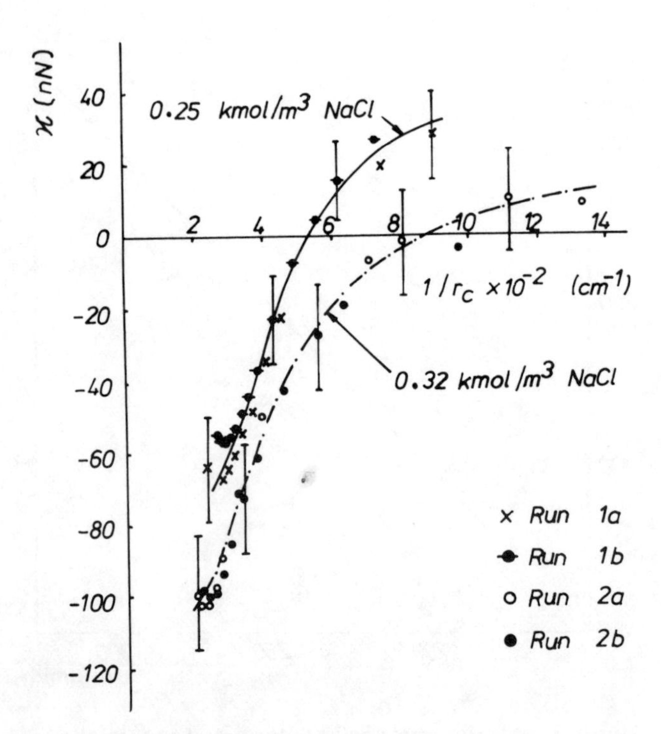

Figure 7. Line tension, κ, vs reciprocal radius r_c^{-1} of the contact line.

POLYVINYL ALCOHOL AS A SUSPENDING AGENT FOR PVC PRODUCTION

Dilva Padovan and Donald R. Woods ■ Dept. of Chem. Engrg., McMaster University, Hamilton,ON, Canada

Water soluble polymers, especially polyvinyl alcohol, are very useful steric stabilizers for the production via suspension polymerization of PVC. These stabilizers vary in their molar mass, the degree of hydrolysis. Often mixtures of such stabilizers are used. To assess the surface phenomena role that these might be playing in stabilizing the monomer drops, film thinning studies were done for a model liquid/liquid system of butyl chloride and water. The film studied was the water trapped between a 700 m diameter butyl chloride drop rising to a planar water/butyl chloride interface. The focus was on the effect of varying the bulk phase concentration of a commercial PVA sample obtained from Esso Chemical: KZ 04 with a viscosity average molar mass of 23,000, a degree of hydrolysis of 72 mol.% and a degree of polymerization of 400. The bulk phase concentrations ranged from 3 to 1200 ppm. Two grades of butyl chloride were used. Photographs were taken of the light interference patterns generated by the trapped film as the film thinned to either an equilibrium thickness or to rupture. For the film studies, the thickness stabilized at 120 to 180 nm for concentrations of 60 and 300 ppm; for impure butyl chloride and for pure butyl chloride at concentrations 300, 600 and 1200 ppm, the film eventually ruptured at about 30 to 45 nm or stabilized about 50 nm. A bulk concentration of 3 ppm was not sufficient to create a stable film. The water film ruptured. The purity of the butyl chloride affected the results. Impure liquid tended to cause coalescence. To complement this work, small batch tests with emulsions for the same nominal concentrations of stabilizer were done. The emulsion coalesced at nominal bulk concentrations of 3 and 60 ppm. It was stable for concentrations of 300 ppm or greater. Estimates suggest that at the low concentrations insufficient stabilizer is present to saturate all of the surface area. (In the small, film cell, saturation was not a problem because the total surface area was small.) An analysis of the rates of film thinning showed that all of the films thinned following an uneven drainage pattern, the Liem model with partial surface mobility was applicable and that the effective viscosity of the water film into which the stabilizer extends was about 6 to 8 times that of water. This is very qualitative because the surface tension varies with concentration, and elapsed time during the thinning process. These estimates represent the minimum increase in nominal viscosity that would occur if the minimum surface tension was attained during the process. The surface mobility for the 60 and 300 ppm system was essentially constant at 1.5 mobile surfaces. For the higher concentrations, the mobility changed from about 0.2 to 1.9. The overall results highlight the importance of the stabilizer on film thinning—and not just equilibrium thickness—as being important for stability.

For the production of PVC, the reactor conditions can use monomer drops about 120μm diameter (called suspension polymerization); monomer drops about 10μm diameter (called emulsion polymerization) and bulk polymerization. For the suspension polymerization, the drops are stabilized so that they do not coalesce by the addition of water soluble polymers, called the stabilizer or protective colloid. A function is to adsorb to the monomer water interface and prevent other interfaces from approaching because of steric repulsion forces. For emulsion polymerization, the drops are stabilized by the addition of surfactants, such as sodium dodecyl sulfate. They prevent coalescence by electrochemical double layer interaction/repulsion. In suspension polymerization, the commonly used stabilizer is polyvinyl acetate that has been partially hydrolyzed to polyvinyl alcohol. One such material is a commercial product PVA-KZ-04. This has a viscosity average molar mass of 23 300, a degree of polymerization of 400, and a degree of hydrolysis of 72% mol. (Thus it is 28 mol% acetate). The purpose of this research was to determine the optimum concentration of PVA-KZ-04 to yield the most stable film of water trapped between two approaching oil/water interfaces and to identify the reasons why the material fufills its role. The butyl chloride/water was used to represent the monomer water system.

BACKGROUND

A few have studied the use of polyelectrolytes as stabilizers for films. Van Vliet (17) used various concentrations and types of PVA to stabilize a film of glycerin/water trapped beteen two liquid/air surfaces. The film was held in a circular, horizontal glass ring. The thickness of the film was determined by light intensity. From the equilibriun thickness estimates could be made of the polymer adsorption, conformation and steric repulsion. Sonntag et al (16) used a similar approach and studied the adsorption at a water/oil surface and the effect on the stability of the water or oil layer trapped between the two surfaces held in a glass ring.

Nilsson et al (13) studied the interfacial tension and the effect of type stabilizer on the drop size distribution and resin porosity in PVC suspension polymerization. They did not consider the film thickness or measure of the stability of the drops. They worked with the vinyl chloride/water system.

Theories to relate the polymer, its conformation and the resulting steric interaction have been developed by Hesselink, (6), Hesselink, Vrij and Overbeek (5), Scheutjens and Fleer (15), Dolan and Edwards (1,2), and Napper (11,12).

EXPERIMENTAL APPROACH:

Drops of n-butyl chloride were formed in water and brought into a cell where they could be "aged" so that the drop would have polyelectrolyte adsorbed to it and then released to rise through water to nestle up against a horizontal oil/water interface. Alternatively, the drop, once it was brought into the cell, could be released immediately so that a relatively clean drop would rise through the water. The water in the cell contained an known, intial concentration of PVA. The horizontal oil water interface could be cleaned via a probe designed especially to ensure that both sides of the interface are cleaned. Thus, the adsorption of polyelectrolyte to the bulk surface could also be controlled by adjusting the time between the surface cleaning and the release of the drop. For this study the drop was not aged and the bulk interface was cleanedimmediately before each drop was released. The elapsed time from cleaning until drop release was in the order of 3 to 50s. Hence, neither the drop nor the bulk interface should have much adsorbed polymer when the film thinning started. The thin film of water trapped between the rising drop and the bulk layer was viewed through a micropscope; the intensity and colour of the light reflected from the surfaces bounding the film are uniquely related to the thickness of the film. Thus, from slides taken of the film thinning and the equilibrium behaviour, the thickness of the film could be estimated. The variables studied included the size of the drop, (from volumes 0.6 to 3.6μL), the initial concentration of PVA in the water phase (from 3 to 1200 mg/L), the grade of the butyl chloride and the age of the bulk interface between cleaning and drop release. The fixed variables included temperature (at 23.5 ± 1.5°C), cleaning procedure and type and characteristics of the PVA.

A sketch of the apparatus, developed by Liem (10) and modified slightly for this work, is given in Fig 1. Details are available (15) (10). The apparatus was all glass and teflon; mounted on a heavy frame that was supported on vibration free pads. Great care was taken in cleaning the apparatus before it was assembled for a run. The cleaning procedure took about a week and included numerous washes and the removal of static charges from the teflon parts that resulted from the cleaning procedure. The butyl chloride was Baker Analyzed reagent of density 880 kg/m^3 at 25°C, refractive index of 1.4021 at 20°C; some experiments were done with Baker TM Grade (98% minimum assay). The interfacial tension between water and butyl chloride was 32 ± 13 mN/m at 25°C as measured by the spinning drop technique, 36 mN/m as measured by the drop weight method when the forming time was about 2 min. and 30.5 ± 3 mN/m via the Wilhelmy plate method. At least three known systems were used to calibrate and confirm operating procedures for each method. The water used in the study was doubly distilled, deionized and carbon filtered to produce a final conductivity of less than 1 micromho. Only trace quantities of organics were detected by liquid chromatography.

The PVA was supplied by Imperial Oil Ltd, Sarnia. We determined the intrinsic viscosity to be 0.229 dL/g from which the root-mean square, end-to-end distance of the polymer in water we estimated to be $(r^2)^{1/2}$ at 25°C = 13.6 nm. This size agrees reasonably well with Van Vliet's estimate of 17.2 nm for a similar polyelectrolyte (17).

The interfacial tension between butyl chloride/water was measured as a function of time and concentration of PVA by the spinning drop and Wilhemy plate methods.

The PVA concentration reported in this work was the initial concentration in the water phase before the phases are contacted. For the film stability studies in the cell, the surface area was small so that a negligible change in bulk concentration occurred. Furthermore, the surfaces would be relatively easy to saturate. Table 1 summarizes the work of Fleer and Lyklema (4) and Lankveld (8) on estimates of the coverage of surfaces by polyelectrolytes. However, in the actual process where many drops exist and the surface area is large, the equilibrium bulk concentration of PVA will be much less than the initial concentration of PVA. To appreciate the impact of this and to relate the cell studies to plant conditions, the same chemicals were used in emulsion studies. Ten mL of PVA solution and 10 mL of n-butyl chloride were placed in a 2.7 cm wide test tube. A high speed propeller mixer emulsified the mixture for about 60 s. The behaviour of the resulting emulsions was observed. The initial concentrations of PVA used were 60, 300 and 600 mg/L.

RESULTS AND DISCUSSION

The first hypothesis was that the "best stabilization" would correspond to concentrations that give the thickest stable films. Figs 2 and 3 show the results of the

Figure 1 A schematic diagram of the apparatus used to determine film thickness.

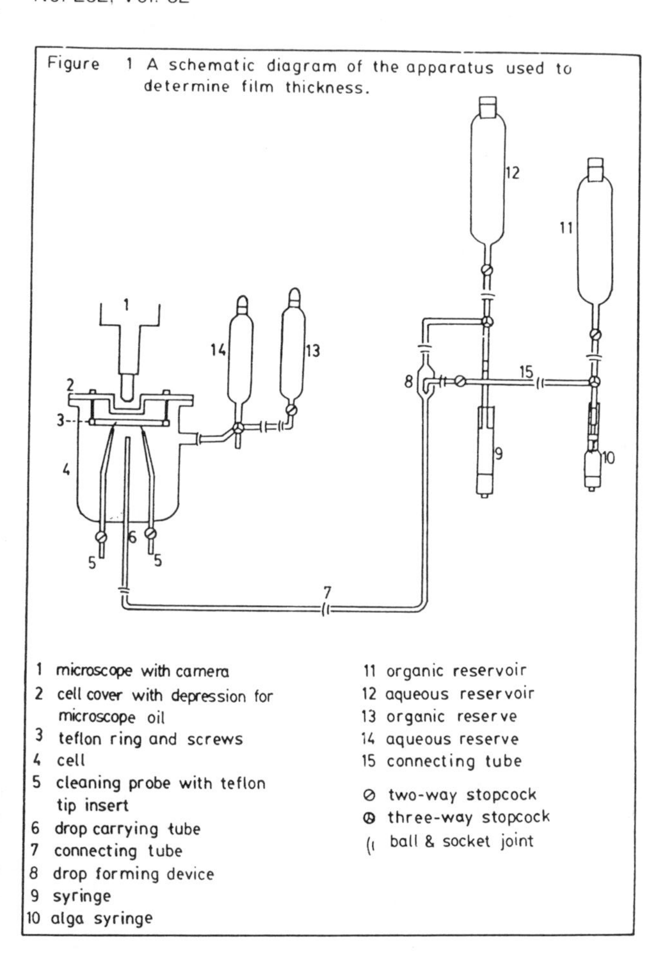

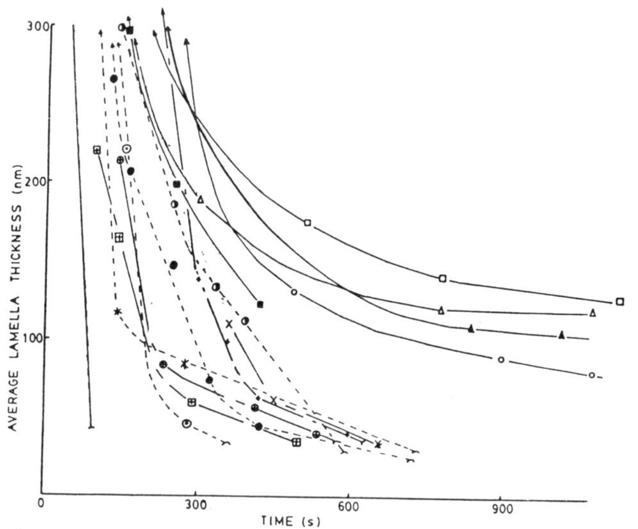

Figure 2 Average lamella thickness as a function of time from drop release. (⌒ =coalescence point)

PVA Concentration (ppm)	Drop Volume (μL)	Symbol
3.11	1.58 to 1.98	——
62.0	1.00	○
	1.17	□
	1.33 & 1.42	△
	1.74	▲
	2.78	■
310	2.46	⊙
	0.64	*
	1.90	●
	3.02	◐
620	1.41	⊕
	3.42	+
	1.49	×
	1.01	⊞

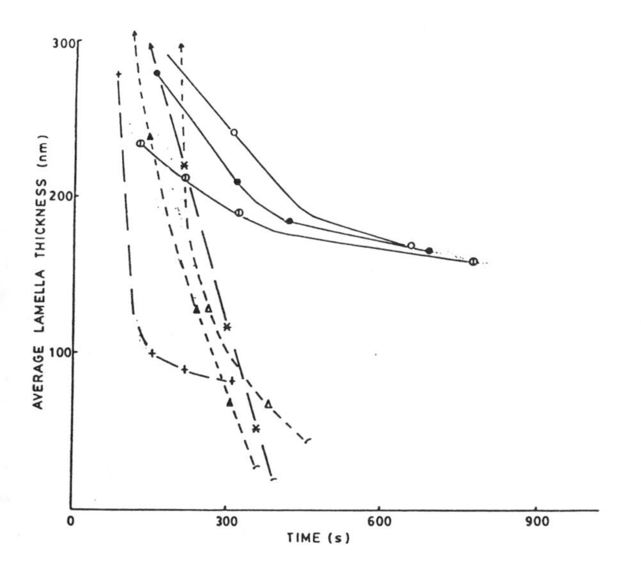

Figure 3 **Average lamella thickness as a function of time from drop release. (n-butyl chloride - Baker Analyzed)**

PVA Concentration (ppm)	Drop Volume (μL)	Symbol
300	2.13	●
	1.40	⦶
	2.45	○
600	1.00	+
	3.58	*
1200	2.05	▲
	3.42	△

⌒ =coalescence point

film thickness measurements as a function of initial polymer concentration for the two different qualities of n-butyl chloride. No trends were noted in the sequence in which the runs were performed, with the total elapsed time from when the cell was first assembled and when the data were taken. This suggested that the interface cleaning procedure used before each drop was released was successful. The data cluster into three groups: the "almost immediate" coalescence (that occurs when the concentration is 3 ppm); some films thin to about 50 nm and stabilize at this thickness or slowly thin further and eventually coalesce, and some films tend to stabilize in the range 120 to 180 nm. The films that eventually coalesce or stabilize at about 50 nm occur when the initial concentration of PVA is 300 mg/L or higher. The initial concentrations of PVA that give stability about 120 to 180 nm are 60 mg/L and 300 mg/L for "pure" butyl chloride. One might conclude from this that a) the quality of the butyl chloride makes a relatively dramatic difference in the stable film thickness and b) the best concentration to use to give the thickest stable film is in the range 60 to 300 mg/L. This assumes that the surface is saturated with polymer. Estimates can be made to establish how much polymer is needed to saturate a given amount of drop area. Table 1 summarizes previous research on polymer adsorption. In general, about 80% of the surface is covered with trains of polymer. About 10% of the polymer segments reside as the trains and each segment occupies between 0.15 and 0.30 nm^2. Alternatively, polymer adsorption tends to require about 2 to 4.5 mg of polymer/m^2. Thus, either approach can be used to estimate how much polymer must be added to the system to saturate all of the area in a suspension polymerizer. If we assume that all of the polymer is adsorbed so that none remains in solution, then for PVA-KZ-04 the maximum area that could be covered for different initial polymer concentrations is given in Table 2. For the cell for the film studies, the 3.1 mg/L is more than enough to saturate the surface. For a polymerization reactor with a suspension made from 50 v/v water/monomer and agitation to produce 10μm, then the surface area generated would be about 300 m^2/L. Thus, from Table 2 the initial concentration of over 600 mg/L would be needed. If less is added, then the surface is not saturated and the tails do not project into the film to stabilize it.

The batch emulsion tests, shown in Table 3, produced drops estimated to be in the general range 20 to 50μm. They confirmed the

Table 1. Estimates of the Surface Coverage by Polymers

Concept	PVA at AgI solid surface, Fleer and Lyklema	PVA at paraffin oil/water surface, Lankveld
Coverage	max = 0.8	
Fraction of train surface	0.1 to 0.2	0.08 to 0.10* 0.15 to 0.28* 0.05 to 0.11** 0.008 to 0.06**
Area/molecule in surface		10 to 70 nm^2 1 to 15 nm^2
Average thickness of layer, nm	10 to 20	8 to 15* 10 to 20**
mass adsorbed, mg/m^2	1.5 to 2	2 to 4.5* 2 to 4.5**

*low molar mass polymer,
** high molar mass polymer

Table 2. Estimated Surface Area that can be Saturated by Polymer Adsorption

Polymer Concentration, mg/L	Estimated Maximum Area, m^2/L segments	2 to 4.5 mg/L
3.	1.2	0.7 to 1.5
62	24	14 to 31
300	115	67 to 150
600	230	135 to 300
620	240	138 to 310
1200	465	270 to 600

Table 3. Stability of Emulsion

Concentration, mg/L	Qualitative Statement about Stability
60	Coalesced to larger drops; unstable
300	Extremely stable
600	Extremely stable

calculations because the emulsions for 3 and 60 mg/L were unstable and coalesced, whereas those for greater than 300 mg/L were stable.

The interfacial tension for different concentrations of PVA was sensitive to the elapsed time from when the surface is created. However, for the higher concentrations,ie above 15 mg/L where most of the interesting data were taken, the interfacial tension was relatively constant after several minutes. This is illustrated in Fig 4. For the smaller concentrations of polymer, the interfacial tension (as measured by the spinning drop method, after an elapsed time of 15 h,and by the Wilhemy plate method, after the interface had rested 30 min after it was formed) are given in Fig 5. The data are in fair agreement, and show that the interfacial tensions decrease to less than 5 mN/m for concentrations greater than 60 ppm. Nilsson et al (13), working with vinyl chloride/distilled water under pressure and at 50oC and with PVA 71.5% hydrolyzed (Rhodoviol 5/270), found that for 100 ppm the interfacial tension was about 4 mN/m.

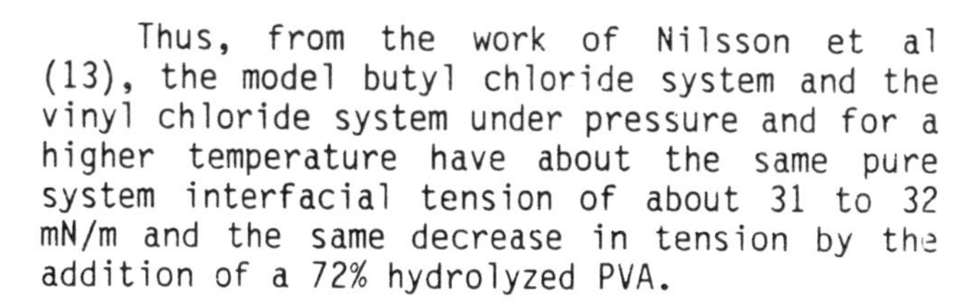

Thus, from the work of Nilsson et al (13), the model butyl chloride system and the vinyl chloride system under pressure and for a higher temperature have about the same pure system interfacial tension of about 31 to 32 mN/m and the same decrease in tension by the addition of a 72% hydrolyzed PVA.

Although the focus of this research is on "stable equilibrium" films, data were taken to show the rate of approach to equilibrium or the rate of film thinning. These data can be used to help understand more about the coalescence process for drops stabilized by polymers.

Various models have been proposed to describe the rate of film thinning; the appropriate model depends on which mechanism prevails in the thinning process. For the polymer concentrations greater than 3 mg/L used in this research, the prevailing mechanism was uneven drainage. Under this mechanism, the fliud is removed from the film by local mobility (rather than by a pressure gradient); the film is of approximately

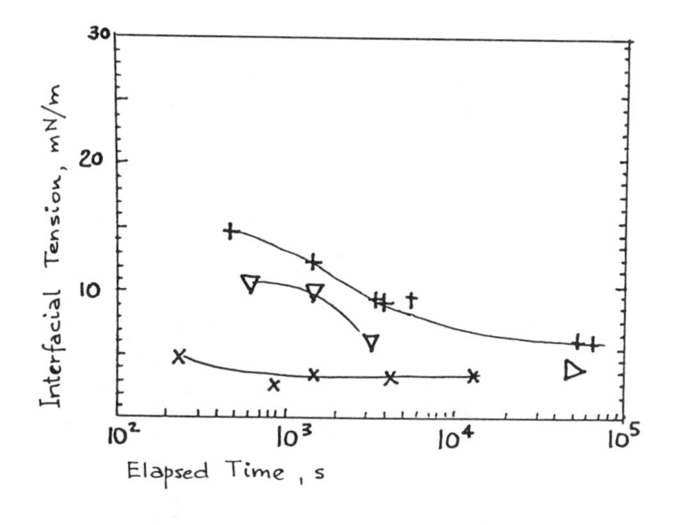

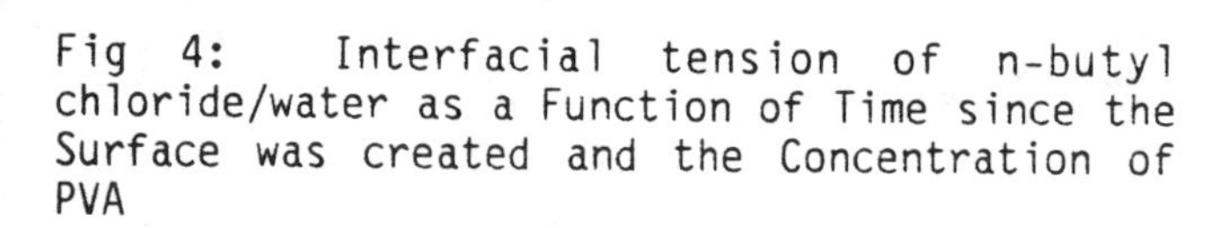

Fig 4: Interfacial tension of n-butyl chloride/water as a Function of Time since the Surface was created and the Concentration of PVA

Concentration, mg/L: ▽ 6 ; + 9 ; ▷ 21 ; × 150 .

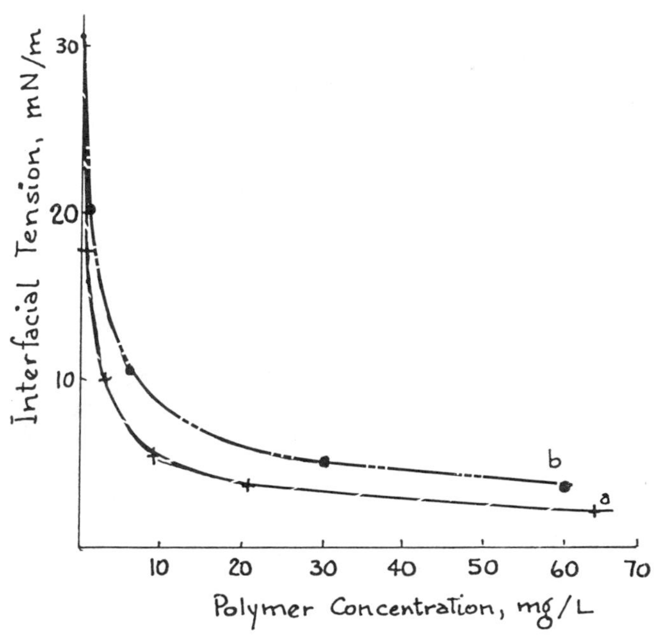

Fig 5: Interfacial tension of n-butyl chloride/water as a function of PVA concentration in the water phase.

a) As measured by the spinning drop method (15 hours after the interface was formed) +.b) As measured by the Wilhelmy plate method (30 min after the interface was formed) ●.

uniform thickness (as opposed to being highly dimpled). Liem and Woods (9) studied films that thinned by this mechanism and proposed the modified parallel disks model of the form:

$$\tau^+ = \{1/h^+\}^2 - 1 \tag{1}$$

where τ^+= dimensionless time

$$\tau^+ = \frac{16}{4\phi}\left(\frac{\gamma^2}{\mu_c \Delta\rho g}\right)\frac{1}{b^3}\left[\frac{h_1}{b}\right]^2 \tau \tag{2}$$

γ = surface tension, mN/m

μ_c = viscosity of the continuous phase in the film,

$\Delta\rho$ = density difference between the phases,

g = acceleration due to gravity,

h_1 = initial thickness of the film "immediately after release from the nozzle" assume 2000 nm,

ϕ = function that varies according to the local mobility
= $(4 - 2\ n_m)/(1 + n_m)$

n_m = number of locally mobile surfaces. This ranges from 0 to 2. Value 0 means that no mobility occurs on either of the two surfaces bounding the film. A value of 2 means that both surfaces can move freely, at least locally.

Fig 6 shows Equation 1 for n_m = 0 (or both surfaces immobile). Also shown are the data for anisole/water and toluene/water

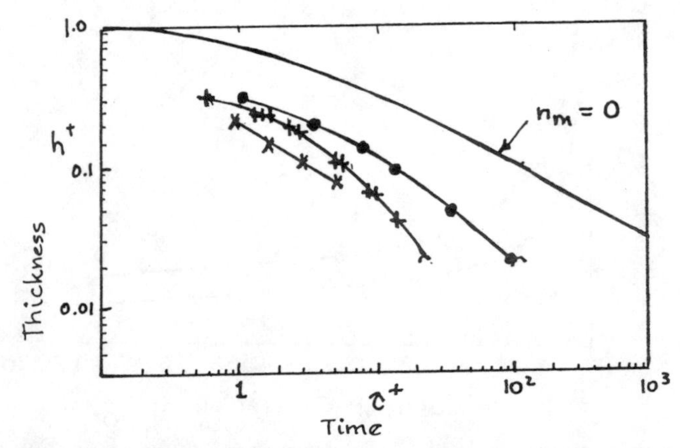

Fig 6: Comparison between film thinning data for Uneven film drainage and the modified parallel discs model of Liem and Woods for zero mobile interfaces.

anisole/water stabilized by SDS 0.015 mL volume drop ●.toluene/water stabilized by SDS 0.0025 mL volume drop x, + .

systems with sodium dodecyl sulfate as the surfactant. The data are for uneven draining only. These results suggest that some local mobility occurs. The amount of mobility depends on the elapsed time and on the amount of surfactant present in the bulk phase. The resulting mobility, expressed as the number of mobile surfaces, is shown in Fig 7. However,

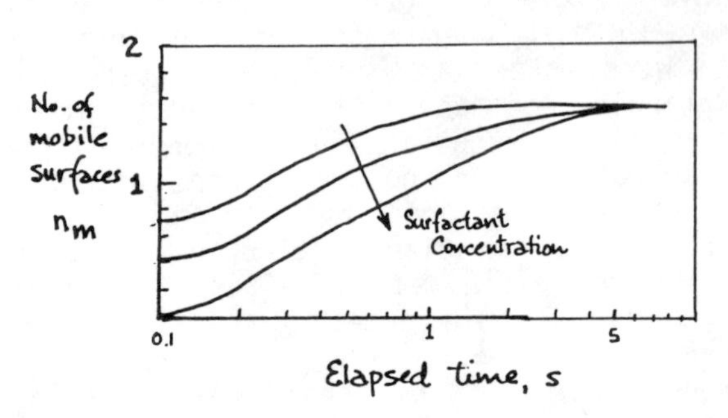

Fig 7: Surface mobility for Uneven Drainage as a Function of Elapsed Time and concentration of SDS (From Liem and Woods(9))

the modified parallel disks model of Equation 1 , which showed a +5 dependency of rate on drop radius, could not adequately correlate the drop diameter dependency for the anisole water uneven drainage (which showed a +1.6 to 2.5 power dependency) or for the cyclohexanol/water uneven drainage (which showed a +2.9 power dependency)(1)(18). Despite these limitations of the modified parallel disks model in predicting uneven drainage behaviour, it might be useful to apply this approach to the polymer stabilized, butyl chloride data. Equation 1 and the current data from Figs 2 and 3 are plotted in Fig 8 (a) to (d) for the individual concentrations and Fig 9 for all the data. There is uncertainty as to what value is most appropriate for the interfacial tension for the bulk and drop interfaces and what is the most appropriate value for the viscosity. Because both interfaces should be clean at the start of the thinning process the argument might be that an interfacial tension of about 25 mN/m would be reasonable. On the other hand, some of the thinning experiments are in the time range 80 to 1000 s which, from Fig 4, suggests that accessible interfaces should have interfacial tensions in the order of 4 to 5 mN/m. Thus, there is uncertainty as to the "best" value of the interfacial tension to use.

Concerning just the data for the same concentration but with different drop diameters, we can estimate the diameter dependence needed to make all the lines

Fig 8: Film Thinning of the Water film between a drop of n-butyl chloride and a planar butyl chloride/water interface. Film thickness vs dimensionless time for different bulk concentrations of PVA and for different drop sizes.

a) PVA concentration: 62 mg/L
1.00 μL
1.17 μL
1.40 μL
1.74 μL
2.78 μL

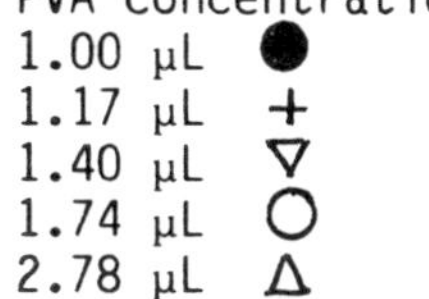

b) PVA concentration: 300 mg/L
1.40 μL
2.13 μL

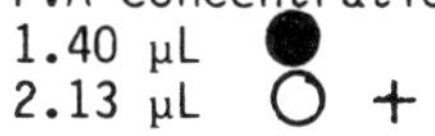

c) PVA concentration: 600 mg/L
1.0 μL ● ○
1.4 μL +
1.5 μL ×
3.4 μL Δ
3.6 μL ∇

d) PVA concentration: 1200 mg/L
2.00 μL ○
3.42 μL +

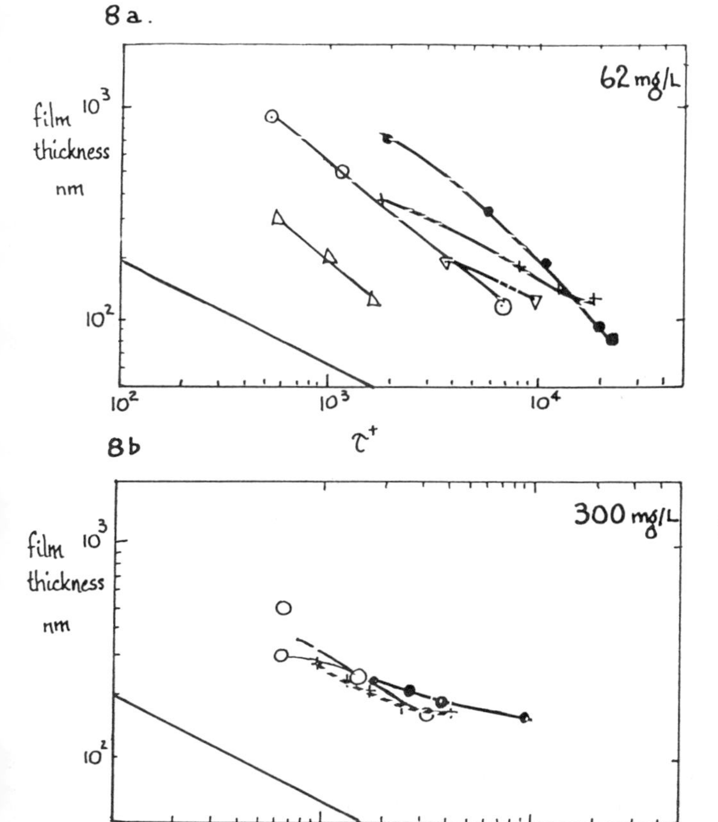

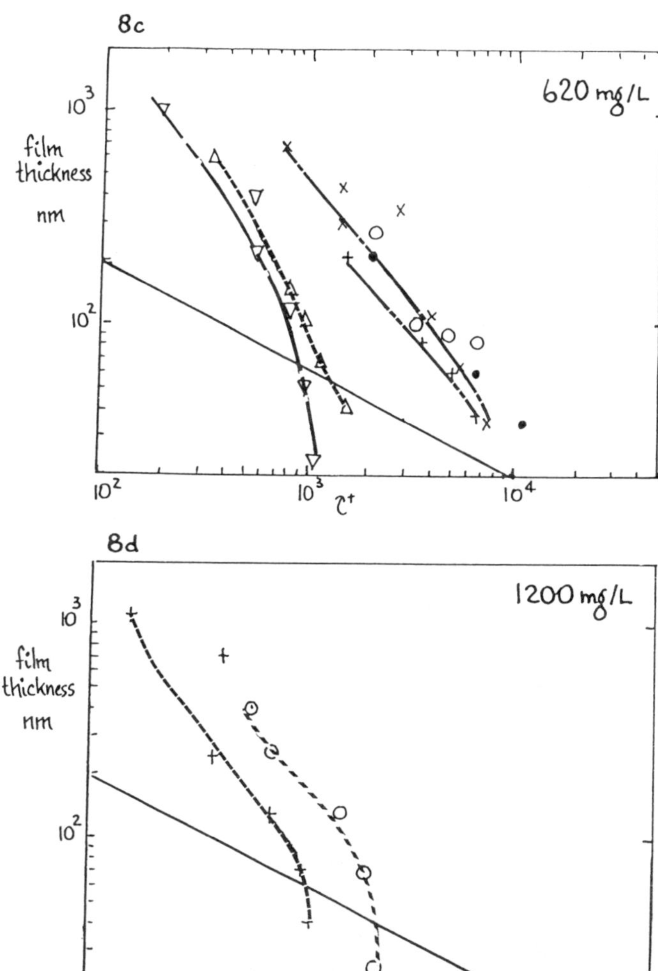

coincident (on the assumption that the cleaning was equally effective and that the effect of the thinning on the adsorption is independent of diameter). From Figs 8- (a) to (d) we conclude that the power dependency is not +5 but is rather in the range n= -2.35 to 5 depending on the concentration. For 62 ppm, n is in the range 0 to -0.15 and for the small drops -2.35; for 300 ppm, n = +5; for 600 ppm n = 0.15 to 1.2 and for 1200 ppm n = 1.3. Except for the low concentration data that show negative dependency, these trends are what we would expect based on previous work with this model.

Because the current data are average thicknesses and because of the uncertainty as to the appropriate value of interfacial tension to use, one must be careful not to read too much into the data. Nevertheless, bounding regions of behaviour can be established (based on the assumption that the interfacial tension is somewhere between 4 and 25 mN/m during the film thinning process) and, subject to those restrictions, some interesting conclusions can be drawn from the data in Fig 9:

1. most of the data are to the right of the zero mobility line (as represented by the plot of Equation 1 for nm = 0). This occurs regardless of the value of the interfacial tension used. Since one cannot have negative mobility, one might interprete this to mean that the polymers have the effect of increasing the viscosity in the film-increasing it up to 8 to 200 times the value of water. Such a hypothesis would then be consistent with the data. The range occurs because we cannot attribute for certain how much of the adjustment occurs because of a decrease in interfacial tension and how much, because of a viscosity increase. The important ratio is (γ^2/μ_c). This ratio varies from 6 to 200. For 3 ppm the ratio increase is about 6 times; for this concentration we might attribute this primarily as a viscosity effect. For all other concentrations the increase is between 70 to 200 times. The physical property ratio accounts primarily for the shift of the data beyond the model. The difference in deviation as τ^+ increases we suspect is because of mobility.

2. two different types of behaviour seem to occur with respect to mobility. For concentrations 620 and 1300 ppm, the mobility changes rapidly starting at film thicknesses of about 800 to 1000 nm. The mobility levels off at about 1.8 to 1.9 locally mobile interfaces. This value is slightly higher than that shown for surfactant stabilized systems. For these systems, the drop diameter power dependency is about what is observed for surfactant stabilized systems. The power dependency for the diameter is about 1.0 to 1.5.

3. for concentrations 60 and 300 mg/L the surface mobility seems to be constant about 1.5 mobile interfaces, although some of the dilute data deviate from this behaviour. This is illustrated in Fig 10.

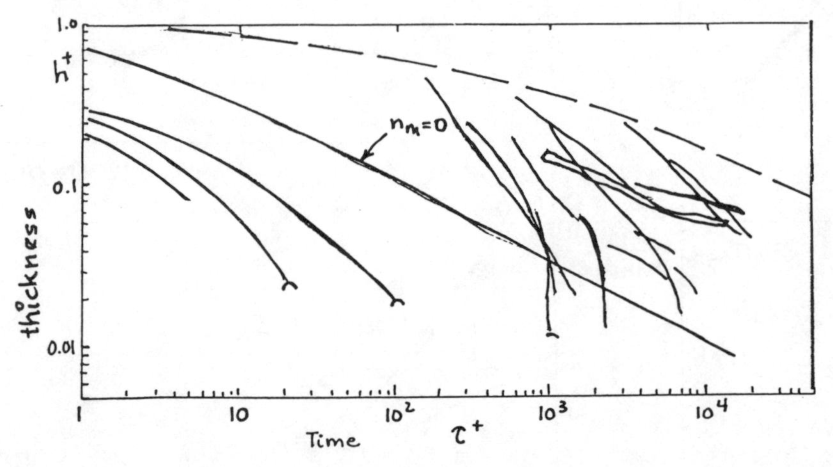

Fig 9: Film thinning of the butyl chloride/water system for Uneven thinning together with the modified parallel disks model for zero mobile surfaces

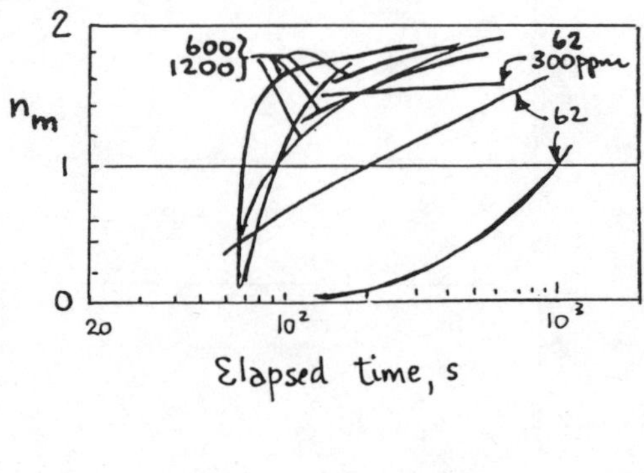

Fig 10: Surface Mobility for Uneven drainage in the butyl chloride/water system for different concentrations of PVA

value depends on the thickness of the film. The lower values of the ratio refer to film thicknesses in the order of 1000nm whereas the factor of 200 refers to behaviour of films of thickness less than 600 nm. If it was all an interfacial tension reduction caused by adsorption into the interfaces bounding the film, for this system this could account for only a factor of 25 difference; the remaining factor of 8 difference would have to be accounted for by something other than the interfacial tension; we would expect it to be the viscosity.

CONCLUSIONS

1. For free films of water trapped between a butyl chloride drop and a planar interface, the most stable thickness occurs at 60 and 300 mg/L as measured in a small coalescence cell where the surface to volume ratio is small.

2.The purity of the n-butyl chloride affects the stable thickness. Inpurities tend to reduce the stable thickness.

3. Contrary to the findings reported in conclusion #1, for an emulsion created with the same chemicals but where the surface area per unit volume is large, the concentrations 60 to 300 mg/L do not yield stable dispersions. Estimates suggest that the reason this occurs is because insufficient polymer is present ot saturate the surfaces. A concentration of about 600 mq/L is needed. This work emphasizes the care that is needed in interpreting film studies and relating these to industrial applications.

4. For concentrations of 60 mg/L and greater all the films thinned by uneven drainage. A model that matches these conditions approximately is the partial mobility model. Although this is a challenge to apply to this system because the interfacial tension is decreasing during the thinning process, nevertheless some general, tentative conclusions are:

a) within the subset of uneven drainage, the lower concentration systems (60 and 300 mg/L) show different diameter dependence and mobile surface behaviour than do the higher concentration systems.

b) for the higher concentration systems, the diameter dependence and the mobility patterns are similar to the behaviour of anisole and toluene /water systems that are stabilized by surfactants.

c) for all the systems, the adsorbed PVA seems to increase the local viscosity of the water in the film by a factor of about 6 to 8. This magnitude is very qualitative because we do not know the interfacial tension accurately throughout the film thinning process. The viscosity could be as large as 200 times that of water but we think this is unlikely.

ACKNOWLEDGEMENTS

We are pleased to thank Imperial Oil Ltd and Natural Sciences and Engineering Research Council of Canada for financially supporting this work. We thank Dr James Wallace of Esso Chemical Co Ltd, Sarnia for his advice during this project and for supplying the PVA.

REFERENCES

1. Burrill, K.A. PhD thesis, McMaster University, Hamilton Ont, (1970).
2. Dolan, A.K. and Edwards, S.F. Proc Roy Soc. London, A 337, p 509 to 516 (1974)
3. Dolan, A.K. and Edwards, S.F.. Proc Royal Soc. London, A 343 427 to 443 (1975)
4. Fleer, G.J. and Lyklema, J. Proc 5th International Congress on Surface Activity, Barcelona, 2, p 247 to 256 (1968)
5. Hesselink, F.Th., Vrij,A. and Overbeek, J.Th. G, J Phs Chem 75 p 2094 (1971)
6. Hesselink, F.Th. J Polymer Sci, 61 p 439 to 449 (1977)
7. Hodgson, T.D. and Woods, D.R. J Colloid and Int Sci 30, P 429 to 446 (1969)
8. Lankveld, J.M.G., "Adsorption of Polyvinyl Alcohol on Paraffin-water Interfaces and the Properties of Paraffin in water Emulsions Stabilized by Polyvinyl Alcohol" PhD Thesis, Agricultural University, Wageningen, The Netherlands (1970)
9. Liem, A.J.S. and Woods, D.R. Can J of Chemical Engng p 225 (1974)
10. Liem, A.J.S. "Film Thinning in Oil/Water Coalescence" PhD Thesis, McMaster University, Hamilton, Canada (1975)
11. Napper, D.H. J Colloid and Interface Sci 32 p 106 (1970)
12. Napper, D.H. in Colloidal Dispersions J.W. Goodwin, ed The Royal Soc of Chemistry, London p 99 (1982)
13. Nilsson et al "Suspension Stabilizers for PVC Production I" J of Vinyl Technology 7, No. 3, p 112 to 118 (1985)
14. Padovan, Dilva "Polyvinyl Alcohol as a Steric Stabilizer" M.Eng Thesis, McMaster University, Hamilton, Ont Canada.(1985)
15. Scheutjens, J.M.H.M. and Fleer, G.J. Advances in Colloid and Interface Science, 16, p 361 to 380 (1982).
16. Sonntag, H., et al "Experimentelle Untersuchung der Sterischen Stabilisierung von Emulsionen durch Polyvinylalkohol" Colloid and Polymer Sci 257, p 286 to 291 (1979)
17. van Vleit, T. "Interactions Between Adsorbed Macromolecules" PhD Thesis, Agricultural University, Wageningen, The Netherlands, (1977).
18. Woods, D.R. and Burrill, K.A.,"The Stability of Emulsions" J of Electroanalytical Chemistry, 37, pp 191 to 312 (1972).

EFFECTS OF SURFACE VISCOSITIES ON THE THINNING AND RUPTURE OF A DIMPLED LIQUID FILM AS A SMALL BUBBLE APPROACHES A LIQUID-GAS INTERFACE

Pil-Soo Hahn and John C. Slattery ■ Department of Chemical Engineering
Northwestern University, Evanston, Illinois 60201

When a small bubble approaches a liquid-gas interface, a thin liquid film forms between them and begins to drain. Lin and Slattery (*1*) and Hahn et al., (*2*) described the drainage of this film in the limit where the surface tension gradients were sufficiently large that the interfaces could be considered to be immobile (the tangential components of velocity at the phase interfaces are zero).

Here the effects of the surface viscosities are included by allowing the interface to be partially mobile. Given the bubble radius and the required physical properties including the surface viscosities, we are able to predict both the configuration of the film as a function of time and the coalescence time (the time at which the film becomes unstable and ruptures).

SCOPE

The rate at which bubbles coalesce is important to the preparation and stability of foams, to the separation of minerals by froth flotation, to the displacement of an unstable foam used for mobility control in a tertiary oil recovery process, and to a broad class of operations including gas adsorption and sparging in which a gas or vapor dispersion is formed. On a smaller scale, when the two bubbles are forced to approach one another in a liquid phase or when a bubble is driven through a liquid phase to a liquid-gas interface, a thin dimpled liquid film forms between the two interfaces and begins to drain. As the thickness of the draining film becomes sufficiently small (about 1000 Å), the effects of the London-van der Waals forces and of any electrostatic double layer become significant. Depending on the sign and the magnitude of the disjoining pressure attributable to the London-van der Waals forces and to the repulsive force of any electrostatic double layer, there may be a critical thickness at which the film becomes unstable, ruptures, and coalescence occurs.

In prior analyses for the effects of the surface viscosities upon the coalescence of small bubbles at a liquid-gas interface (3 to (5) the dimpling of the liquid film has been neglected. The film has been assumed to be bounded by parallel, plane interfaces.

The analysis of Hahn et al. (2; see also 1) account for the dimpling of the liquid film as a function of time, but it assumes that the interfaces are immobile (the tangential components of velocity at these interfaces are zero) and that the surface viscosities play no role in the drainage process. In what follows, we extend their discussion to include the effects of the surface viscosities.

CONCLUSIONS AND SIGNIFICANCE

When the bubble radius and the required physical properties are specified, the coalescence time can be determined for a small bubble approaching a liquid-gas interface. Our predicted coalescence time is an upper bound in the sense that it does not allow for the development of asymmetric drainage and of instabilities leading to premature rupture as observed by some experimentalists. It would not necessarily give an upper bound for systems in which electrostatic double-layer forces played an important role. It should also not be expected to given an upper bound, when a poor estimate of the strength of the London-van der Waals force is employed.

The stability of a draining liquid film is a strong function of the surface viscosities, the bubble radius, and the viscosity of the film liquid for a large intermediate range of these variables. If the surface viscosities are sufficiently large and if the bubble radius and the viscosity of the liquid film are sufficiently small, the liquid-gas interfaces are immobile, our prediction for the coalescence time approaches that of Hahn et al. (2), and the coalescence time is insensitive to increases in the surface viscosities.

There are significant differences between the dependence of the coalescence time predicted by our current analysis and that found by our previous discussion (5), in which the film was assumed to be bounded by parallel planes. We conclude that it is important to recognize the dimpling of the liquid film formed as a small bubble approaches a liquid-gas interface.

INTRODUCTION

The effects of a surfactant on the coalescence time or rest time (the time during which a bubble appears to rest at a phase interface before coalescence occurs) has been studied experimentally by Allan et al. (6), Hartland (7), Hodgson and Lee (8), Hodgson and Woods (9), Komasawa and Otake (10), Lang and Wilke (11), and Burrill and Woods (12, 13). In the absence of surfactant, the coalescence times were very small. The addition of a small amount of surfactant dramatically increases the coalescence time.

Some previous studies suggest that the stability of foams and eumulsions is strongly affected by the surface viscosities (3, 14 to 21). These studies are not entirely convincing, either because they are not sufficiently systematic or because the technique used to measure the surface shear viscosity is open to question. They do not include measurements of the surface dilatational viscosity, which is likely to be the dominant of the two (22 to 24).

Barber and Hartland (3) included the effects of the surface viscosities in describing the thinning of a liquid film bounded by partially mobile parallel planes. Flumerfelt et al. (4) extended this analysis to include the effects of the London-van der Waals forces. The impact of these discussions is diminished by the manner in which the effects of the surface viscosities were combined with those of the surface tension gradient.

Hahn and Slattery (5) also adopted the development of Barber and Hartland (3; see also 4), but with three important modifications. First, the rate of thinning of the liquid film was argued to be sufficiently small that the effects of any surface tension gradient could be neglected rather than combined with those of the surface viscosities as did Barber and Hartland (3; see also 4). Second, since critical film thicknesses measured or predicted are normally larger than 120 Å, they suggested that the London-van der Waals forces were retarded rather than non-retarded as assumed by Flumerfelt et al. (4). Third, they adopted the linear stability analysis suggested by Chen et al. (25), in order to predict the critical time or coalescence time at which the film becomes unstable and presumably begins to rupture. They found that, for a large intermediate range of the surface viscosities, the coalescence time is a strong function of these parameters.

In what follows, we extend the analysis of Hahn et al. (2; see also 1) to include the effects of the surface viscosities, which leads us to the prediction of the dimpled film profile as a functon of time and to (what we would expect to be) a better estimation of the coalescence time.

STATEMENT OF PROBLEM

Figure 1 shows the draining liquid film formed as a small bubble approaches a liquid-gas interface. We will adopt a number of same assumptions made by Lin and Slattery (1; see also 2), who did not include the effects of surface viscosities.

i) Viewed in the cylindrical coordinate system of Figure 1, the two interfaces bounding the draining liquid film are axisymmetric (i=1,2):

$$z^* = h_i^*(r^*, t^*) \tag{1}$$

(The superscript $...^*$ denotes a dimensional variable.)

ii) The dependence of h_i^* (i=1,2) upon r^* is sufficiently weak that

$$\left(\frac{\partial h_i^*}{\partial r^*}\right)^2 << 1 \tag{2}$$

iii) Introducing

$$h^* \equiv h_1^* - h_2^* \tag{3}$$

let R^* be the rim radius of the bubble such that

at $r^* = R^*(t^*)$:

$$\frac{\partial h^*}{\partial r^*} = 0 \tag{4}$$

The Reynolds lubrication theory approximation applies in the sense that, if

$$h_0^* \equiv h^*(0,0) \tag{5}$$

and

$$R_0^* \equiv R^*(0) \tag{6}$$

we will require

$$\left(\frac{h_0^*}{R_0^*}\right)^2 << 1 \tag{7}$$

iv) The effect of mass transfer on the velocity distribution is neglected.

v) The pressure p_0^* within the bubble is independent of time and position. The pressure within phase B is equal to the local hydrostatic pressure p_h^*, which is also assumed to be a constant.

vi) The liquid is an incompressible, Newtonian fluid, the viscosity of which is a constant.

vii) All inertial effects are neglected.

viii) The effects of gravity and of electrostatic double-layer forces are neglected within the liquid film. London-van der Waals forces are taken into account.

ix) The pressure within the draining film approaches its local hydrostatic value beyond the rim where the Reynolds lubrication theory approximation (assumption iii) is still valid. At this point $(r^* = R_h^*)$, the two principal curvatures of the bubble are constants independent of time,

at $r^* = R_h^*$:

$$\frac{\partial h^*}{\partial r^*} = \left(\frac{\partial h^*}{\partial r^*}\right)_{t^*=0} \tag{8}$$

at $r^* = R_h^*$:

$$\frac{\partial^2 h^*}{\partial r^{*2}} = \left(\frac{\partial^2 h^*}{\partial r^{*2}}\right)_{t^*=0} \tag{9}$$

x) Experimental observations (Allan et al., 6, 9) suggest that there is a time at which the thinning rate at the rim is equal to the thinning rate at the center. At time $t^* = 0$ in our computations, the thinning rate is independent of radial position. We will assume that for $t^* > 0$ the thinning rate at the center is always greater than the thinning rate at the rim, so long as the effects of any disjoining pressure are negligible.

xi) The bubble is sufficiently small that it may be assumed to be spherical. This is equivalent to assuming that the Bond number

$$N_{Bo} \equiv \frac{\Delta\rho^* g^* R_b^{*2}}{\gamma^*} << 1 \qquad (10)$$

Here $\Delta\rho^*$ is the density difference between the continuous phase and the gas phase, g^* the magnitude of the acceleration of gravity, R_b^* the radius of the bubble, and γ^* the surface tension.

xii) Within the draining liquid film, the mutual force per unit mass $\underset{\sim}{b}_m^*$ known as the London-van der Waals force is representable in terms of a scalar potential ϕ^*:

$$\underset{\sim}{b}_m^* = -\nabla\phi^* \qquad (11)$$

At a planar liquid-gas interface (26, 27)

$$\rho^* \phi^* (h_i^*) = \Phi_B^* + \frac{B^*}{h^{*m}} \qquad (12)$$

where ρ^* is the mass density of the liquid film, $\phi^*(h_i^*)$ is the London-van der Waals interaction potential per unit mass evaluated at the liquid-gas interface, and Φ_B^* is the interaction potential per unit volume of a semi-infinite film liquid in the limit as the liquid-gas interface is approached. When the film thickness is less than 120 Å, $m \doteq 3$; when the film thickness is larger than 400 Å, $m \doteq 4$ (28, 29). We speak of

$$\Pi^* \equiv -\frac{B^*}{h^{*m}} \qquad (13)$$

as the disjoining pressure of a flat film of thickness h^*. Note that B^* is always positive when two homophases approach each other. In this case, the interaction potential per unit volume of the continuous phase at the interface is larger than it would be if the continuous phase were semi-infinite. This corresponds to a negative disjoining pressure that acts to draw the two liquid-gas interfaces together. A stable film of uniform thickness will never be formed in this case.

xiii) Because the dependence of h^* upon r^* is weak (assumption ii), we will assume that the local value of the interacton energy per unit volume of the film liquid at the liquid-gas interface is equal to that of a flat film of the same thickness.

xiv) Since the critical film thicknesses measured or predicted by Allan et al. (6), MacKay and Mason (30), Vrij (31), Ivanov et al. (32), Burrill and Woods (12), and Chen et al. (25) are normally larger than 120 Å, we expect that m = 4 in Eq. 12 is more appropriate. Several authors (26, 28, 33 to 35) reported the approximate values of B^* between two flat surfaces as

for $m = 4$: $|B^*| \sim 10^{-19}$ erg cm

Since we lack more detailed information, we will use this value for B^* in our analysis.

In addition to these assumptions, we will use (5):

xv) The liquid-gas interfacial stress-deformation behavior can be represented by the linear Boussinesq surface fluid model (36, 37)

$$\underset{\approx}{S}^{(\sigma)*} \equiv \underset{\approx}{T}^{(\sigma)*} - \gamma^{*}\underset{\approx}{P}$$

$$= [(\kappa^{*} - \varepsilon^{*})\mathrm{div}_{(\sigma)}\underset{\sim}{v}^{(\sigma)*}]\underset{\approx}{P} + 2\varepsilon^{*}\underset{\approx}{D}^{(\sigma)*} \quad (14)$$

Here $\underset{\approx}{S}^{(\sigma)*}$ is the viscous portion of the surface stress tensor, $\underset{\approx}{T}^{(\sigma)*}$ the surface stress tensor, $\underset{\approx}{P}$ the projection tensor that transforms vectors defined on the dividing surface into their tangential components, κ^{*} the surface dilatational viscosity, ε^{*} the surface shear viscosity, $\underset{\sim}{v}^{(\sigma)*}$ the surface velocity vector,

$$\underset{\approx}{D}^{(\sigma)*} \equiv \frac{1}{2}[\underset{\approx}{P}\cdot\nabla_{(\sigma)}\underset{\sim}{v}^{(\sigma)*} + (\nabla_{(\sigma)}\underset{\sim}{v}^{(\sigma)*})^{T}\cdot\underset{\approx}{P}] \quad (15)$$

the surface rate of deformation tensor, $\nabla^{(\sigma)}$ the surface gradient operator, and $\mathrm{div}_{(\sigma)}$ the corresponding surface divergence operation (38, 39).

xvi) The rate of thinning of the liquid film is sufficiently small that surfactant concentration in each of the dividing surfaces bounding the film may be considered to be nearly independent of position. Since the surface tension and the two surface viscosities are determined by the local surfactant concentration, they may be treated as being independent of position as well. In order to establish the validity of this assumption, repeat the argument that Giordano and Slattery (40, their Appendix A; see also 5) developed for displacement in a capillary. It is because we assume that the surfactant concentration in each of the dividing surfaces is independent of position that we neglect the effect of the surface tension gradient with respect to the effect of the surface viscosities in the jump momentum balance or the force balance at each of the interfaces.

xvii) We expect that a freshly formed bubble will have a different surfactant concentration than the older interface that it is approaching and that this surfactant concentration is changing rapidly as a function of time as the result of adsorption. For simplicity, we will assume that both interfaces have the same surfactant concentrations, surface tension, and surface viscosities and that all of these properties are independent of time.

xviii) Viscous effects in the gas phase are neglected with respect to those of the liquid phase in the jump momentum balance at each of the interfaces.

SOLUTION

The fundamental equations describing this problem are identical with those presented by Hahn et al. (2) except in one respect: the jump mementum balance or the force balance at the phase interfaces. With assumptions iv, v, vii, viii, and xv through xviii, the r and z components of the jump momentum balance assume the forms at $z = h_1$

$$-\frac{2}{N_{ca}} k H_1 \frac{\partial h_1}{\partial r} - k(p - p_h)\frac{\partial h_1}{\partial r} - \frac{1}{k}\frac{\partial v_r}{\partial z} + \frac{(\kappa+\varepsilon)}{k}\left(\frac{\partial^2 v_r^{(\sigma)}}{\partial r^2} + \frac{1}{r}\frac{\partial v_r^{(\sigma)}}{\partial r} - \frac{1}{r^2} v_r^{(\sigma)}\right) = 0 \qquad (16)$$

$$\frac{2}{N_{ca}} H_1 + (p - p_h) - 2\frac{\partial v_z}{\partial z} + \frac{\partial v_r}{\partial z}\frac{\partial h_1}{\partial r} + (\kappa+\varepsilon)\left[\frac{1}{r}\frac{\partial}{\partial r}\left(r\frac{\partial v_r^{(\sigma)}}{\partial r}\right)\frac{\partial h_1}{\partial r} + \frac{\partial v_r^{(\sigma)}}{\partial r}\frac{\partial^2 h_1}{\partial r^2}\right] + (\kappa-\varepsilon)\left[\frac{1}{r}\frac{\partial v_r^{(\sigma)}}{\partial r}\frac{\partial h_1}{\partial r} + \frac{1}{r} v_r^{(\sigma)}\frac{\partial^2 h_1}{\partial r^2}\right] = 0 \qquad (17)$$

Here and in what follows, we define

$$r \equiv \frac{r^*}{R_0^*} \qquad z \equiv \frac{z^*}{h_0^*}$$

$$h_i \equiv \frac{h_i^*}{h_0^*} \qquad H_i \equiv H_i^* R_0^* \quad (i=1,2)$$

$$h \equiv h_1 - h_2 \qquad p \equiv \rho^*\left(\frac{R_0^*}{\mu^*}\right)^2 (p^* - p_0^*)$$

$$v_r \equiv \frac{\rho^* R_0^*}{\mu^*} v_r^* \qquad v_z \equiv \frac{1}{k}\frac{\rho^* R_0^*}{\mu^*} v_z^*$$

$$t \equiv \frac{\mu^*}{\rho^*(R_0^*)^2} t^* \qquad \phi \equiv \left(\frac{\rho^* R_0^*}{\mu^*}\right)^2 \phi^*$$

$$\mathcal{P} \equiv p + \phi \qquad (18)$$

$$\kappa \equiv \frac{\kappa^*}{\mu^* R_0^*} k \qquad \varepsilon \equiv \frac{\varepsilon^*}{\mu^* R_0^*} k$$

$$N_{ca} \equiv \frac{\mu^{*2}}{\rho^* \gamma^* R_0^*} \qquad k \equiv \frac{h_0^*}{R_0^*} \qquad (19)$$

and H_i^* $(i=1,2)$ are the mean curvatures of the interfaces. The θ component of the jump momentum balance is satisfied identically. Adding $k\frac{\partial h}{\partial r}$ times Equation (17) to Equation (16) and recognizing assumptions ii and iii, we have at $z = h_1$:

$$-\frac{\partial v_r}{\partial z} + (\kappa+\varepsilon)\left[\frac{\partial^2 v_r^{(\sigma)}}{\partial r^2} + \frac{1}{r}\frac{\partial v_r^{(\sigma)}}{\partial r} - \frac{1}{r^2} v_r^{(\sigma)}\right] = 0 \qquad (20)$$

and Equation (16) implies

$$\text{at } z = h_1: \quad 2H_1 + N_{ca}(p - p_h) = 0 \qquad (21)$$

In a similar fashion, we can also see that

at $z = h_2$:

$$\frac{\partial v_r}{\partial z} + (\kappa+\varepsilon)\left[\frac{\partial^2 v_r^{(\sigma)}}{\partial r^2} + \frac{1}{r}\frac{\partial v_r^{(\sigma)}}{\partial r} \frac{1}{r^2} v_r^{(\sigma)}\right] = 0 \qquad (22)$$

at $z = h_2$: $2H_2 - N_{ca}p = 0$ (23)

Hahn et al. (2) assumed that the interfaces were immobile. We recognize here that they are mobile and require

$$\text{at } z = h_1, h_2: \quad v_r = v_r^{(\sigma)} \tag{24}$$

Integrating the r component of the Navier-Stokes equation twice consistent with Equation (24), we find

$$v_r = \frac{k^2}{2}\frac{\partial \mathcal{P}}{\partial r}[z^2 - (h_1 + h_2)z + h_1 h_2] + v_r^{(\sigma)} \tag{25}$$

Substituting this into the equation of continuity and integrating once, we have

$$-v_z = \frac{k^2}{2}\left(\frac{1}{r}\frac{\partial \mathcal{P}}{\partial r} + \frac{\partial^2 \mathcal{P}}{\partial r^2}\right)\left[\frac{1}{3}z^3 - \frac{1}{2}(h_1 + h_2)z^2 + h_1 h_2 z\right] - \frac{k^2}{2}\frac{\partial \mathcal{P}}{\partial r}\left[\frac{1}{2}z^2\left(\frac{\partial h_1}{\partial r} + \frac{\partial h_2}{\partial r}\right) - \left(h_1\frac{\partial h_2}{\partial r} + \frac{\partial h_1}{\partial r}h_2\right)z\right] + \left(\frac{1}{r}v_r^{(\sigma)} + \frac{\partial v_r^{(\sigma)}}{\partial r}\right)z + C(r) \tag{26}$$

in which $C(r)$ is an as yet undetermined function of r.

Since we neglect the effect of mass transfer on the velocity distribution (assumption iv)

$$\text{at } z = h_i: \quad -\frac{\partial h_i}{\partial t} = -v_z^{(\sigma)} + \frac{\partial h_i}{\partial r}v_r^{(\sigma)} \tag{27}$$

Taking the difference between these two expressions and employing Equations (25) and (26), we find

$$\frac{\partial h}{\partial t} = \frac{k^2}{12}\left(\frac{1}{r}\frac{\partial \mathcal{P}}{\partial r} + \frac{\partial^2 \mathcal{P}}{\partial r^2}\right)h^3 + \frac{k^2}{4}\frac{\partial \mathcal{P}}{\partial r}\frac{\partial h}{\partial r}h^2 - \frac{1}{r}\frac{\partial(r\, v_r^{(\sigma)})}{\partial r}h - v_r^{(\sigma)}\frac{\partial h}{\partial r} \tag{28}$$

Taking the difference between Equations (21) and (23), recognizing that

$$\frac{\partial \mathcal{P}}{\partial z} << \frac{\partial \mathcal{P}}{\partial r} \tag{29}$$

and applying the appropriate expressions for the dimensionless mean curvatures, we see

$$\mathcal{P} = \frac{1}{2N_{ca}}\left[N_{ca}\, p_h + 2\,\Phi_B + \frac{2kB}{h^m} - k\left(\frac{1}{r}\frac{\partial h}{\partial r} + \frac{\partial^2 h}{\partial r^2}\right)\right] \tag{30}$$

in which

$$\Phi_B \equiv \frac{R_0^*}{\gamma^*}\Phi_B^*$$

$$B \equiv \frac{R_0^{*2}}{\gamma^* h_0^{*m+1}}B^* \tag{31}$$

Inserting Equations (12) and (30) into Equation (28), we discover

$$-\frac{\partial h}{\partial t'} = \frac{h^3}{3}\left(\frac{1}{r^3}\frac{\partial h}{\partial r} - \frac{1}{r^2}\frac{\partial^2 h}{\partial r^2} + \frac{2}{r}\frac{\partial^3 h}{\partial r^3} + \frac{\partial^4 h}{\partial r^4}\right) + h^2\frac{\partial h}{\partial r}\left(-\frac{1}{r^2}\frac{\partial h}{\partial r} + \frac{1}{r}\frac{\partial^2 h}{\partial r^2} + \frac{\partial^3 h}{\partial r^3}\right)$$

$$+ \frac{8}{3} B \left[\frac{1}{r} \frac{1}{h^2} \frac{\partial h}{\partial r} + \frac{1}{h^2} \frac{\partial^2 h}{\partial r^2} - \frac{2}{h^3} \left(\frac{\partial h}{\partial r}\right)^2\right]$$

$$+ \frac{8}{r} \frac{\partial (r v_r^{(\sigma)'})}{\partial r} h + 8 v_r^{(\sigma)'} \frac{\partial h}{\partial r} \qquad (32)$$

in which we have found it convenient to introduce

$$t' \equiv \frac{h_0^* \gamma^*}{8 R_0^{*2} \mu^*} t^*$$

$$v_r^{(\sigma)'} \equiv \frac{\mu^*}{k^3 \gamma^*} v_r^* \qquad (33)$$

Given Equations (25) and (30), we can rearrange either Equation (20) or (22) as

$$r^2 \frac{\partial^2 v_r^{(\sigma)'}}{\partial r^2} + r \frac{\partial v_r^{(\sigma)'}}{\partial r} - v_r^{(\sigma)'} = - M(r) \quad (34)$$

with

$$M(r) \equiv \frac{r^2 h}{4(\kappa+\varepsilon)} \frac{\partial}{\partial r} \left(\frac{1}{r} \frac{\partial h}{\partial r} + \frac{\partial^2 h}{\partial r^2} - \frac{2B}{h^4}\right) \qquad (35)$$

Applying the method of variation of parameters, we find that the solution to this equation has the form

$$v_r^{(\sigma)'} = C_1 r + \frac{C_2}{r} + r A_1(r) + \frac{1}{r} A_2(r) \qquad (36)$$

where

$$A_1(r) \equiv - \int_0^r \frac{M}{2r^2} dr$$

$$A_2(r) \equiv \int_0^r \frac{M}{2} dr \qquad (37)$$

We require

$$\text{at } r = 0: \quad v_r^{(\sigma)'} = 0 \qquad (38)$$

and (assumption ix)

$$\text{as } r \to R_h: \quad \frac{\partial v_r}{\partial z} \to 0 \qquad (39)$$

Equation (38) demands

$$C_2 = 0 \qquad (40)$$

For $r > R_h$, either Equation (20) or (22) reduce to

$$\frac{\partial^2 v_r^{(\sigma)'}}{\partial r^2} + \frac{1}{r} \frac{\partial v_r^{(\sigma)'}}{\partial r} - \frac{1}{r^2} v_r^{(\sigma)'} = 0 \qquad (41)$$

in view of Equation (39). This last has a solution of the form

$$v_r^{(\sigma)'} = \frac{C_3}{r} \qquad (42)$$

when we observe that the r component of the surface velocity must be positive and it must decrease as r increases. Since the velocities and the velocity gradients calculated from Equations (36) and (42) should be identical at $r = R_h$, we have two equations for the two unknowns C_1 and C_3 and we conclude

$$C_1 = - A_1(R_h) \qquad (43)$$

Substituting Equations (36) and (40) into Equation (32), we conclude

$$- \frac{\partial h}{\partial t'} = \frac{h^3}{3} \left(\frac{1}{r^3} \frac{\partial h}{\partial r} - \frac{1}{r^2} \frac{\partial^2 h}{\partial r^2} + \frac{2}{r} \frac{\partial^3 h}{\partial r^3} + \frac{\partial^4 h}{\partial r^4}\right)$$

$$+ h^2 \frac{\partial h}{\partial r} \left(- \frac{1}{r^2} \frac{\partial h}{\partial r} + \frac{1}{r} \frac{\partial^2 h}{\partial r^2} + \frac{\partial^3 h}{\partial r^3}\right)$$

$$+ \frac{8}{3} B \left[\frac{1}{rh^2} \frac{\partial h}{\partial r} + \frac{1}{h^2} \frac{\partial^2 h}{\partial r^2} - \frac{2}{h^3} \left(\frac{\partial h}{\partial r}\right)^2\right]$$

$$+ 8 \frac{\partial h}{\partial r} \left[r(C_1 + A_1) + \frac{A_2}{r}\right]$$

$$+ 16h(C_1 + A_1) \qquad (44)$$

Our first objective is to calculate the initial dependence of h upon radial position consistent with assumption x. Note that, after an application of L'Hospital's rule [1, their Equation (45); 2, their Equation (18)], we obtain

limit $r \to 0$:

$$-\frac{\partial h}{\partial t'} = \frac{8}{9} h^3 \frac{\partial^4 h}{\partial r^4} + \frac{16}{3} \frac{B}{h^2} \frac{\partial^2 h}{\partial r^2} + 16hC_1 \qquad (45)$$

Recognizing that the rate of thinning is independent of radial position at the initial time, we can use Eqs. 44 and 45 to say at $t' = 0$

$$\frac{8}{3} \left(\frac{\partial^4 h}{\partial r^4}\right)_{r=0} + 16B\left(\frac{\partial^2 h}{\partial r^2}\right)_{r=0} + 48C_1$$

$$= h^3\left(\frac{1}{r^3} \frac{\partial h}{\partial r} - \frac{1}{r^2} \frac{\partial^2 h}{\partial r^2} + \frac{2}{r} \frac{\partial^3 h}{\partial r^3} + \frac{\partial^4 h}{\partial r^4}\right)$$

$$+ 3h^2 \frac{\partial h}{\partial r}\left(- \frac{1}{r^2} \frac{\partial h}{\partial r} + \frac{1}{r} \frac{\partial^2 h}{\partial r^2} + \frac{\partial^3 h}{\partial r^3}\right)$$

$$+ 8B\left[\frac{1}{rh^2} \frac{\partial h}{\partial r} + \frac{1}{h^2} \frac{\partial^2 h}{\partial r^2} - \frac{2}{h^3} \left(\frac{\partial h}{\partial r}\right)^2\right]$$

$$+ 24 \frac{\partial h}{\partial r} \left[r(C_1 + A_1) + \frac{A_2}{r}\right]$$

$$+ 48h(C_1 + A_1) \qquad (46)$$

As discussed by Lin and Slattery (1), we require that the result be consistent with

$$\text{at } r = 0: \quad \frac{\partial h}{\partial r} = 0 \qquad (47)$$

$$\text{at } r = 0: \quad \frac{\partial^3 h}{\partial r^3} = 0 \qquad (48)$$

$$\text{at } r = 1: \quad \frac{\partial h}{\partial r} = 0 \qquad (49)$$

$$\text{as } r \to R_h: \quad \frac{1}{r} \frac{\partial}{\partial r} \left(r \frac{\partial h}{\partial r}\right) \to \frac{2R_0^*}{R_b^* k} \qquad (50)$$

In order to integrate a finite-difference form of Equation (46), we replace Equation (50) by (1, 2)

$$\text{at } r = 1: \quad \frac{\partial^2 h}{\partial r^2} = C \qquad (51)$$

where C, which is the difference between the sum of the principal curvatures for interface 1 and the sum of the principal curvatures for interface 2, is a free parameter, the value of which will be determined shortly.

For each value of C, we can determine for given values of B and $\kappa + \varepsilon$ a tentative initial configuration of the film by integrating Equation (46) consistent with Equations (47) through (49) and (51). The dimensionless radial position at which the pressure gradient becomes negligible is tentatively identified as R_h, subject to later verification that assumption ii is still satisfied at this point.

Equation (44) can be integrated consistent with each of these tentative initial configurations, Equations (47), (48), (8), and (9), the latter two boundary conditions first having been made dimensionless. We employed the Crank-Nicolson technique (41); accuracy was checked by decreasing the time and space intervals. We used $\Delta r = 0.02$ and $\Delta t' = 0.003 - 0.05$.

In addition to requiring that at time $t' = 0$ the thinning rate is independent of radial position, assumption x demands that for $t' > 0$ the thinning rate at the center is always greater than the thinning rate at the rim, so long as the effects of any disjoining pressure are negligible. Our numerical computations indicate that, for every value of $\kappa+\varepsilon$, there is a mimimum value of C (Figure 9) such that the thinning rate at the center is always greater than the thinning rate at the rim in the early stage of the thinning process. Since Lin and Slattery (1) and Hahn et al. (2) found that the effects of disjoining pressure can be neglected in the early stage of thinning, we examined only the case of $B = 0$ to determine the mimimum value of C as a function of $\kappa+\varepsilon$. For each mimimum value of C , there is a corresponding maximum value of h_0^* for which the thinning rate at the center is always greater than the thinning rate at the rim for $t' > 0$. We will choose this maximum value of h_0^* as our initial film thickness at the center for a given value of $\kappa+\varepsilon$.

Equation (49) permits us to identify R as a function of time. By R_f we mean the value of R either as $t^* \to \infty$ in the case of a positive disjoining pressure or as $t^* \to t_c^*$ in the case of a negative disjoining pressure, where t_c^* is the time at which the film ruptures and coalescence occurs. The value of R_f can be obtained from our numerical computation. For a bubble freely approaching a liquid-gas interface under the influence of gravity, R_b^* is measured and R_f^* is determined by (1, 42, 43)

as $t' \to \infty$:

$$R^* \to R_f^* = \left(\frac{4}{3}\frac{\Delta\rho^* g^*}{\gamma^*}\right)^{1/2} R_b^{*2} \qquad (52)$$

We identify

$$R_0^* = \frac{R_f^*}{R_f} \qquad (53)$$

From Equations (50) and (52), h_0^* is given by

$$h_0^* = \frac{8}{3KR_f^2}\frac{\Delta\rho^* g^* R_b^{*3}}{\gamma^*} \qquad (54)$$

in which

$$K \equiv \left[\frac{1}{r}\frac{\partial}{\partial r}\left(r\frac{\partial h}{\partial r}\right)\right]_{r=R_h} \qquad (55)$$

Values of K , R_f , and R_h are obtained from our numerical computation and tabulated in Table 1.

RESULTS

Figures 2 through 8 show the dimensionless film thickness h as a function of r and t' for $B = 10^{-7}$ and varying values of $\kappa+\varepsilon$. For different values of B , the trends are similar. The initial profiles are functions of $\kappa+\varepsilon$ through their dependence upon C as shown in Figure 9. For $0 < \log(\kappa+\varepsilon) < 1.5$, C is sensitive to small changes in $\kappa+\varepsilon$. As $\kappa+\varepsilon$ approaches 0 , our computation predicts an almost flat initial film profile. For $\kappa+\varepsilon = 10^4$ (Figure 2), all profiles are identical to those of Hahn et al. (2, their figure 4), suggesting that the phase interfaces are nearly immobile for sufficiently large values of $\kappa+\varepsilon$. This is consistent with the conclusions of Hahn and Slattery (5).

In Figure 10, the dimensionless film thickness at the center is plotted against t' and $\kappa+\varepsilon$ for $B = 10^{-8}$. Even after taking into account the relatively small variation of initial film thickness at the center for different $\kappa+\varepsilon$ [see Equation (54) and Table 1; note that γ^* decreases as surfactant concentration increases], Figure 10 shows that the film drains much faster as $\kappa+\varepsilon$ decreases.

The dimensionless thinning rate at the rim as a function of h and $\kappa+\varepsilon$ for $B = 10^{-4}$ and 10^{-6} is shown in Figures 11 and 12. It is the London-van der Waals forces that lead the film to rupture. The effect of $\kappa+\varepsilon$ is to retard the rate of thinning to the thickness from which the London-van der Waals forces are dominant. As B increases, the thickness at which the effects of London-van der Waals forces dominate increases and the coalescence time decreases. As $\kappa+\varepsilon$ decreases, the time required to reach the thickness, from which the effects of London-van der Waals forces dominate, decreases. In contrast with Figures 11 and 12 as well as Equations (32) and (35) through (38), Traykov and Ivanov (44) argue that the thinning rate is independent of surface shear viscosity.

The dimensionless surface velocity as a function of r and $\kappa+\varepsilon$ for $B = 10^{-6}$ and $t' = 0$ is shown in Figure 13. Our numerical computations show that the character of these profiles is preserved as a function of time, until the film reaches the thickness at which the London-van der Waals forces become dominant.

For a system of fully immobile phase interfaces, Hahn et al. (2) obtained

$$t^*_{c(H)} \equiv 0.79 \frac{\mu^* R_b^{*4.06} (\Delta\rho^* g^*)^{0.84}}{\gamma^{*1.38} B^{*0.46}} \quad (56)$$

with which the coalescence time can be predicted. We will find it convenient to define

$$t_c \equiv \frac{t^*_c}{t^*_{c(H)}} \quad (57)$$

Figure 14 shows t_c as a function of $\kappa+\varepsilon$ and B. The small circles represent the results of our numerical computations. Our argument assumes that the motion is sufficiently slow that the surface concentration of surfactant is nearly independent of position in the phase interfaces. This would certainly not be true as $\kappa+\varepsilon \to 0$. Our computation also neglects inertial effects, which may become important as $(\kappa+\varepsilon) \to 0$. For these reasons as well as the increasing computation time required as $\kappa+\varepsilon$ was decreased, we limited our computations to $2 < \kappa+\varepsilon < 10^4$. For $\kappa+\varepsilon > 10^4$, the system can be safely regarded as fully immobile and our numerical results for coalescence times are identical with those calculated using Equation (56).

Hahn et al. (2) were relatively unsuccessful in using Equation (56) to describe the rest times for nitrogen bubbles rising in various liquids observed by Allan et al. (6) and by MacKay and Mason (30). Those data could be explained by the present theory using plausible values for the interfacial viscosities of those systems.

Hahn and Slattery (5), who assumed in their analysis that the film was bounded by parallel planes, presented their results in terms of

$$t_{c(M)} \equiv \frac{t^*_c}{t^*_{c(M)}} \quad (58)$$

and

$$\alpha^{1/4} \equiv \frac{B^{*1/4}\gamma^{*3/4}(\kappa^{*}+\varepsilon^{*})}{8\mu^{*}\Delta\rho^{*}g^{*}R_b^{*15/4}} \tag{59}$$

Here

$$t^{*}_{c(M)} \equiv 1.046\,\frac{\mu^{*}R_b^{*9/2}\Delta\rho^{*}g^{*}}{\gamma^{*3/2}B^{*1/2}} \tag{60}$$

is the corresponding coalescence time calculated by Chen et al. (25) starting from the thinning rate equation derived by MacKay and Mason (30) for a film bounded by immobile parallel planes. From Equations (19), (31), (52) through (54), and (59), we see that

$$\kappa+\varepsilon = 5.04\ R_f^2 K^{1/4}\alpha^{1/4}B^{-1/4} \tag{61}$$

Using Equation (61) and Table 1, we have replotted the results of Hahn and Slattery (5) in Figure 14 as the solid lines.

The comparison in Figure 14 between the present results for a dimpled film and those of Hahn and Slattery (5) for a film bounded by parallel planes is distorted, since the coalescence times are made dimensionless in different fashions. From Equations (33), (52) through (54), (56) through (58), and (60) as well as Table 1, we can say

$$\frac{t_c}{t_{c(M)}} = \frac{t^{*}_{c(M)}}{t^{*}_{c(H)}}$$

$$= 1.10\ R_f^{0.32}K^{0.2}B^{-0.04} \sim 1.87\ B^{-0.04} \tag{62}$$

This allows us to replot in Figure 15 the results of Hahn and Slattery (5) in such a manner that they can be compared more directly with the current results. The plane parallel film model predicts that the coalescence time will be about three times larger than the current prediction [or the prediction of Hahn et al., (2)], when $\kappa+\varepsilon = 10^4$. As $\kappa+\varepsilon$ decreases, the coalescence time predicted by the plane parallel film model decreases much faster than the current results indicate. The plane parallel filmmodel is too superficial to give an accurate prediction for the coalescence time.

Figures 14 and 15 indicate that, for a given system, t^{*}_{c} decreases as the bubble size increases [$\kappa+\varepsilon$ decreases as R^{*}_{b} increases according to Equations (19), (52), and (53) together with Table 1]. This may suggest why the comparison of theoretical predictions of coalescence time shows increasing divergence between theory and experiment as the bubble size increases (2, 25, 45).

ACKNOWLEDGMENT

This work was sponsored in part by grant no. CBT-8511207 from the National Science Foundation. One of the authors (PSH) would also like to thank the Korean Government for financial support during a portion of this work.

NOTATION

A_1, A_2	Defined by Equation (37)
$\underset{\sim}{b}^{*}_{m}$	London-van der Waals mutual force per unit mass
B^{*}	London-van der Waals constant in Equation (12)
B	Dimensionless London-van der Waals constant defined by Equation (31)
C	Defined by Equation (51)

C_1, C_2, C_3 Constants of integration in Equations (36) and (42)

$\underset{\approx}{D}^{(\sigma)*}$ Surface rate of deformation tensor defined by Equation (15)

g^* Magnitude of acceleration of gravity

h^* film thickness, defined by Equation (3)

h Dimensionless film thickness defined by Equation (18)

h_0^* Film thickness at $t^* = 0$ and $r^* = 0$

h_1^*, h_2^* Configurations of the liquid-gas interfaces. See Figure 1 and Equation (1)

h_1, h_2 Dimensionless configurations of the bubble-liquid interfaces, defined by Equation (18). See Figure 1 and Equation (1)

H_1^*, H_2^* Mean curvatures of liquid-gas interfaces

H_1, H_2 Dimensionless mean curvature of liquid-gas interfaces, defined by Equation (18)

k Defined by Equation (19)

K Defined by Equation (55)

m Parameter in Equation (12)

$M(r)$ Defined by Equation (35)

N_{Bo} Bond number defined by Equation (10)

N_{ca} Capillary number defined by Equation (19)

p^* Pressure

p Dimensionless pressure, defined by Equation (18)

p_h^* Local hydrostatic pressure in phase B (see Figure 1)

p_h Dimensionless local hydrostatic pressure in phase B

p_0^* Pressure within the bubble

$\underset{\approx}{P}$ Projection tensor that transforms vectors defined on the dividing surface into their tangential components

$\mathcal{P}$ Dimensionless modified pressure, defined by Equation (18)

r^* Cylindrical coordinate

r Dimensionless cylindrical coordinate, defined by Equation (18)

Δr Grid size used in numerical computation

R^* Rim radius of bubble, defined by Equation (4)

R_b^* Radius of the bubble

R_f^* Rim radius as $t \to \infty$ or just prior to the development of an instability and coalescence

R_h^* Radial position where the pressure p^* within the draining film approaches the local hydrostatic pressure in the neighborhood of the bubble

R_0^* Rim radius of the bubble at $t^* = 0$

$\underset{\approx}{S}^{(\sigma)*}$ Viscous portion of the surface stress tensor, defined by Equation (14)

t^* Time

t Dimensionless time, defined by Equation (18)

t' Dimensionless time, defined by Equation (33)

t_c^* Coalescence time

t_c Dimensionless coalescence time defined by Equation (57)

$t^*_{c(H)}$	Predicted coalescence time defined by Equation (56)
$t^*_{c(M)}$	Predicted coalescence time defined by Equation (60)
$t_{c(M)}$	Dimensionless coalescence time defined by Equation (58)
$\Delta t'$	Time step used in numerical computation
$\underset{\approx}{T}^{(\sigma)*}$	Surface stress tensor
$\underset{\sim}{v}^{(\sigma)*}$	Surface velocity vector
v^*_r, v^*_z	Components of velocity
v_r, v_z	Dimensionless components of velocity defined by Equation (18)
$v^{(\sigma)}_r, v^{(\sigma)}_z$	Dimensionless components of surface velocity, defined by Equation (18)
$v^{(\sigma)'}_r$	Dimensionless r-component of surface velocity, defined by Equation (33)
z^*	Cylindrical coordinate
z	Dimensionless cylindrical coordinate, defined by Equation (18).

Greek Letters

α	Defined by Equation (59)
γ^*	Surface tension
ε^*	Surface shear viscosity
ε	Dimensionless surface shear viscosity defined by Equation (19)
κ^*	Surface dilatational viscosity
κ	Dimensionless surface dilatational viscosity defined by Equation (19)
μ^*	Viscosity of the film liquid
Π^*	Disjoining pressure defined by Equation (13)
ρ^*	Density of the liquid film
$\Delta\rho^*$	Density difference between the continuous phase and the gas phase
ϕ^*	London-van der Waals potential energy per unit mass of the liquid film
ϕ	Dimensionless London-van der Waals potential, defined by Equation (18)
Φ^*_B	Interaction potential energy per unit volume of a semi-infinite liquid film in the limit as the liquid-gas interface is approached
Φ_B	Dimensionless interaction potetial of a semi-infinite film defined by Equation (31)

Others

$\nabla_{(\sigma)}$	Surface gradient operator (38 39)
$\ldots^T$	Superscript denoting transpose operation.

REFERENCES

1. Lin, C. Y. and J. C. Slattery, AIChE J., 28, 786 (1982).
2. Hahn, P. S., J. D. Chen, and J. C. Slattery, AIChE J., 31, 2026 (1985).
3. Barber, A. D. and S. Hartland, Can. J. Chem. Eng., 54, 279 (1976).
4. Flumerfelt, R. W., J. P. Oppenheim, and J. R. Son, AIChE Symp. Ser., 78 (212), 113 (1982).
5. Hahn, P. S. and J. C. Slattery, AIChE J., 31, 950 (1985).
6. Allan, R. S., G. E. Charles and S. G. Mason, J. Colloid Sci., 16, 150 (1961).
7. Hartland, S., Trans. Inst. Chem. Eng., 46, T275 (1968).
8. Hodgson, T. D., and J. C. Lee, J. Colloid Interface Sci., 30, 94 (1969).

9. Hodgson, T. D. and D. R. Woods, J. Colloid Interface Sci., 30, 429 (1969).

10. Komasawa, I. and T. Otake, J. Chem. Eng. Japan, 3, 243 (1970).

11. Lang, S. B. and C. R. Wilke, Ind. Eng. Chem. Fundam., 10, 341 (1971).

12. Burrill, K. A. and D. R. Woods, J. Colloid Interface Sci., 42, 15 (1973).

13. Burrill, K. A. and D. R. Woods, J. Colloid Interface Sci., 42, 35 (1983).

14. Brown, A. G., W. C. Thuman, and J. W. McBain, J. Colloid Sci., 8, 491 (1953).

15. Davies, J. T., "A Study of Foam Stabilizer Using a New Surface Viscometer," in Second International Congress of Surface Activity, vol. 1, Schulman, J. H. (Ed.), p. 220, Academic Press, New York (1957).

16. Kanner, T. and J. E. Glass, Ind. Eng. Chem., 61, 31 (1969).

17. Bikerman, J. J., Foams, Springer-Verlag New York, New York (1973).

18. Joly, M., "Rheological Properties of Monomolecular Films Part II: Experimental Results, Theoretical Interpretation, Applications," in Surface and Colloid Science, vol. 1, Matijevic, E. (Ed.), p. 1, Academic Press, New York (1972).

19. Ivanov, I. B. and D. S. Dimitrov, Colloid and Polymer Sci., 252, 982 (1974).

20. Sagert, N. H. and M. J. Quinn, J. Colloid Interface Sci., 65, 415 (1978).

21. Sagert, N. H. and M. J. Quinn, J. Colloid Interface Sci., 65, 415 (1978).

22. Marv, H. C., V. Mohan, and D. T. Wasan, Chem. Eng. Sci., 34, 1283 (1979).

23. Djabbarah, N. F. and D. T. Wasan, Chem. Eng. Sci., 37, 175 (1982).

24. Stoodt, T. J., and J. C. Slattery, AIChE J., 30, 564 (1984).

25. Chen, J. D., P. S. Hahn, and J. C. Slattery, AIChE J., 30, 622 (1984).

26. Sheludko, A., D. Platikanov, and E. Manev, Discuss. Faraday Soc., 40, 253 (1965).

27. Ruckenstein, E. and R. K. Jain, Faraday Trans. II, 70, 132 (1974).

28. Churaev, N. V., Colloid J. USSR (Engl. Transl.), 36, No. 2, 283 (1974).

29. Churaev, N. V., Colloid J. USSR (Engl. Transl.), 36, No. 2, 287 (1974).

30. MacKay, G. D. M. and S. G. Mason, Can. J. Chem. Eng., 41, 203 (1963).

31. Vrij, A., Dis. Faraday Soc., 42, 23 (1966).

32. Ivanov, I. B., B. Radoev, E. Manev, and A. Sheludko, Trans. Faraday Soc., 66, 1262 (1970).

33. Kitchener, J. A. and A. P. Prosser, Proc. R. Soc. Ser. A., 242, 403 (1957).

34. Black, W., J. G. V. De Jongh, J. Th. G. Overbeck, and M. J. Sparnaay, Trans. Faraday Soc., 56, 1597 (1960).

35. Derjaguin, B. V., Y. I. Rabinovich, and N. V. Churaev, Nature (London), 265, 520 (1977).

36. Boussinesq, J., Compt. Rend. Hebd. Seances Acad. Sci., 156, 983 (1913).

37. Scriven, L. E., Chem. Eng. Sci., 12, 98 (1960); R. Aris, Vectors, Tensors, and the Basic Equations of Fluid Mechanics, Prentice-Hall, Englewood Cliffs, NJ(1962). For a list of typographical errors, see Slattery,J. C., Chem. Eng. Sci., 19, 379 (1964).

38. Wei, L. Y., W. Schmidt, and J. C. Slattery, J. Colloid Interface Sci., 48, 1 (1974).

39. Briley, P. B., A. R. Deemer, and J. C. Slattery, J. Colloid Interface Sci., 56, 1 (1976).

40. Giordano, R. M. and J. C. Slattery, AIChE J., 29, 483 (1983).

41. Myers, G. E., Analytical Methods in Conduction Heat Transfer, p. 274, McGraw-Hill, New York (1971).

42. Chappelear, D. C., J. Colloid Sci., 16, 186 (1961).

43. Princen, H. M., J. Colloid Sci., 18, 178 (1963).

44. Traykov, T. T. and I. B. Ivanov, Int. J. Multiphase Flow, 3, 471 (1977).

45. Manev, E. D., S. V. Sazdanova, and D. T. Wasan, J. Colloid Interface Sci., 97, 591 (1984).

TABLE 1: K, R_h AND R_f AS FUNCTIONS OF k+e

k+e	10^4	10^2	50	20	10	5	2
K	12.29	12.10	11.89	11.38	10.61	9.74	7.49
R_h(x50)	1.68	1.68	1.68	1.70	1.74	1.76	1.88
R_f(x50)	1.10	1.10	1.12	1.12	1.14	1.16	1.20

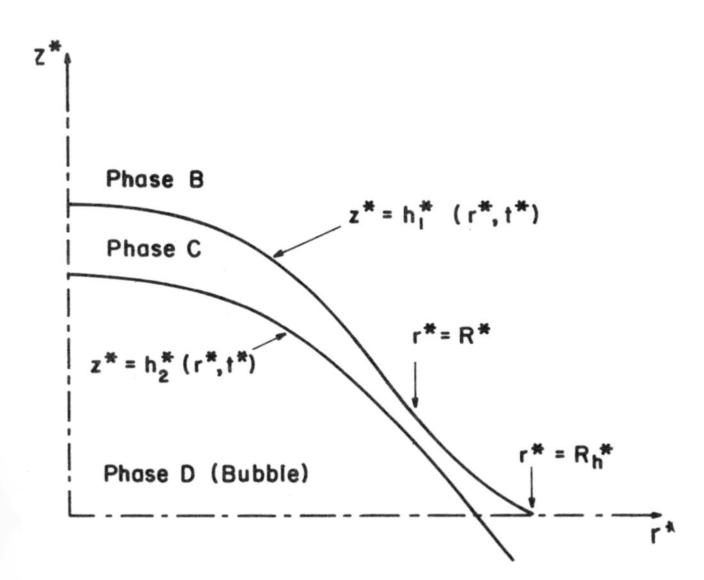

Figure 1: A symmetric bubble (phase D) moves through a liquid (phase C) as it approaches a liquid-gas interface (between phases C and B). The configuration of the bubble-liquid interface is given by $z^* = h_2^*(r^*,t^*)$; that of the liquid-gas interface by $z^* = h_1^*(r^*,t^*)$.

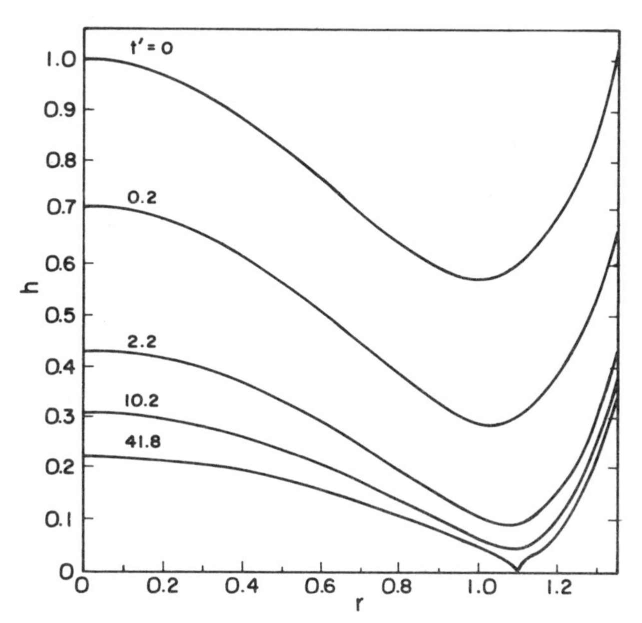

Figure 2: Dimensionless film thickness h as a function of dimensionless radial position and dimensionless time for $B = 10^{-7}$, $\kappa+\varepsilon = 10^4$, $R_h = 1.68$, $C = 4.88$.

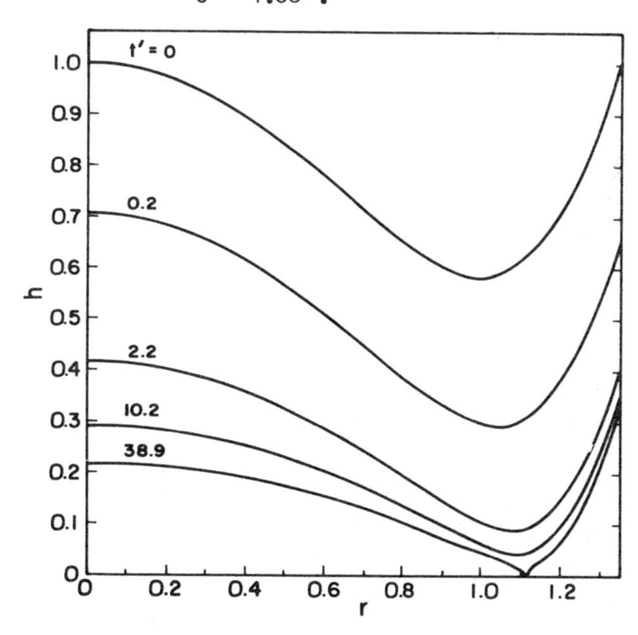

Figure 3: Dimensionless film thickness h as a function of dimensionless radial position and dimensionless time for $B = 10^{-7}$, $\kappa+\varepsilon = 10^2$, $R_h = 1.68$, $C = 4.76$.

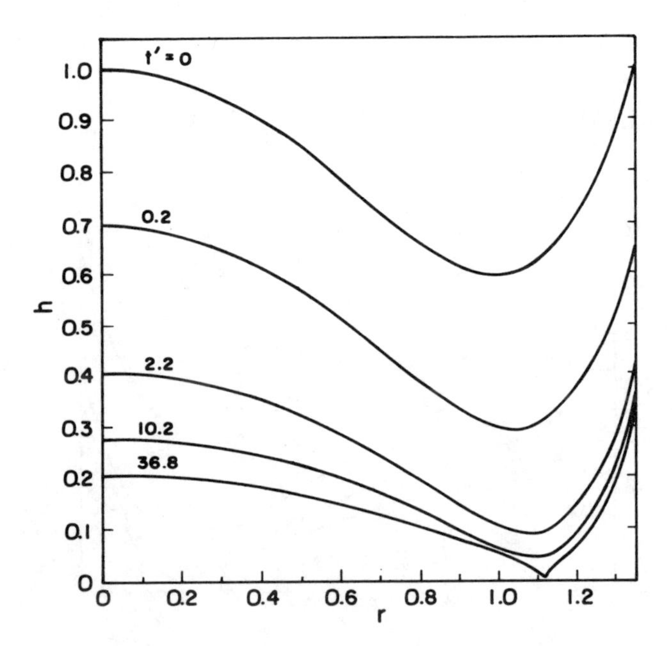

Figure 4: Dimensionless film thickness h as a function of dimensionless radial position and dimensionless time for $B = 10^{-7}$, $\kappa+\varepsilon = 50$, $R_h = 1.68$, $C = 4.64$.

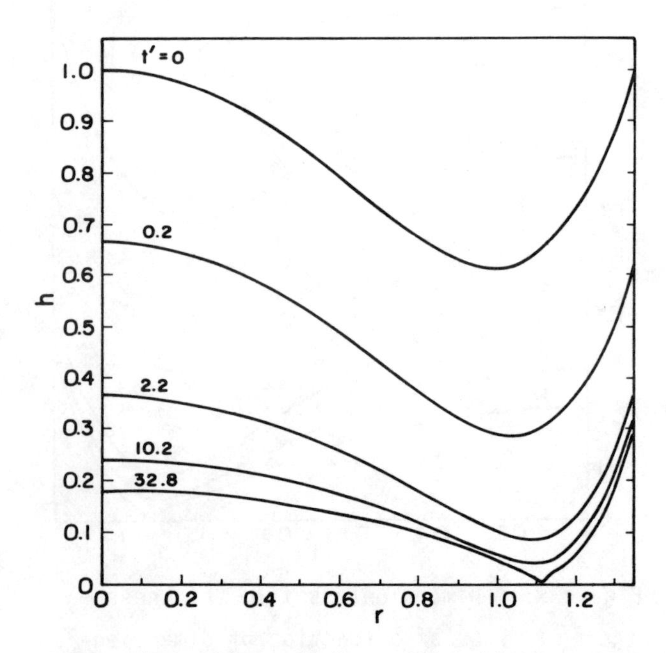

Figure 5: Dimensionless film thickness h as a function of dimensionless time for $B = 10^{-7}$, $\kappa+\varepsilon = 20$, $R_h = 1.70$, $C = 4.32$.

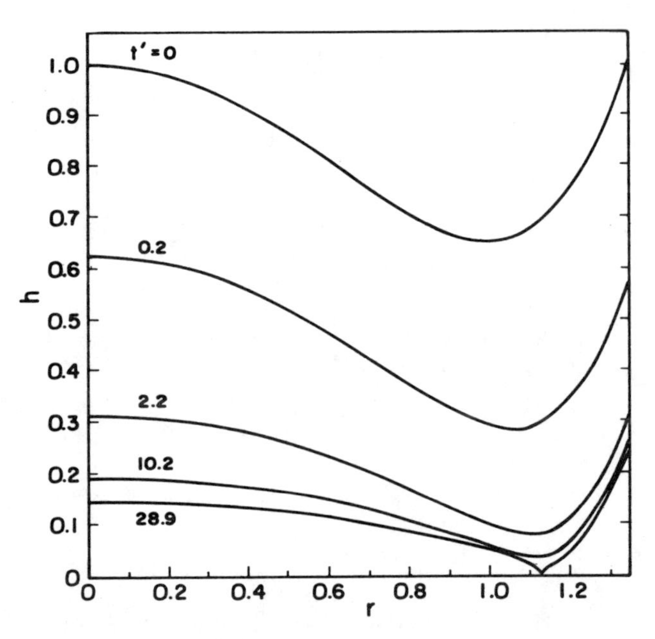

Figure 6: Dimensionless film thickness h as a function of dimensionless radial position and dimensionless time for $B = 10^{-7}$, $\kappa+\varepsilon = 10$, $R_h = 1.74$, $C = 3.86$.

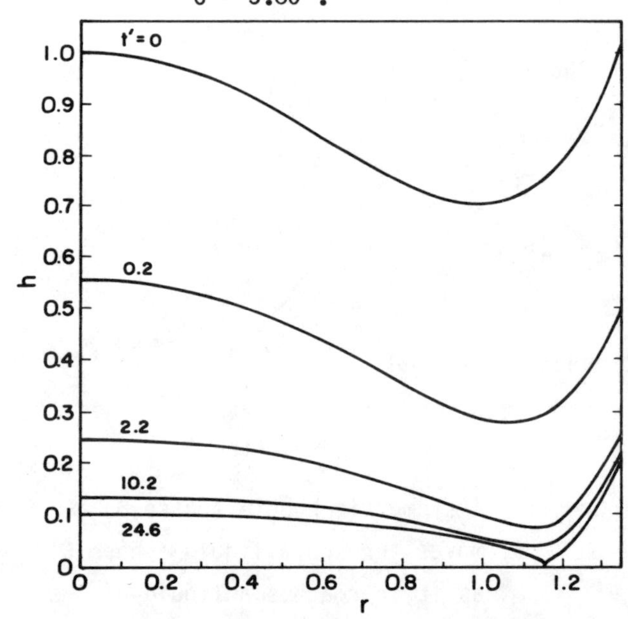

Figure 7: Dimensionless film thickness h as a function of dimensionless radial position and dimensionless time for $B = 10^{-7}$, $\kappa+\varepsilon = 5$, $R_h = 1.76$, $C = 3.31$.

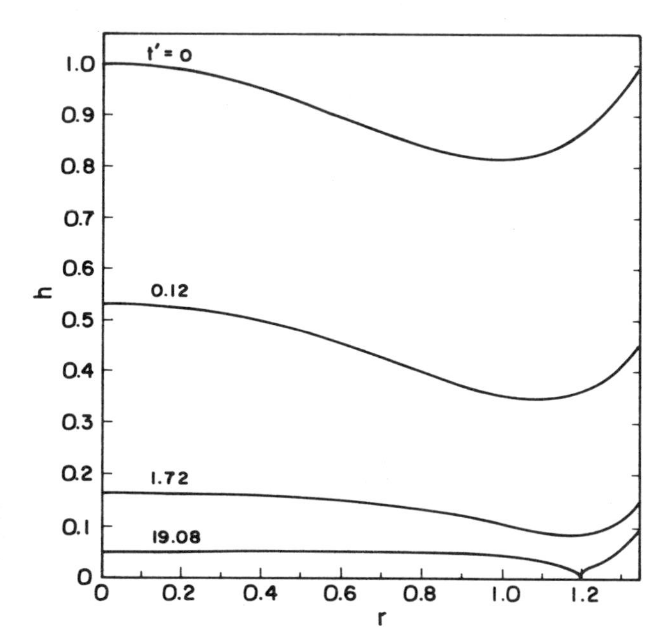

Figure 8: Dimensionless film thickness h as a function of dimensionless radial position and dimensionless time for $B = 10^{-7}$, $\kappa+\varepsilon = 2$, $R_h = 1.88$, $C = 2.04$.

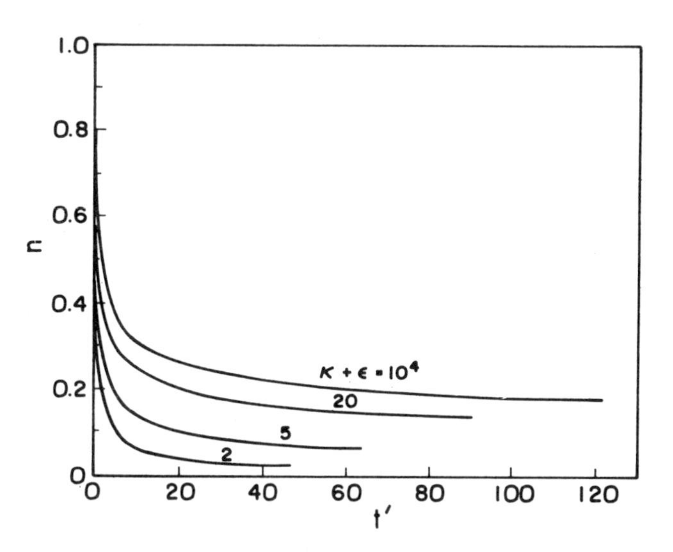

Figure 10: Dimensionless film thickness at the center as a function of dimensionless time and $\kappa+\varepsilon$ for $B = 10^{-8}$.

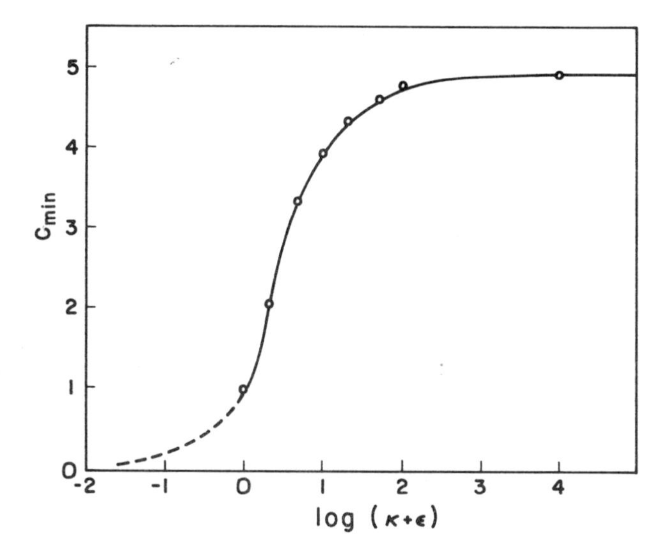

Figure 9: Variation of C_{min} as a function of $\kappa+\varepsilon$ (dashed line is an expected trend).

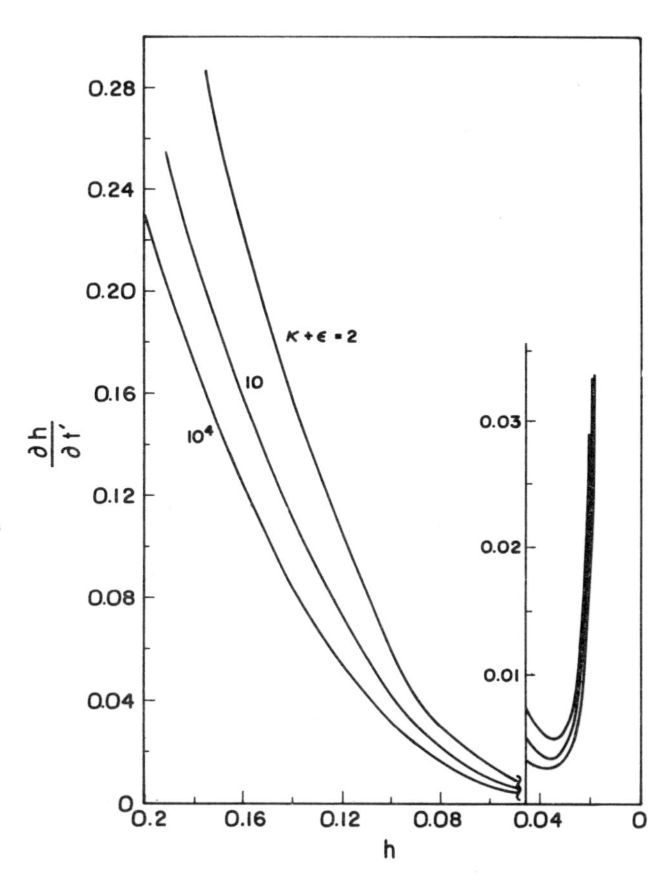

Figure 11: Dimensionless thinning rate at the rim as a function of dimensionless film thickness and $\kappa+\varepsilon$ for $B = 10^{-6}$.

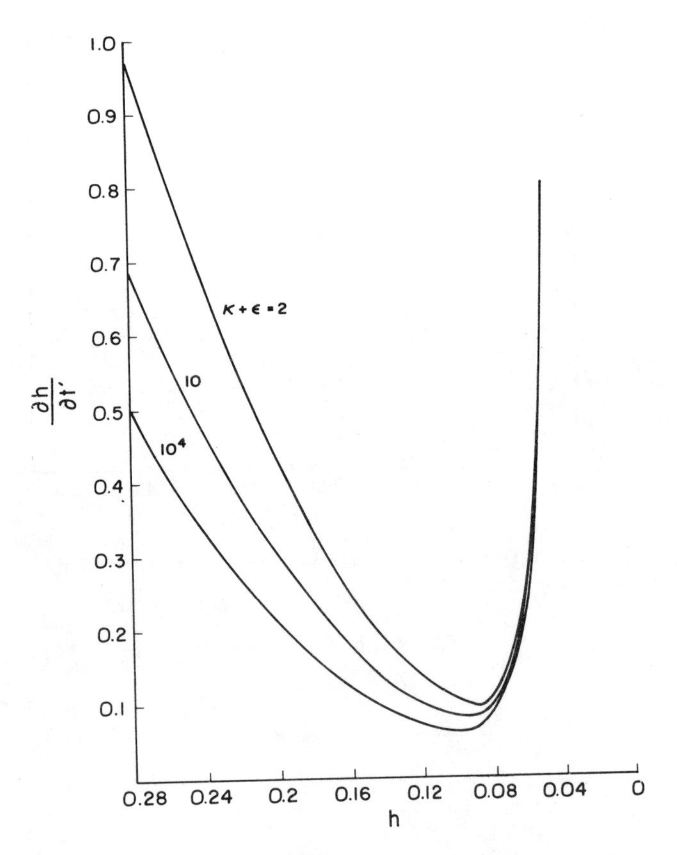

Figure 12: Dimensionless thinning rate at the rim as a function of dimensionless film thickness and $\kappa+\varepsilon$ for $B = 10^{-4}$.

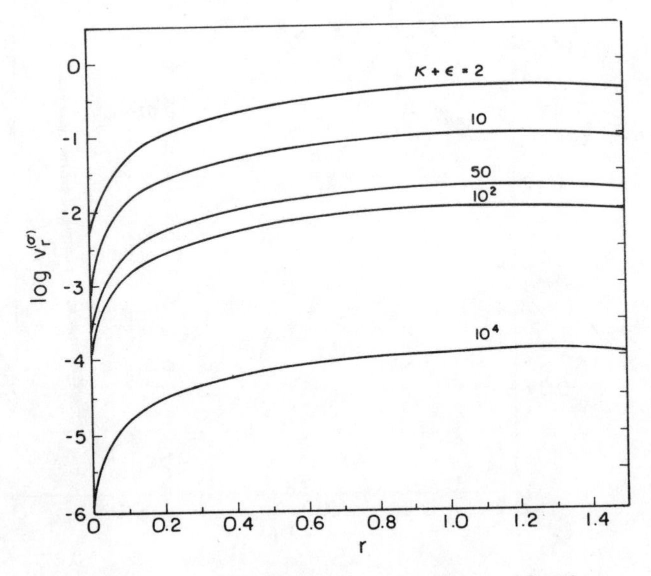

Figure 13: Dimensionless surface velocity at $t' = 0$ as a function of dimensionless radial position and $\kappa+\varepsilon$ for $B = 10^{-6}$.

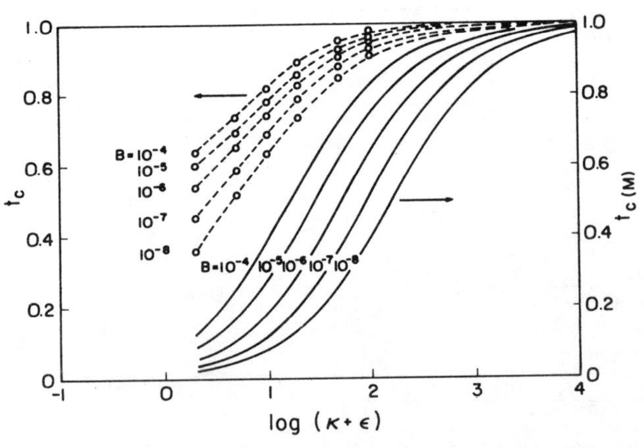

Figure 14: Dimensionless coalescence time predicted by the present theory (dashed curves) and by Hahn and Slattery (1985; solid curves) as functions of $\kappa+\varepsilon$ and B. In this case, the coalescence times are referred to different bases.

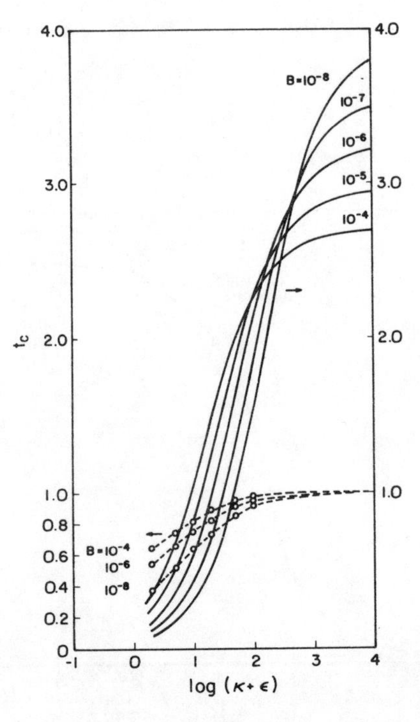

Figure 15: Dimensionless coalescence time predicted by the present theory (dashed curves) and by Hahn and Slattery (1985; solid curves) as functions of $\kappa+\varepsilon$ and B. In this case, the coalescence times are referred to the same basis.

THE EFFECT OF ADSORPTION OF LIQUID CRYSTALLINE LAYERS ON THE VAN DER WAALS INTERACTION IN FOAMS

Stig E. Friberg and Hironobu Kunieda ■ Chemistry Department, University of Missouri-Rolla
Rolla, Missouri 65401

Calculations were made of the Van der Waals potential of a multi-layered structure of alternating oil and water layers on a central oil film to imitate the conditions in an oil film covered by a lamellar liquid crystal.

The results showed an enhanced Van der Waals potential compared to that of a homogeneous oil film of a thickness equal to the central oil film plus the layered structures.

The reason for the stability of liquid crystal covered phase oil films is not a change of colloidal forces; the reason must be sought in the mechanical properties of the surface layer.

INTRODUCTION

The stability of thin films has traditionally been treated as a problem involving one phase liquid films with an adsorbed monomolecular layer of surfactant (1-6). However, the research during the last 15 years (7-13) has shown many foams to contain multimolecular layers (9,11) or that the condensed part may consist of two separate phases (7,8,12,13). As a matter of fact, for hydrocarbon foams, the presence of an additional spreading layered phase appears to be a sine qua non for their stability.

A multitude of factors contribute to the stability of such multi-phase foams and the final clarification of them may be rather distant in the future.

However, the Van der Waals potential is always an essential part of the force spectrum involved in the stability and we found a calculation of the influence on the Van der Waals energy by a multilayered structure to be of some interest. We have used the Vold approach (14) for spherical systems in our derivation of a general formula for a film with planar parallel walls.

THE MODEL CALCULATION

The model shown in Figure 1, includes a central oil layer, alternating layers of water and oil as a model for the liquid crystal lamellar structure and air surrounding the structure.

Successive addition gives:

Hironobu Kunieda's permanent address is: Department of Applied Chemistry, Yohohama National University, Hodogaya, Yohohama, JAPAN.

$$\overline{V} = -\frac{1}{12\pi}\Bigg[\sum_{q=0}^{n-1}\sum_{p=1}^{n} \frac{A_w}{[h+(p+q-1)(d_w+d_o)]^2} - \frac{2A_w}{[h+d_w+(p+q-1)(d_w+d_o)]^2}$$

$$+ \frac{A_w}{[h+2d_w+(p+q-1)(d_w+d_o)]^2} + \frac{2A_{ow}}{[h+d_w+(p+q-1)(d_w+d_o)]^2} - \frac{2A_{ow}}{[h+(p+q)(d_w+d_o)]^2}$$

$$- \frac{2A_{ow}}{[h+2d_w+(p+q0-1)(d_w+d_o)]^2} + \frac{2A_{ow}}{[h+d_w+(p+q)(d_w+d_o)]^2} + \frac{A_o}{[h+2d_w+(p+q-1)(d_w+d_o)]^2}$$

$$- \frac{2A_o}{[h+d_w+(p+q)(d_w+d_o)]^2} + \frac{A_o}{[h+(p+q+1)(d_w+d_o)]^2}$$

$$- 2 \times \sum_{p=1}^{n}\Bigg[\frac{A_{ow}}{[h+(p-1)(d_w+d_o)]^2} - \frac{A_{ow}}{[h+d_w+(p-1)(d_w+d_o)]^2} + \frac{A_o}{[h+d_w+(p-1)(d_w+d_o)]^2}$$

$$- \frac{A_o}{[h+p(d_w+d_o)]^2}\Bigg] + \frac{A_o}{h^2}\Bigg]$$

in which h is the thickness of the oil layer, d_w and d_o of the aqueous and oil layers in the liquid crystal. A_w, A_{ow} and A_o are the Hamaker constants for water/vacuum/water, oil vacuum/water and oil/vacuum/oil systems and n is the number of layers in the liquid crystal.

Figure 2 gives an overview of the change in Van der Waals potential, when a multilayer structure covers the central oil film. The results are given in the form of the ratio between Van der Waals potential of the central oil film and the structure with layers.

The addition of layers causes a reduction of the Van der Waals potential from the value for the central oil film, Fig. 2.

DISCUSSION

The result in Fig. 2 may, at first, be taken as an indication that the layered structure per se will cause a reduction of the Van der Waals potential.

However, it is essential to realize that addition of the layered structure changes not only the Van der Waals potential but also results in an extended total thickness of the film. With this in mind, a more appropriate comparison may be between a central oil film covered with a layered structure and an oil film with dimensions equal to the total thickness of the combined structure.

The results of such a comparison, Fig. 3, show the Van der Waals potential actually to be enhanced by the layered structure and for the small extension of the thicker films the enhanced potential is linear with the extension.

The reason for this enhanced Van der Waals potential is obviously based on the difference between the potential for a single layer, which is the last term in Equation [1] and the entire expression. However, the potential for a many-layered structure is complex and hence, is difficult to relate to the difference in the Van der Waals potential in a straight forward manner.

Limiting the expression to the case of a double layer on each side of the film provides the essential information to understand the difference. The expression reads

$$- 12\bar{V}\pi = (A_w + A_o - 2A_{ow}) \left[\frac{1}{h^2} - \frac{2}{(h+d_w)^2} + \frac{1}{(h+2d_w)^2} \right]$$

$$+ 2(A_o - A_{ow}) \left[\frac{1}{(h+d_w+d_o)^2} - \frac{1}{(h+2d_w+d_o)^2} \right]$$

$$+ A_o/(h+2d_w+2d_o)^2$$

The third term is identical to the potential over an oil film with the total thickness of the film covered with the layers and the difference (Δ) may be written

$$\Delta = k_1 \left[1 - \frac{1}{(1+\alpha)^2} - \left(\frac{1}{(1+\alpha)^2} - \frac{1}{(1+2\alpha)^2} \right) \right]$$

$$+ k_2 \left[1 - \frac{1}{(1+\alpha)^2} \right]$$

Realizing that all terms are differences between squares and assuming $\alpha \ll 1$, the difference may be approximated

$$\Delta \cong k_3 \cdot \alpha$$

A linear relation between the potential ratio of the non-layered and layered structure as found for thick films and thin layers in Fig. 3.

CONCLUSIONS

The conclusion from these results is evident. The layered structure will increase the Van der Waals potential in comparison with a liquid film of identical thickness. With this in mind, the reason for the stability of hydrocarbon films covered with a liquid crystal should rather be sought in the changed mechanical properties of the film surface.

ACKNOWLEDGMENTS

This research was supported by a grant from the Clorox Company.

LITERATURE CITED

1. Kitchener, J. A., Recent Prog. Surf. Sci. 1, 51 (1964).

2. Clunie, J. S., J. F. Goodman, and B. T. Ingram, "Surface and Colloidal Sci.", Vol. 3 (E. Matijevic, ed.), Wiley-Inter-Science, New York, 1971.

3. Bikerman, J. J., "Foams", Springer, New York, 1973.

4. Lange, H. and B. Kurzendörfer, Fetter, Siefen Anstrichm. 76, 120 (1974).

5. Mysels, K. J., K. Shinoda, and S. Frankel, "Soap Films", Pergamon, New York 1959.

6. Ross, S., "Foams", in Encyclopedia of Chemical Technology, Vol. 2, 3rd Ed., Wiley, New York 1980.

7. Friberg, S. E. and S. I. Ahmad, J. Colloid Interface Sci., 35, 175 (1971).

8. Jederström, G., L. Rydhag and S. Friberg, J. Pharm. Sci. 62, 1979 (1973).

9. Lyklema, J. and H. Bruil, Nature (London) Phys. Sci. 233, 19 (1971).

10. Manev, E., A. Scheludko and D. Exerow, Colloid Polym. Sci. 252, 586 (1974).

11. Manev, E. D., S. V. Sazdanova, A. A. Rao and D. T. Wasan, J. Dispersion Science and Technology 3(4), 435 (1985).

12. Friberg, S. E., in "Advances in Liquid Crystals", (G. H. Brown, ed.), Vol. 3, Academic Press, (1978), p. 149.

13. Friberg, S. E., C. S. Wohn, B. Greene and R. Van Gilder, J. Colloid Interface Sci. 101, 593 (1984).

14. Vold, M. J., Ibid, 16, 1 (1961).

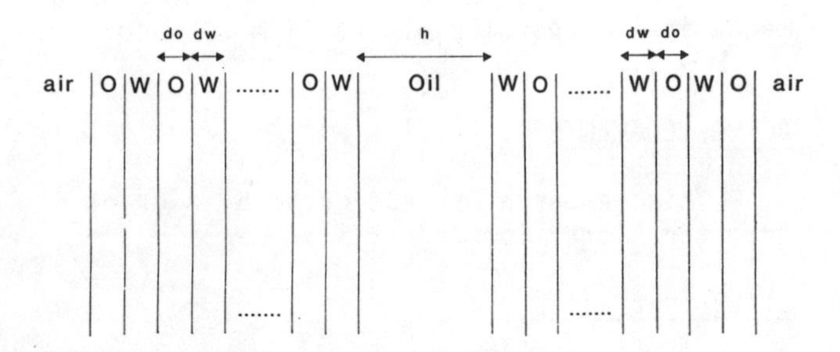

Figure 1. The model used in the calculations consisted of a central oil film of thickness h. The film was covered by alternate water (W) and (O) layers of thicknesses d_w and d_0.

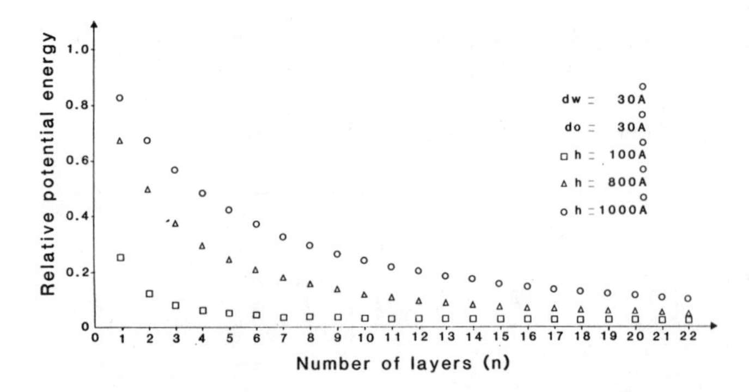

Figure 2. With increased number of layers, the Van der Waals potential over the entire film was reduced in comparison to the one over the central oil layer.

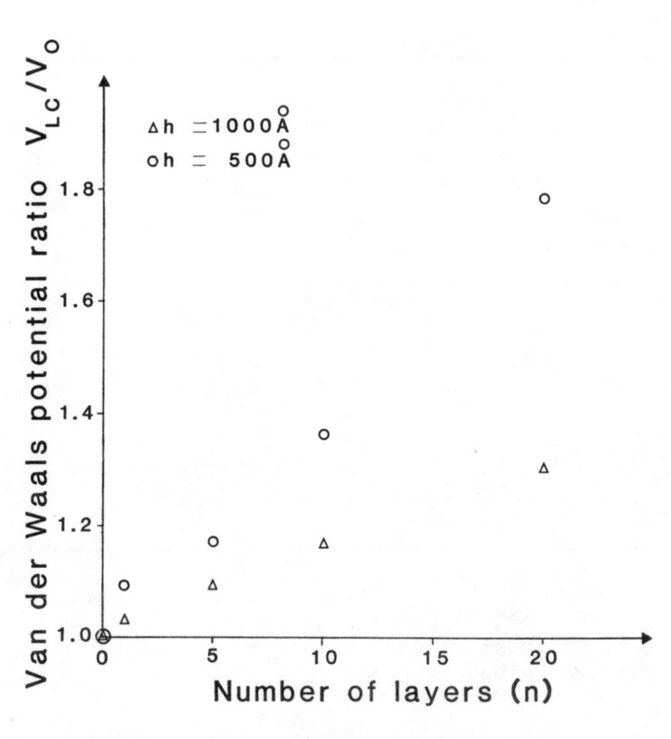

FIGURE 3

Figure 3. A comparison with an oil film of identical thickness to the entire layered film showed an increase of the Van der Waals potential.

NONLINEAR STABILITY OF THIN FREE LIQUID FILMS: RUPTURE AND MARANGONI EFFECTS

M. Prevost and D. Gallez ■ Universite Libre de Bruxelles, Faculte des Sciences
Unite de Conformation des Macromolecules Biologiques, Bruxelles, Belgique

Thin free liquid films, i.e., liquid sheets sandwiched between two bulk phases, are encountered in many colloid and biocolloid systems. Rupture of soap films, fusion of lipid bilayers, coalescence of emulsions and pattern formation all involve unstable processes of the liquid film.

In the present theory, we examine nonlinear effects on the stability of a free thin liquid film in a simplified situation where only the influence of Van der Waals interaction is taken into account. First, in the limit of long wavelength of the perturbation, we obtain a nonlinear evolution equation describing the temporal variation of the film thickness along the lateral space dimension in the case of tangentially immobile surfaces. We determine numerically the time of rupture for the film through the amplification of the squeezing mode, this latter being the more appropriate mode leading to rupture.

Secondly, we analyze the stability of a film submitted to solutal Marangoni effects at the two surfaces which allows predicting bifurcations to new flow patterns.

Instability and rupture of free liquid films occurs in many colloid and biocolloid systems : rupture of soap films (1), coalescence of emulsions (2), or in the biological domain : fusion of vesicles (3) or cells (4), tear film rupture (5), all involve instability of a free liquid film.

We will consider here the stability of an aqueous film with insoluble surfactants sandwiched between two fluid phases, which is an adequate model for the examples quoted above. Until now, much attention has been devoted to the linear stability analysis with respect to infinitesimal perturbations of the two surfaces of the film (6-7) : this analysis has shown the existence of two modes of vibration, uncoupled for symmetrical films (8) : the bending mode (the two surfaces tranversally in phase) and the squeezing (SQ) mode (the two surfaces 180° out of phase), both of which can be detected experimentally by laser light scattering (9). The instability of the SQ mode, which will be considered here leads to the film rupture.

However a linear theory ceases to be valid at finite amplitude perturbations aswell as the character of the flow induced by inhomogeneous distributions of surfactants remains undetermined in a linear theory.

So in the present paper, we examine nonlinear effects on the stability of free films, in the situation where van der Waals attractive forces are operative. We consider first the simplified case of tangentially immobile surfaces (10). The appropriate smallness parameter is, in that case, related to the long-wavelength compared to the film thickness. We derived a non-linear evolution equation for the surfaces, which is solved numerically. This allows to determine the true rupture time of the film and to compare it to the rupture time obtained by the linear theory. Such deviation was also previously performed for a film on a solid substrate (11).

On the other hand, when there exists an inhomogeneous distribution of insoluble surfactants, surface tension gradients will develop (Marangoni effects). The generating mechanism of the concentration gradient may be chemical reactions or adsorption-desorption. Linear stability criteria for one interface have been extensively developed (12). The interplay between chemical reactions and a fluid motion was recently analysed in a non-linear treatment, for a thin film supported by a rigid planar wall (13). In the present paper, we defined similarly a long lateral scale related to the diffusion of surfactants along the surface, and obtain the velocity field at the two free surfaces. The linear approximation for the stationary normal modes is then recovered in a simplified case and the role of surfactants on the rupture time assessed.

HYDRODYNAMIC MODEL

The free liquid film is always submitted to perturbations originating from external or internal fluctuations. The system whose stability is analysed is defined by a nonthinning, infinite and planar film sandwiched between two fluid bulk phases. Two interfaces are respectively located at $z=h_0/2$ and $z=-h_0/2$ (see Fig.1). Moreover, the film is described as an isotropic, incompressible and viscous fluid. The dynamic behaviour of each phase is given by the Navier-Stokes equations :

$$\rho\,[u_t + uu_x + vu_z] = \mu\nabla^2 u - p_x - W_x \qquad (1)$$

$$\rho\,[v_t + uv_x + vv_z] = \mu\nabla^2 v - p_z - W_z \qquad (2)$$

and the conservation of mass is

$$u_x + v_z = 0 \qquad (3)$$

where u and v are respectively the tangential and normal velocities, ρ the mass density, μ the viscosity and p the mechanical pressure. The van der Waals potential W reads :

$$W = A\,h^{-3} \qquad (4)$$

with A the Hamaker constant and h the film width.
The dynamic boundary conditions are the continuity of velocities at the two surfaces, together with a normal stress balance (Laplace condition) :

$$-p + \mu/(1+h_x^2)[\,2v_z(1-h_x^2) - h_x(v_x+u_z)] = \sigma\, h_{xx}/(1+h_x^2)^{3/2} \quad \text{at } z=\pm h/2 \qquad (5)$$

where σ is the surface tension, and a tangential stress balance (Marangoni condition) :

$$\mu(u_z+v_x)(1-h_x^2) + 2h_x\,(v_z-u_x) = \sigma_x(1+h_x^2)^{1/2} \quad \text{at } z=\pm h/2 \qquad (6)$$

Moreover, a conservation equation for the surface active substance is needed :

$$\Gamma_t + (\dot{a}/2a)\Gamma + \partial_x(\Gamma u) = D_s\,.\,\Gamma_{xx} + F(\Gamma) \qquad (7)$$

where Γ is the surfactant concentration, D_s the surface diffusion coefficient, and $F(\Gamma)$ the reaction function. The second term in l.h.s. of Equation (7) represents the rate of change of concentration due to the dilatation of the surface, and ∂_x denotes differentiation with respect to the longitudinal coordinate along the surface. A linear dependence is supposed between the variation of surface tension and surface concentration.
The kinematic equation expresses the temporal evolution of the surface $z=h(x,t)/2$:

$$v = h_t/2 + h_x u/2 \qquad (8)$$

For the squeezing mode (180° out-of-phase normal velocities) leading to film rupture, we have the additional conditions at the centre of the film :

$$v(0) = 0 \quad \text{and} \quad u_z(0) = 0 \qquad (9)$$

Because of the symmetry, hereafter only the liquid sheet between $z=h/2$ and $z=0$ is taken into consideration.

NONLINEAR THEORY FOR TANGENTIALLY IMMOBILE SURFACES

Experimental evidence on soap films shows that disturbances begin to grow to a large size and that films rupture for a critical value of the film thickness h_c. Therefore, linear theories cannot follow the evolution of the film up to the rupture. We consider then small but finite amplitude waves and take into account the nonlinearities in Equations (1), (2), (5)-(8). For simplicity, we shall restrict this treatment to nearly neutral films with a quantity of surfactants sufficient to render the surfaces tangentially immobiles (the tangential condition (6) reads simply u=0).

These surface waves are characterized by wavelengths long compared to the film thickness h. Indeed, long waves are efficient in extracting the potential energy due to van der Waals forces while minimizing the work done against surface tension. The longwavelength approximation will be used in this nonlinear theory to obtain successive approximations for the solutions of the equations of motion (1)-(2). The solutions obtained for each component of the velocity field will be substituted in the kinematic equation (8). We formalize this longwave theory by defining a small parameter $\varepsilon=h/\lambda$ directly related to the small wavenumber $k=2\pi/\lambda$ of the disturbance. For a viscous fluid having a kinematic viscosity ν, we first "nondimensionalize" the governing equations (1)-(3) and boundary conditions using the Stokes scale which is convenient for slow motion : length $\sim h_0$, time $\sim h_0^2/\nu$, velocity $\sim\nu/h_0$ and pressure $\sim\rho\nu^2/h_0^2$. Moreover, we introduce a stretching transformation appropriate to the longwave expansion : $\tilde{x}=\varepsilon x$, $\tilde{z}=z$, $\tilde{t}=\varepsilon t$. Approximate solutions of the system of equations can be obtained by developing u, v and p in regular pertubation expansion in ε. At zero order the velocity field reads (10) (with ~

suppressed) :

$$u^0 = \phi_x(z^2/2 - (h/2)^2/2) \quad (10)$$

$$v^0 = -\phi_{xx}(z^3/6 - (h/2)^2 z/2) + \phi_x h_x hz/4 \quad (11)$$

where $\phi = \bar{A}h^{-3} - 3\bar{S}h_{xx}$ with $\bar{A}$ related to the Hamaker constant A by $\bar{A}=\varepsilon A/6\pi\rho\nu^2 h_0$ and $\bar{S}$ related to the surface tension σ by $\bar{S}=\varepsilon^3 h_0\sigma/3\rho\nu^2$. By substituting the derived solutions in the kinematic Equation (8), we obtain at lower order in ε the following evolution equation :

$$4h_t - 1/3(\phi_x h^3)_x = 0 \quad (12a)$$

This evolution equation is similar (apart from the factor 4 which is due to the free film geometry) to the equation derived by Williams and Davis (11) for a film without surfactants on a solid substrate. It was also derived in the lubrication approximation by Dimitrov (14) for the general case of two different interfaces. We are interested to solve explicitly Equation (12a) for tangentially immobile surfaces where the velocity field is undetermined. When replacing ϕ_x by its value the rescaled Equation (12a) reads :

$$4h_T + (h_{XXX}.h^3)_X + (h^{-1}.h_X)_X = 0 \quad (12b)$$

with an initial condition

$$h(X,0) = g(X)$$

The evolution equation (12b) shows already that the nonlinearities will accelerate the rupture process as the capillary forces (3rd term) increase with decreasing h. This equation governs infinitesimal as well as finite perturbations. We can thus recover the linear situation as a test of the theory. Introducing the development h = 1 + 2H for an interface at z=1/2 and linearizing Equation (12b) in H, we obtain :

$$H_T + H_{XXXX}/4 + H_{XX}/4 = 0 \quad (13)$$

Substituting the normal modes in the linearized Equation (13), we recover the following dispersion relation :

$$\omega = -k^2/4\,(k^2 - 1) \quad (14)$$

Moreover, the dimensionless marginal wavenumber obtained for $\omega = 0$, which separates stable and unstable states can be written as $k_0 = 1$. In addition, the dispersion relation (14) displays a fastest growth rate corresponding to a maximal wavelength $(k_{max})^{-1}$ (see Fig.2) :

$$k_{max} = \sqrt{2}/2$$

$$\text{and} \quad \omega_{max} = 1/16 \quad (15)$$

The results (15) are identical to those found by Ruckenstein and Jain (6) for tangentially immobile surfaces. Our analysis describing the evolution of infinitesimal perturbations gives thus the same results as those derived in the linear analysis under the same approximations.

Let us solve now the general nonlinear Equation (12b) and compare it to the resolution of the linear Equation (13). As concerns the nonlinear theory, it is necessary to solve Equation (12b) numerically, using a finite difference method (15). Forward differences in space are used to obtain the successive derivatives with respect to x, while an explicit scheme in time is used as concerns the time derivative h_t. The difference equations obtained are then solved iteratively. We want to derive the time characteristics of the rupture process T_N, i.e. the time for which the thickness h(X,T) vanishes at a point along the space lateral dimension X. The solution h(X,T) will depend on the parameters included in the initial condition of (12b). We consider initial periodic values in space of the form :

$$g(X) = 1/2 + B \sin kX \quad (16)$$

Let us consider the evolution of the most probable wavenumber $k=k_{max}$. The solutions are then periodic in λ_{max} and the evolution of the most amplified wavelength given by the linear theory is examined. We choose $\Delta X=\lambda_{max}/10$ and $\Delta T=(50\omega_{max})^{-1}$ where ΔX and ΔT symbolise the space and time intervals along the X and T axes. The value ΔX is such that the numerical stability criterion $0 < \Delta T/(\Delta X)^2 < 0.5$ is verified (15).

The figure 3 shows the evolution of one interface obtained from the numerical solution of the linear Equation (13) (Fig.3a) and the nonlinear Equation (12b) (Fig.3b) when the amplitude B in the initial condition (16) is equal to 0.1 and for $k=k_{max}$. The interface reaches the film centre (z=0) at $T_L=26.2$ (for the linear case) and at $T_N=2.24$ (for the nonlinear case). It clearly shows that the nonlinearities in the evolution equation contribute significantly to the instability of the free liquid film and accelerate the rupture process. The T_N/T_L ratio is independent of the different scales and measures the effect of the nonlinearities. It is equal here to 0.18. With the following values of the parameters : Hamaker constant $A = 10^{-13}$ ergs, h = 200 Å, $\mu = 10^{-2}$ g/(cm.sec), $\rho = 1$ g/cm^3 and $\sigma = 30$ dyn/cm, the case $T_N=2.24$ corresponds to a di-

-mensional time of the order of 2. 10^{-3} sec.; the linear time T_L equals 26.2 which leads to a time of the order of 2. 10^{-2} sec.. The T_N/T_L ratio depends on the initial condition (16). It has been shown that this ratio tends to 1 as the amplitude B of the initial condition tends to 0.

However, an important question remains open : as the instability begins to grow for $h < h_c$, an infinity of wavelengths ($0 \leq k \leq 1$) appears, one of which has the fastest growth rate. But the latter is not necessarily the experimentally observed wavelength. Radoev et al. (16), using a complete description of the process of thinning together with growth of surface fluctuations, deduced an optimal wavelength $\lambda_{opt}=\sqrt{8\pi\sigma/3A}$ which yields for the smallest thickness h = 240 Å a value of λ_{opt} of about 5. 10^{-4}cm. By comparing this criterion with the marginal case of the linear dispersion Equation (13), we obtain (see Fig.2) $k_{opt}=0.87$ k_0. On the other hand, an upper limit for λ (i.e. a lower limit for k) can be obtained by taking $\lambda/2=d$ with d the diameter of the black spots formed in the film (16). As d is slightly larger than 5. 10^{-3}cm this gives a lower adimensional limit $k \sim 0.01$. Our estimate for the true nonlinear rupture time T_N using $k=k_{max}$ falls thus within the plausible range of wavenumbers.

SOLUTAL MARANGONI EFFECTS

In this paragraph, we will consider the more general case where an inhomogeneous distribution of insoluble surfactants is established at the surfaces of the free film, as a result for instance of a surface chemical reaction (17). In that case, a large diffusional length scale can be introduced when the characteristic reaction time far exceeds the time for diffusion across a distance comparable with the film thickness (13). We define thus a new scaling with the distance x scaled by the corresponding diffusion length $L=\sqrt{D_s t_r}$, t_r being the characteristic reaction time. We define the dimensionless variables (primed) :

$$(t', x', z') = (D_s/L^2 t,\ L^{-1}x,\ h_0^{-1}z)$$
$$p' = pL^2/\mu D_s$$
$$\sigma' = \sigma_0^{-1}.\sigma$$
$$\Gamma' = \Gamma_0^{-1}.\Gamma$$

with σ_0 the tension of the clean surface, h_0 the mean film thickness, and Γ_0 the mean surfactant concentration. The presence of a new small parameter $\varepsilon = h_0/L$ allows, as in the preceding section, a long-wave expansion procedure. At lower order the system of equations reads (with primed suppressed) :

$$u^o_{zz} = p_x + W_x \quad (17a)$$

$$0 = p_z + W_z \quad (17b)$$

$$u^o_x + v^o_z = 0 \quad (17c)$$

$$-p = \sigma T\, h_{xx} \quad \text{at } z=h/2 \quad (17d)$$

$$u^o_z = -M\,\Gamma_x \quad \text{at } z=h/2 \quad (17e)$$

$$\text{and} \quad \Gamma_t + \partial_x(\Gamma u^o) = \Gamma_{xx} + F(\Gamma) \quad (17f)$$

with the fluid crispation number

$$T = \sigma_0 h_0/\mu D_s \quad (18a)$$

and the Marangoni number

$$M = (\partial\sigma/\partial\Gamma)\Gamma_0\, h_0/\mu D_s \quad (18b)$$

and the van der Waals potential

$$W = A' / h^3 \quad \text{with } A'=AL^2/\mu D_s h_0^3 \quad (18c)$$

The variation of the surface concentration is assumed to be symmetric at the two surfaces. We retain here the effect of the surface tension by considering $T \sim O(1/\varepsilon^2)$ in relation (17d). This case corresponds to a moderate surface tension and is similar to the value parameters used in the previous section. Two other extreme cases are possible, which will not be considered here :
$T \sim O(1)$: surface highly deformable and
$T \gg 1/\varepsilon^2$: undeformable surface
The velocity field at zero order reads :

$$u^o = \phi_x z^2/2 + C_3 z + C_2 \quad (19)$$

$$v^o = -\phi_{xx} z^3/6 - C_{3x} z^2/2 - C_{2x} z + C_1 \quad (20)$$

where $\phi = -\sigma T\, h_{xx} + A'\, h^{-3}$. The constants C_3 and C_1 can be eliminated by using the specific conditions (9) for the SQ mode :

$$u^o_z(0) = 0 \text{ so that } C_3 = 0$$

$$v^o(0) = 0 \text{ so that } C_1 = 0$$

By introducing the velocity field (19) in the Equation (17f) for the conservation of surface concentration, one obtains :

$$\Gamma_t + \Gamma_x\,[\phi_x (h/2)^2/2 + C_2] + \Gamma\,[\phi_{xx}(h/2)^2/2 + C_{2x}\,] = \Gamma_{xx} + F(\Gamma) \quad (21)$$

Equation (21) must be complemented by the evolution equation at the 2 surfaces :

$$h_t/2 = v_0 - h_x u^0 /2 \qquad (22)$$

where the velocity fied u^0 and v^0 is given by Equations (21) and (22) together with the tangential condition (17e) is a rather tedious task which will be considered in further investigations. Similar equations have been solved for films on a solid substrate (13) combining idea of the catastrophe theory and longscale expansion. In the case of free films, we intend to obtain stationary waves solutions (patterns of the SQ mode) and progressive waves solutions (propagation of solitary waves, as black and white spots in soap films (18)). Interesting qualitative features, however, are already seen by taking the linear limit of Equation (21), and neglecting the specific chemical source $F(\Gamma) = 0$. Introducing the linear development $\Gamma = 1 + \partial\Gamma$ and linearizing (21) in Γ, one obtains

$$\delta\Gamma_t + u_x^0 = \delta\Gamma_{xx}$$

which for normal modes $\delta\Gamma = Ge^{ikx}e^{\omega t}$ gives

$$G^{\sim} = -iku^0 / (\omega + k^2)$$

and $$\delta\Gamma_x = k^2 u^0 e^{ikx} e^{\omega t} / (\omega + k^2) \qquad (23)$$

When this value is introduced in the tangential condition (17e) at z=h/2, it allows to determine C_2 :

$$C_2 = -\phi_x [(h/2)^2/2 + (h/2)/\tilde{M}k^2] \,(24a)$$

with $\tilde{M} = M/(\omega + k^2)$. Introducing the previous linear development h=1+2H and linearizing in H, leads to :

$$C_2 = -ikH(k^2\sigma T - 3A)(1/4 + 1/\tilde{M}k^2) \quad (24b)$$

and $$C_{2x} = k^2 H(k^2\sigma T - 3A)(1/4 + 1/\tilde{M}k^2) \quad (24c)$$

At this stage, several interesting features may already be seen, by replacing the value of C_2 given by Equation (24a) in the velocity field (19) and (20). The tangential velocity u^0 tends to the value given by Equation (10) for tangential immobile surfaces as $\tilde{M} \to \infty$; in particular it tends to zero at the surfaces; similarly, the normal velocity v^0 tends to the value given by Equation (11) as $\tilde{M} \to \infty$. For a finite density of surfactants, the velocity field at the surfaces is inversely proportional to the surfactant concentration. For the symmetric variation of surfactants concentrations considered here, the convective flow pattern of the SQ mode is thus stabilized by the surface tension gradient. The kinematic equation $v^0 = h_t/2 + u^0 h_x/2$ can also be linearized and gives the following dispersion relation by replacing the constant C_2 and C_{2x} in the expression for the velocity field :

$$\omega = -k^2/12(k^2\sigma T - 3A)(1 + 6/k^2\tilde{M}) \qquad (25)$$

By rescaling this equation, one obtains :

$$\omega = -k^2/4(k^2 - 1)(1 + 1/k^2 M^{\cdot\cdot}) \qquad (26)$$

with $M^{\cdot\cdot} = A'\tilde{M}/2\sigma T$. For a high concentration of surfactants ($M^{\cdot\cdot} \to \infty$), the dispersion relation (14) of the preceding section for tangentially immobile surfaces is recovered. The maximal wavelength k_{max}^{-1} and the corresponding fastest rate of growth ω_{max} reads respectively :

$$k_{max} = (1/2 - 1/2M^{\cdot\cdot})^{1/2}$$
$$\omega_{max} = 1/16(M^{\cdot\cdot} + 1)^2/M^{\cdot\cdot 2} \qquad (27)$$

The role of the surfactants on the dispersion relation is illustrated in Fig.4. It is clearly seen that increasing the surfactants concentration decreases the fastest rate of growth (i.e. increases the rupture time) and shifts the dominant wavelength to smaller values. These stabilizing effects of the surfactants are similar to those predicted by Ruckenstein and Jain (6).

CONCLUSIONS

We performed a nonlinear stability analysis of a free liquid film with insoluble surfactants subjected to van der Waals attractive forces. We focused on the instability of the squeezing mode which leads to film rupture.

Two cases were considered : firstly, the case of tangentially immobile surfaces. In that case, it is shown that the nonlinearities accelerates the rupture process, essentially by increasing the long range attractive forces with respect to the stabilizing capillary forces.

Secondly, the case of gradients of concentration at the surfaces (solutal Marangoni effects). Preliminary results indicate the stabilizing role of the surfactants on the rate of growth of surface fluctuations. For the squeezing mode, symmetric gradients of concentration at the two surfaces, are shown to stabilize the convective flow pattern. These results must be extended to more general solutions (stationary or progressive waves) for free films and compared to experiments with specific chemical reactions.

LITERATURE CITED

1. Scheludko, A. D., Adv. Coll. and Interf. Sci., 1, 391 (1967).

2. Marucci, G., Chem. Eng. Sci., 24, 975 (1969).

3. Papahadjopulos, D., G. Poste, B. E. Schaeffer and W. J. Vail, Biochim. Biophys. Acta, 352, 10 (1974).

4. Coackley, W. T., L. A. Hewison and D. Tilley, Eur. Bioph. J., to appear.

5. Sharma,A. and E. Ruckenstein, J. Coll. Interf. Sci., 106, 12 (1985).

6. Ruckenstein,E. and R. K. Jain, J. Chem. Soc. Far. Trans. II, 70, 132 (1974).

7. Prevost,M., D. Gallez and A. Sanfeld, J. Chem. Soc. Far. Trans. II, 79, 961 (1983). Prevost,M. and D. Gallez, Ibid., 80, 517 (1984).

8. Felderhof,B. U., J. Chem. Phys., 49, 44 (1968).

9. Joosten,J.G.H., J. Chem. Phys., 80, 2363 (1984); Ibid., 80, 2383 (1984).

10. Prevost,M. and D. Gallez, J. Chem. Phys., 84, 4043 (1986). Gallez,D. and M. Prevost, Phys. Chem. Hydr. J., 6, 731 (1985).

11. Williams, M.B. and S.H. Davis, J. Coll. Interf. Sci., 90, 220 (1982).

12. Dalle Vedove, W. and A. Sanfeld, J. Coll. Interf. Sci.,84, 318 (1981); Ibid., 84, 328 (1981).

13. Dagan, Z. and L. M. Pismen, J. Coll. Interf. Sci., 99, 215 (1984). Pismen, L.M., Ibid., 102, 237 (1984).

14. Dimitrov, D. S., Progress in surface Science, 14 (4), 295 (1984).

15. Lederman, W., " Handbook of Applicable Mathematics ", vol.III, Churchhouse, R. F. (Ed.), Wiley Interscience, New York (1981).

16. Radoev, B. P., A. D. Scheludko and E. D. Manev, J. Coll. Interf. Sci., 95, 254 (1983).

17. Avnir, D. and M. Kagan, Nature, 307, 717 (1984).

18. Joosten, J. G. H., J. Chem. Phys., 82, 2427 (1985).

RUPTURE OF THIN FREE FILMS WITH INSOLUBLE SURFACTANTS: NONLINEAR ASPECTS

Ashutosh Sharma and Eli Ruckenstein ■ Department of Chemical Engineering, State University of New York at Buffalo, Buffalo, New York 14260

While the linear analyses of the thin film instabilities have been rather extensive, their results are quantitative only for infinitesimal disturbances. In practice, however, the thin films that occur in flotation, foams and emulsions experience large mechanical perturbations. The nonlinearities associated with Marangoni-motion and intermolecular interactions cannot be ignored in such an event. A simple analytical formalism is proposed here for studying the stability of thin films that are subjected to finite amplitude initial perturbations. The essential idea of the proposed formalism is to investigate the stability of spatially nonuniform stationary solution of the governing equations. This is to be contrasted with the linear stability analysis, where the base state is assumed to be the same as the plane parallel configuration of the thin film. Owing to its finite amplitude, the spatially nonuniform steady state is "closer" to the initial disturbance as compared to the planar configuration and hence, more suitable as a base state. It is only in the limit of vanishingly small amplitudes that the spatially nonhomogeneous steady state reduces to the trivial solution (planar interface) and thus the results of the linear stability analysis are recovered. For a nondraining, thin foam film with insoluble surfactants, the time of rupture is derived as a function of the thin film parameters and amplitude of the initial disturbances. The condition under which the film is neutrally stable is also shown to depend on the amplitude of perturbations. The nonlinear theory predicts a significantly faster growth of the perturbations as compared to the linear theory. It is deduced that the destabilizing influence of the Van der Waals interactions is underestimated by the linear theory and that the stabilizing influences of the surface tension, the Marangoni-effect and surface viscosity are overestimated.

Numerous studies (Scheludko (1), Vrij and Overbeek (2), Felderhof (3), Ruckenstein and Jain (4), Ivanov and Dimitrov (5), Gumerman and Homsy (6)) of thin film stability have been carried out with a view towards predicing the lifetime and critical thickness of thin film that occur in dispersed systems such as flotation, foams and emulsions. An understanding of such diverse biological phenomena as the onset of microvilli in neoplastic cells (Jain et al. (7), Maldarelli et al. (8), Maldarelli and Jain (9)), the adhesion and fusion of membranes (Dimitrov and Zhelev (10)), the tear film breakup (Sharma and Ruckenstein (11), Sharma and Ruckenstein (12)) and the chemical reaction induced instability of membranes (Dalle Vedove and Sanfeld (13)) have also motivated the study of thin film instabilities.

A principal difficulty in solving the thin film hydrodynamic equations stems from the nonlinearities of the governing equations, the molecular interaction potentials and the boundary conditions, as well as from the movement of the boundaries. Thus, the solution methodology underlying all of the studies of the thin film stability has been to employ the linear stability analysis for obtaining estimates of the time of rupture. The strategy has been to linearize the governing equations around the unperturbed state (also referred to as the trivial state) of the thin film and then seek the normal mode solutions to these linearized set of equations. If at least one eigenvalue (the growth rate) so determined is positive for a set of wave-numbers, the planar configuration of the thin film is unstable to disturbances with certain wave-numbers. The time of rupture is then usually interpreted as the inverse of the dominant (maximum) growth rate. While this analysis is relatively straightforward, its results are expected to be quantitatively correct only if the interfacial disturbances are vanishingly small and therefore, the initial state of the thin film is closely approximated by the trivial steady-state, i.e., an undisturbed planar film. This is confirmed by the numerical simulations of Williams and Davis (14) for the long wavelength instability of a thin film that is devoid of solutes and is in contact with a solid. Large discrepancies were observed between the times of rupture as computed from the simulations and the linear theory when the disturbances had a large amplitude. The simulations also show that the linear theory is satisfactory for amplitudes that are very small compared to the mean thickness of the film, which, for instance, is the case with perturbations of purely thermal origin. This fact explains the apparent success of linear stability results in predictions concerning the isolated thin films in vibrationless environments. The design of flotation, foam and emulsion systems, however, requires the prediction of the lifetimes of thin films in an environment where the mechanically generated external disturbances are often quite large. A recent experimental study (Desai and Kumar (15)) of foam collapse concluded that the observed critical holdups were one to two orders of magnitude higher than that arrived at by the linear analysis, i.e., the breakup times were substantially smaller. The authors also attributed this discrepancy to the finite amplitude mechanical disturbances that clearly dominate over the thermal perturbations. The extremely low interfacial tension systems such as proteinaneous coatings (e.g. mucous coating of the corneal epithelium) and oil-water interfaces with surfactants encountered in the oil recovery, are even more susceptible to both large mechanical and thermal perturbations. It is in view of these considerations, and a diversity of existing thin film models, that it is desirable to develop an analytical, albeit simple, method of analysis that yields information about the set of parameter values for which a film is unstable, as well as about the growth rate of disturbances when they have a finite amplitude. We have earlier proposed such a formalism (Sharma and Ruckenstein (16)) that predicts the response of a nondraining thin film to the finite amplitude external disturbances. This formalism is outlined in the next secion. We then apply it to a model of free thin film (prototype for a foam film) with insoluble surfactants. The information sought is the wave-numbers of the neutrally stable and the fastest growing perturbations and the growth rate of the latter, all as functions of the thin film parameters and the amplitude of the initial disturbance. We are then able to delineate the effects of nonlinearities associated with the surface tension restoring force, van der Waals interactions, the Marangoni-effect and the surface viscosity, on the rupture of a thin, nondraining free film. The last section is devoted to the possible extensions of the present methodology and its limitations.

THE METHODOLOGY

The present goal is to develop a perturbative analysis for the finite amplitude instability of thin films (typically less than or of the order of 1000 Å) for which the long range van der Waals dispersion forces are important. The problem to be considered is this: a thin film is subjected to a finite amplitude external perturbation and hence, its free interfaces have spatially nonuniform Shapes h(x,0) at the initial moment (Figure 1). The

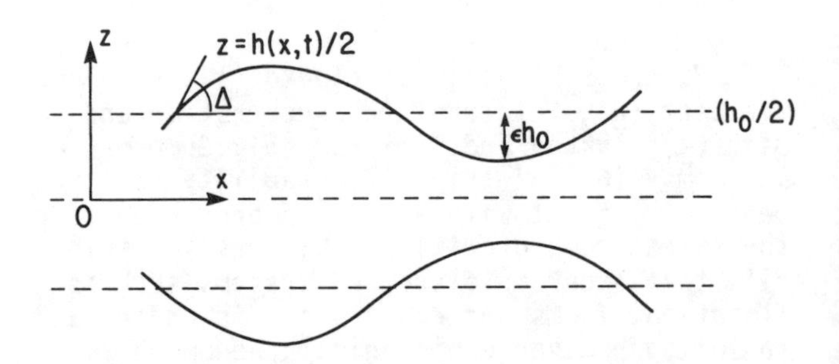

Figure 1. A thin free film with corrugated interfaces. The amplitude of the initial disturbance on each interface of the film is ϵh_0, where $\epsilon < 1/2$. The film surface is described by z=h(x,t).

nondimensional amplitude of the perturbation is denoted by ε, when it is nondimensionalized by the mean thickness of the film, h_o and consequently, ε is less than half. One needs to determine now the space-time evolution of the initial distrubance and from this, the time of rupture of the thin film.

Hitherto, the solution methodology has been to linearize the governing equations around the mean value of the film thickness and then seek the normal mode solutions to these linearized set of equations. The mean value of the film thickness, h_o would be referred to as the "trivial steady state" and the aforementioned procedure as the "linear stability" analysis, if the effect of all but the first order, dominant harmonic are ignored. As is shown by the numerical simulations of Williams and Davis (14), the linear stability theory indeed predicts the correct times of rupture whenever the amplitude of the initial disturbance is vanishingly small compared to the mean thickness of the film. Further, the simulations reveal that the linear theory begins to systematically overestimate the time of rupture as the perturbation amplitude increases. The discrepancy, however, cannot be attributed to the influence of harmonics other than the dominant one, because the numerical step size deemed satisfactory was not small enough to keep track of the higher order harmonics adequately. This contention was analytically proved for a wetting film by us (Sharma and Ruckenstein (16)) and is later shown to hold for a tangentially immobile free film as well.

It may therefore be argued that the dominant growth coefficient for a finite amplitude perturbation is itself different from that inferred by linearizing the equations around the trivial steady state. We now demonstrate that a perturbation solution of the form

$$h(x,t) = H_o(x) + C_1\,\varepsilon\,H_1(x,t) + \sum_2 C_i\,\varepsilon^i H_i\,(x,t) \qquad \text{(A)}$$

obtains the dominant growth coefficient for a finite amplitude perturbation and reduces to the linear analysis as ε goes to zero. Here, $h(x,t)$ is the location of the free interfaces, x and t are the lateral space coordinate and time, respectively and $H_o(x)$ is a time stationary solution of the governing equations. $H_1(x,t)$ is the leading order term that determines the growth/decay rate of disturbances that are envisaged to be imposed upon the stationary solution, $H_o(x)$. C_1, C_i are constants. As is apparent, $H_o(x)$ and $H_1(x,t)$ are determined respectively from the zeroth order and the first order (in small parameter, ε) governing equations. Further, the solution (A) must satisfy the given initial condition $h(x,0)$ at $t=0$. viz.,

$$h(x,0) = H_o(x) + C_1\,\varepsilon\,H_1(x,0) + O(\varepsilon^2) \qquad \text{(B)}$$

The expressions (A) and (B) formally describe the space-time evolution of an initial condition (disturbance) $h(x,0)$ that originates in the neighborhood of the steady-state (or the base state) $H_o(x)$. The effective linearization (expansion (A)) around the base state $H_o(x)$ is expected to work well only if the initial state, $h(x,0)$ is "close" to the base state. From this reasoning, the linear stability analysis is anticipated to be at its best whenever the initial amplitude is vanishingly small, viz., $\varepsilon \to 0$. This is so because in the linear theory, the base state is the mean thickness of the film and the deviation of the initial state from the base state is of the order of ε. Thus, whenever $\varepsilon \to 0$, the initial state coincides with the trivial base state h_o and hence the linearization around the mean film thickness yields correct results.

However, the above scenario does not have an a-priori justification if ε is finite (say 0.3) and thus the deviation of the initial state from the trivial state becomes considerable. As is indeed demonstrated by the simulations, the linearization around the mean film thickness now ceases to be adequate.

The problem besetting the linear theory for a finite ε may be circumvented by noting that the base is not necessarily the trivial state, h_o, but it is a steady state solution of the governing equations. Thus, the possibility of the base state itself being spatially nonuniform with a certain finite amplitude cannot be ruled out. Such a "nontrivial", finite-amplitude, time-stationary solution does indeed exist for a dissipative system (system for which the growth rate is real and hence the temporal oscillations are ruled out). In fact, the leading order term of this nontrivial steady solution corresponds to the neutrally stable wave of the linear stability analysis. As is obvious, a neutrally stable solution satisfies the dual requirements of both being the time stationary and spatially nonuniform and hence, nontrivial. The neutrally stable wave of the linear stability is determined up to a multiplicative constant however and this amplitude is uniquely established by the "matching" condition (B). This condition yields a proportionality relation between the amplitudes of the base state and the perturbation (initial state). It is in view of this that the base state reduces to the trivial solution h_0 in a natural manner whenever $\varepsilon \to 0$. The results of the linear analysis are, of course, recovered in such an event. For any finite ε however, the nontrivial base state, $H_0(x)$ is closer to a finite amplitude disturbance as compared to the trivial state, h_0. The choice of base state as $H_0(x)$ is therefore expected to yield better results while pursuing a perturbative analysis.

In what follows then, we will use expansion (A) in conjunction with condition (B) with the understanding that the base state, $H_0(x)$ is a finite amplitude steady state solution of the governing equations. Up to terms of order ε, $H_0(x)$ will be shown to coincide with the neutrally stable wave as determined from the linear analysis.

FREE THIN FILM WITH INSOLUBLE SURFACTANTS

The thin film model being considered is depicted in Figure 1 with one wavelength of the disturbance. The film is considered to be unbounded (radius is very large) and, therefore, nondraining. The effect of drainage is, of course, important for films of small radii. In this section, we address the issue of how conditions for neutral stability and the growth rate of perturbations change with a variation in the disturbance amplitude. The possible extensions of the present approach to situations where the drainage is important are then indicated.

The governing equations are the continuity and Navier-Stokes equations with an extra body force term that arises due to the long range van der Waals interactions. As has been shown by the linear theories (Ruckenstein and Jain (4), Maldarelli et al. (8)), the dominant growth rate leading to rupture occurs for the long wavelength, symmetric disturbances and thus we confine our attention to these disturbances from the onset. Whenever the wavelengths of disturbances far exceed the mean thickness of the film, the hydrodynamic equations can be simplified without placing any restrictions on the amplitude of the perturbations. The long wavelength reduction procedure was developed and applied by Benny (17) and by Atherton and Homsy (18) to the falling films. Williams and Davis (14) extended the analysis to a thin wetting (in contact with a solid) film that is devoid of solutes and derived a nonlinear equation of evolution for the location of the free interface. While the essentials of a formal perturbative analysis may be found in these references and were also pursued by us (Sharma and Ruckenstein (16)) for a wetting film with solutes, it suffices to note that the angle Δ in Figure 1 is small for a long wavelength perturbation. It is in view of this that the curvature effects may be neglected in all of the boundary conditions. Further, the following simplified Navier-Stokes equations describe the thin films subjected to long wavelength disturbances (Wiliams and Davis (14), Sharma and Ruckenstein (16)). These are, in effect the equations in the "lubrication approximation".

$$\mu u_{zz} = p_x + \phi_x \tag{1}$$

and

$$p_z + \phi_z = 0 , \tag{2}$$

where subscripts denote differentiation, u is the velocity component in x-direction, μ, p and ϕ are the film viscosity, pressure and the van der Waals interaction potential (for a unit volume of fluid), respectively. The differentials of the interaction potentials ϕ_x and ϕ_z describe the intermolecular forces in the x and z directions, respectively. The continuity equation is

$$u_x + v_z = 0 \tag{3}$$

The following simplified boundary conditions hold for long wave disturbances.

$$u_z = v = 0 \text{ at } z = 0 , \tag{4}$$

$$\mu u_z = \sigma_x \text{ at } z = (h/2) \tag{5}$$

and

$$-p + 2\mu v_z = \sigma h_{xx}/2 \text{ at } z = (h/2) , \tag{6}$$

where σ is the surface tension and $h(x,t)/2$ represents the location of the free interface as a function of x and t. The interfacial tension varies due to the uneven distribution of surfactants and also because of the surface viscosity. The contribution of the term $2\mu v_z$ in boundary condition (6) can be shown to be insignificant in the presence of surfactants, both by the long wave reduction procedure (Williams and Davis (14)) and by the linear stability results (Ruckenstein and Jain (4)). This approximation has been employed in many studies of thin film stability (Dimitrov (19)).

The approximation, however, breaks down for a free film devoid of solutes, as the omission of the term $2\mu v_z$ now results in an infinite rate of growth of perturbations. In practice, however, even a small amount of insoluble surfactants ("gaseous" monolayer type adsorption) usually make this term negligible. It is, therefore convenient to consider two separate cases, namely, a film devoid of solutes and the one with an arbitrary concentration of surfactants. The latter case reduces to the case of a tangentially immobile film when either the Marangoni-flow or the surface viscosity are sufficient to render the surface tangentially immobile.

Pure Free Film

The interfacial tension gradient vanishes for a pure film, i.e., $\sigma_x = 0$. The solution of Equation (2) gives

$$\Psi \equiv p(x,t) + \phi(x,t) = p(h/2) + \phi(h/2) \quad (7)$$

The solution of Equation (1) with the help of expression (7) and boundary conditions (4) and (5) leads to

$$\Psi_x = p_x(h/2) + \phi_x(h/2) = 0 \quad (8)$$

and

$$u = C(x,t) . \quad (9)$$

Here, C(x,t) is an integration constant. Consequently, an integration of the continuity equation (3) with boundary condition (4) yields:

$$v = -C_x z . \quad (10)$$

Use of Equation (10) in conjunction with the remaining condition (6) allows the determination of pressure at the interface, substitution of which in Equation (8) then provides an equation for the unknown coefficient C(x,t):

$$4\mu\, C_{xx} = (2\phi_x - \sigma h_{xxx}) . \quad (11)$$

Finally, the kinematic condition

$$h_t + h_x u - 2v = 0 \text{ at } z = h/2 \quad (12)$$

is transformed to

$$h_t + (Ch)_x = 0 . \quad (13)$$

The solution of the coupled nonlinear Equations (11) and (13) provides the evolution of the initial disturbances when the film is devoid of solutes. Before undertaking this, however, evolution equations are derived for the more important case of a film with insoluble surfactants.

Film with Insoluble Surfactants: Role of Marangoni-Flow and the Surface Viscosity

In this case, Equation (1) may be integrated with the help of expression (7) and the boundary conditions (4) and (5). This gives

$$u = (p_x + \phi_x)\frac{z^2}{2\mu} + a(x,t) \quad (14)$$

and

$$\frac{2\sigma_x}{h} = p_x + \phi_x , \quad (15)$$

where a(x,t) is a constant of integration and $p_x = p_x(h/2)$ is given by the boundary condition (6). Assuming $v_z = 0$ in boundary condition (6), one obtains:

$$p_x = \frac{1}{2}(\sigma h_{xx})_x . \quad (16)$$

The continuity equation now yields the normal velocity as:

$$v = -\frac{z^3}{3\mu}\left(\frac{\sigma_x}{h}\right)_x - a_x z . \quad (17)$$

Substitution of u and v in the kinematic condition (12) gives the sought after equation of evolution as

$$h_t + (ah)_x + \frac{1}{12\mu}(h^2\sigma_x)_x = 0 \quad (18)$$

The interfacial tension gradient is determined by substituting for p_x from Equation (16) into Equation (15) and solving for σ_x:

$$\sigma_x = \frac{h}{2}(\phi_x - \frac{1}{2}\sigma h_{xxx})[1 + \frac{h}{4}h_{xx}]^{-1} \quad (19)$$

The determination of the unknown coefficient a(x,t) requires posing the transport equation for the surfactant and a constitutive relation for the variation of the interfacial tension. Since the surfactant is assumed to be insoluble, the transport equation in the long wave limit is

$$\Gamma_t + [u(h/2)\Gamma]_x - D_s\Gamma_{xx} = 0 , \quad (20)$$

where Γ is the surface excess (or surface) concentration of the insoluble surfactant, D_s is the surface diffusion coefficient and

$u(h/2)$ is the x component of velocity at the interface. The constitutive relation for a purely viscous adsorbed layer may be written as:

$$\sigma = \sigma_o(\Gamma_o) + M(\Gamma)\Gamma + \mu_s[u(h/2)]_x , \quad (21a)$$

where $\sigma_o(\Gamma_o)$ is the surface tension for a certain equilibrium surface concentration of Γ_o, $M(\Gamma) = (\partial\sigma/\partial\Gamma)$ and μ_s is the surface viscosity. The parameter M is sometimes constant (for a "gaseous" monolayer, for instance) or it may be evaluated at a certain base case concentration, Γ_o. Equations (18) to (21) constitute a set of nonlinear equations for four unknowns - h, σ_x, Γ and a, as $u(h/2)$ can be eliminated from Equations (20) and (21) by using Equation (14). Equations (14) and (15) determine the tangential velocity at the interface as:

$$u(h/2) = (\sigma_x h/4\mu) + a . \quad (21b)$$

It is advantageous to solve Equation (21a) for u_x and thus rewrite Equation (20) as:

$$\Gamma_t + \Gamma[(\sigma-\sigma_o-M\Gamma)/\mu_s] - D_s\Gamma_{xx} = -\Gamma_x u . \quad (22)$$

Also, Equation (20) can be solved for a_x by combining it with Equation (21b):

$$a_x h = \left[\frac{D_s\Gamma_{xx}-\Gamma_t-\Gamma_x u}{\Gamma}\right]h - \frac{1}{4\mu} h(\sigma_x h)_x . \quad (23)$$

The evolution equation (18) can now be rewritten in a more convenient form with the help of expression (23) as

$$h_t + \frac{1}{12\mu}(\sigma_x h^2)_x - \frac{1}{4\mu}h(\sigma_x h)_x + ah_x$$
$$+ (\frac{h}{\Gamma})(D_s\Gamma_{xx}-\Gamma_t-\Gamma_x u) = 0 , \quad (24)$$

which then may be rearranged with the help of expression (19) as:

$$h_t - \frac{1}{12\mu} [h^3(\phi_x - \frac{1}{2}\sigma h_{xxx})(1 + \frac{h}{4} h_{xx})^{-1}]_x$$
$$+ \frac{1}{4\mu} (hh_x\sigma_x) + ah_x$$
$$+ (\frac{h}{\Gamma})(D_s\Gamma_{xx} - \Gamma_t - \Gamma_x u) = 0 \quad (25)$$

As is shown later, the present formalism requires only the simultaneous solution of Equations (22) and (25) for obtaining the nonlinear dynamics of the film. Finally, the intermolecular potential for a unit volume of fluid located at the interface needs to be specified. If the interactions are nonretarded, the potential for London van der Waals interaction is of the form

$$\phi = A/6\pi h^3 , \quad (26)$$

where A is the Hamaker constant for the interactions between molecules of the thin film. The potential may be suitably modified to include other interactions like double layer repulsion and retardation effects.

An important limiting case of the above equations is a thin film with tangentially immobile surfaces. In this event, Equation (21b) may be solved for the unknown coefficient $a(x,t)$ by setting the tangential velocity at the interface to be zero. This gives

$$a = - \frac{1}{4\mu} (\sigma_x h) . \quad (27)$$

Substitution of this in Equations (20) and (25) leads to the following interfacial equation of evolution for a surface that is rendered tangentially immobile by the surfactants:

$$h_t + \frac{A}{24\pi\mu} (h^{-1}h_x)_x + \frac{\sigma}{24\mu} (h^3 h_{xxx})_x = 0 . \quad (28)$$

The derivation of Equation (28) supposes $D_s = 0$ and $(h/4)h_{xx} << 1$. The latter assumption holds for the long wavelength disturbances that we are considering here, as $h\, h_{xx} \sim (k^2h_o^2)$, where kh_o is the ratio of the film thickness to the disturbance wavelength. It is interesting to note that this equation may be recast into a parameterless form by the following transformations:

$$H = (h/h_o), \xi = h_o^{-2}(A/\pi\sigma)^{\frac{1}{2}}x \text{ and}$$
$$\tau = (A^2/24\pi^2\mu\sigma h_o^5)t ,$$

which transform Equation (28) to

$$H_\tau + (H^{-1} H_\xi)_\xi + (H^3 H_{\xi\xi\xi})_\xi = 0 . \quad (29)$$

This parameterless form is the same as that obtained and numerically solved by Williams and Davis (14) for a thin film in contact with a solid and devoid of solutes. That one obtains the same functional form of the nonlinear equation for two different thin film models is perhaps not surprising in view of the fact that the linear dispersion relations for these two models are also the same up to a multiplicative constant (Ruckenstein and Jain (4). In any event, the equivalence between the two models enables us to compare our analytical results for a free thin film with excess surfactants to the numerical simulations of Williams and Davis (14) for a wetting film that is devoid of surfactants. Having obtained the nonlinear equations describing the evolution of initially corrugated free interfaces, we now turn towards first the linear analysis and then finite amplitude instability of these interfaces.

Linear Stability and the Higher Harmonics

The linear stability of the free interface is easily ascertained by substituting the following perturbative expansions in Equations (22) and (25).

$$h = h_o + 2\varepsilon h_o \sin kx\, e^{\omega t} + O(\varepsilon^2) \qquad (30a)$$

and

$$\Gamma = \Gamma_o + \varepsilon \Gamma_1 h_o \sin kx\, e^{\omega t} + O(\varepsilon^2)\ , \qquad (30b)$$

where ε is the nondimensional amplitude of the initial perturbation at one of the free interfaces of the film, ω is a growth coefficient, k is a wave number and the initial perturbation is assumed to be of the form sin kx. Linearization of the governing equations by this procedure leads to the following dispersion relation:

$$\omega_L = \frac{k^2 h_o^3}{24\mu}\left(\frac{A}{\pi h_o^4} - \sigma_o k^2\right)\left[1 + \frac{6\mu}{h_o k^2\{\mu_s + \Gamma_o |M| (D_s k^2 + \omega_L)^{-1}\}}\right] \qquad (31)$$

where ω_L is the growth coefficient of the linear theory and σ_o is the surface tension evaluated at the equilibrium surface concentration. The above relation reduces to the result for a tangentially immobile film when either surface viscosity (μ_s) or the surface elasticity ($\Gamma_o|M|$) is large and therefore

$$\omega_{L1} = \frac{k^2 h_o^3}{24\mu}\left[\frac{A}{\pi h_o^4} - \sigma_o k^2\right]\ . \qquad (32)$$

As discussed earlier, relation (31) cannot be used for predicint the growth rates in the event the film is devoid of solutes (or impurities), i.e., both μ_s and $\Gamma_o|M|$ vanish. For a pure free film, the linear dispersion relation is obtained with the help of Equations (11) and (13) in conjunction with expression (30a). In this case

$$\omega_{L2} = \frac{h_o}{4\mu}\left[\frac{A}{\pi h_o^4} - \sigma_o k^2\right]\ . \qquad (33)$$

The wave number for which the growth coefficient, Equation (32), is maximum is given by the following solution of $(\partial \omega_L/\partial k) = 0$:

$$k_m^2 = (A/2\pi\sigma_o h_o^4) \qquad (34)$$

and the corresponding growth coefficient is

$$\omega_{L1} = A^2(96\pi^2\mu\sigma_o h_o^5)^{-1}\ . \qquad (35a)$$

There is no dominant wave number for a pure free film and the growth coefficient is maximum for $k \to 0$, viz.

$$\omega_{L2} = A(4\pi\mu h_o^3)^{-1}\ . \qquad (35b)$$

The dominant growth coefficient for a pure film, ω_{L2} is several orders of magnitude higher than the dominant growth coefficient in the presence of an excess of surfactant, ω_{L1}. These linear stability results (32) to (35) were obtained in these forms by Ruckenstein and Jain (4). Finally, if the results of the linear theory are taken seriously up to the point of rupture, the time of rupture may be computed from expansion (30a) by setting h = 0. This, for a finite amplitude perturbation gives

$$\tau_L = \frac{1}{\omega_{Lm}} \ln(1/2\varepsilon)\ , \qquad (36)$$

where ω_{Lm} is the maximum value of the linear theory growth coefficient. That the inclusion of higher order terms in expansions (30) does not change these results significantly may be shown in the following manner for a tangentially immobile thin free film. Assume a perturbation expansion of the following form with a small ordering parameter, μ^*:

$$H = 1 + \mu^* h_1(\xi,\tau) + \sum_{i=2} (\mu^*)^i\, h_i(\xi,\tau)\ . \qquad (37a)$$

Its insertion in Equation (29) yields the most unstable first harmonic as

$$h_1 = \sin q\xi \exp(\omega_m \tau) \tag{37b}$$

where $q^2 = (1/2)$ and $\omega_m = (1/4)$ are the nondimensional, scaled counterparts of expressions (34) and (35a), respectively. The second harmonic is determined by comparing the terms of order μ^{*2} and is given by the solution of the following linear, nonhomogeneous equation:

$$h_{2\tau} + (h_{2\xi}-h_{1\xi}h_1)_\xi + (h_{2\xi\xi\xi} + 3h_1h_{1\xi\xi\xi})_\xi = 0. \tag{37c}$$

The solution of which is given by

$$h_2 = \frac{3}{10} \cos 2q\xi\, e^{2\omega\tau} . \tag{37d}$$

Similarly, the third order equation

$$h_{3\tau} + \{h_{3\xi}-h_1h_{2\xi}+h_{1\xi}(h_1^2-h_2)\}_\xi$$

$$+ \{h_{3\xi\xi\xi}+3h_1h_{2\xi\xi\xi}+3h_{1\xi\xi\xi}(h_1^2+h_2)\}_\xi = 0 \tag{37e}$$

has the solution

$$h_3 = \frac{12}{33} \sin 3q\xi\, e^{3\omega\tau} . \tag{37f}$$

Thus, the time of rupture up to this approximation may be determined by setting H = 0 in expansion (37a), i.e.,

$$1 - \mu^* e^{\tau/4} - \frac{3}{10}\mu^{*2} e^{\tau/2} - \frac{12}{33}\mu^{*3} e^{3\tau/4} = 0 , \tag{37g}$$

where the parameter μ^* is determined by matching the maximum amplitude of expansion (37a) at time t = 0 with the amplitude of the external perturbation, viz.

$$\mu^*(1 + \frac{3}{10}\mu^* + \frac{12}{33}\mu^{*2}) = 2\varepsilon . \tag{37h}$$

If terms of order μ^{*2} and higher are ignored, $\mu^* \simeq 2\varepsilon$ and hence the time of rupture is given by $\tau = 4 \ln(1/2\varepsilon)$. The ratio of the time of rupture as computed from Equation (37g) and that given by the first term, $4\ln(1/2\varepsilon)$, is always more than about 0.85, even for total nondimensional amplitudes as large as 0.8 of the film thickness. These calculations indicate that the effect of higher harmonics on the time of rupture is rather insignificant. So far, we have obtained the linear theory results and have indicated that the effect of the first harmonic is dominant in calculating the time of rupture of a thin free film. With this background, we now derive the expressions for the first harmonic corresponding to a finite amplitude perturbation.

Finite Amplitude Instability

As discussed earlier, we now consider a finite amplitude spatially nonuniform base state. The film thickness and the concentration may thus be expanded as

$$h = H_o(x) + 2\mu^* H_1(x,t) \tag{38a}$$

and

$$\Gamma = \Gamma_o + \Gamma_1 \sin kx\, e^{\omega_n t} , \tag{38b}$$

where μ^* is as yet an undetermined coefficient which is later related to the amplitude. The base state, $H_o(x)$ is obtained by solving the zeroth order governing equations. The calculation of $H_o(x)$ may be simplified somewhat by noting that the velocities corresponding to the steady state (or the base state) are zero in the absence of any imposed velocities. Further, the absence of any fluid motion in the film implies that the surface concentration is uniform and consequently, the steady state gradients of concentration and the surface tension also vanish. Thus, for a film with surfactants, the base state is a solution of the following zeroth order steady-state version of Equation (25):

$$(H_o^{-1} H_{ox})_x + (\pi\sigma_o/A)(H_o^3 H_{oxxx})_x = 0 . \tag{39}$$

An approximate stationary solution may be found by representing H_o as

$$H_o = h_o(1 + 2\mu^* \sin k_1 x) \tag{40}$$

and equating the terms of order μ^* in Equation (39). This gives the compatibility condition

$$k_1^2 = (A/\pi h_o^4 \sigma_o) \tag{41}$$

As is obvious, this is nothing but the condition for neutral stability of the trivial state, viz., the condition for which ω_{L1} from Equation (32) is identically zero. The expression (40) with the nontrivial wave number, k_1, therefore constitutes a time stationary, spatially nonhomogeneous solution of the governing equations. Physically, the surface tension restoring force exactly counters the destabilizing influence of the van der Waals force for the steady state solution (40). While we have simplified the calculation of $H_0(x)$ by anticipating the velocities and the concentration gradient to be zero from the onset, a complete analysis up to terms of order μ^* indeed shows this to be the case. As is obvious, this complete analysis is the same as the linear stability analysis pursued earlier, if the growth coefficient corresponding to the time stationary base state is set equal to zero. With this in view, the condition for the neutral stability (41) and the steady state solution (40) are directly obtained from the general Equations (31) and (30a), respectively. That the gradient of interfacial tension and the velocities corresponding to the base state are indeed zero, may now be verified by a direct substitution of the steady state solution (40) into expressions (19), (23), (14) and (17). The bifurcation of a time dependent solution may now be investigated by perturbing $H_0(x)$ according to expansions (38).

Differentiating Equation (24) with x gives

$$\Gamma_{tx} + \Gamma_x(\sigma-\sigma_0-M\Gamma)/\mu_s + \Gamma(\sigma_x-M\Gamma_x)/\mu_s - D_s\Gamma_{xxx} = -(\Gamma_x u)_x \;. \quad (42)$$

Substituting in it the expansion (38b) and recalling that for the base state, $u_0 = \Gamma_{0x} = 0$ gives

$$\Gamma_1 \cos kx\, e^{\omega_n t} = -(\Gamma_0\, \sigma_x/\mu_s)\left(k\omega_n + \frac{\Gamma_0|M|k}{\mu_s} + D_s k^3\right)^{-1} . \quad (43)$$

Further, the substitution of expansions (38) in the equation of evolution (25) results in the following leading order equation.

$$H_{1t} + \frac{A}{24\pi\mu}(H_0^{-1}H_{1x} - H_{0x}H_0^{-2}H_1)_x + \frac{\sigma}{24\mu}(H_0^3 H_{1xxx} + 3H_0^2 H_1 H_{0xxx})_x - \frac{H_0}{2\mu^*\Gamma_0}(D_s k^2\Gamma_1 + \omega_n\Gamma_1)\sin kx\, e^{\omega_n t} = 0 \quad (44)$$

The first order concentration, Γ_1, is of course determined from Equations (43) and (19). However, at this point it may be observed that Equations (44) involves coefficients (which are functions of $H_0(x)$) that depend on the spatial coordinate x. In order to make any further analytical progress, we confine our attention to the growth of disturbances at a point where the film thickness is the least and, therefore, the rupture first occurs here. This means that we study the time-evolution of disturbance where $\sin k_1 x = -1$ in Equation (40) and, therefore, $H_0 \simeq h_0(1-2\mu^*)$. Based on this reasoning, various derivatives of the base state may be evaluated as follows:

$$H_0 \simeq h_0(1-2\mu^*) \quad (45a)$$

$$H_{0x} \simeq 2\mu^* h_0 k_1 \cos k_1 x = 0 \quad (45b)$$

$$H_{0xx} \simeq -2\mu^* h_0 k_1^2 \sin k_1 x = 2\mu^* h_0 k_1^2 \;, \quad (45c)$$

$$H_{0xxx} \simeq -2\mu^* h_0 k_1^3 \cos k_1 x = 0 \quad (45d)$$

and

$$H_{0xxxx} \simeq 2\mu^* h_0 k_1^4 \sin k_1 x = -2\mu^* h_0 k_1^4 \;. \quad (45e)$$

This procedure allows the construction of a normal mode solution for H_1, as Equation (44) is now reduced to a linear, homogeneous equation with constant coefficients. The solution of Equation (44) therefore is

$$H_1 = \sin kx\, e^{\omega_n t}$$

and it is in view of this that the evolution equation (44) now gives the following expression for the growth rate associated with a finite amplitude perturbation:

$$\omega_n = \left[\frac{A}{\pi H_0^4} - \sigma_0 k^2\right]\left[\frac{k^2 H_0^3}{24\mu} + \frac{H_0^2}{4}\,\frac{1}{(\mu_s + \Gamma_0|M|(D_s k^2 + \omega_n)^{-1})}\right] + 2\mu^*\left(\frac{A^2}{8\pi^2\mu\sigma_0 H_0^5}\right)\left[\frac{1}{3}(1-2\mu^*)^3 + (1-2\mu^*)^7\right]. \quad (46)$$

In arriving at the above relation, the definition of k_1^2 from Equation (41) has been used in evaluating Equations (45b) and (45e).

The unknown small parameter μ^* may be

related to the nondimensional amplitude of the initial disturbance by matching the location of the interface, $h(x,t)$ at time $t = 0$ with the amplitude of the imposed perturbation. Therefore, with this approximation

$$h_o(1-4\mu^*) \simeq (1-2\varepsilon)h_o$$

or

$$2\mu^* \simeq \varepsilon < \frac{1}{2} \qquad (47)$$

It may be recalled that ε is the nondimensional amplitude of the disturbance at each face of the free film. The dispersion relation may thus be rewritten as

$$\omega_n(\varepsilon) = \frac{k^2h_o^3(1-\varepsilon)^3}{24\mu}\left[\frac{A}{\pi h_o^4}(1-\varepsilon)^{-4} - \sigma_o k^2\right]$$

$$\left[1 + \frac{6\mu(1-\varepsilon)^{-1}}{h_o k^2\{\mu_s+\Gamma_o|M|(D_s k^2+\omega_n)^{-1}\}}\right]$$

$$+ \left(\frac{A^2\varepsilon}{8\pi^2\mu\sigma_o h_o^5}\right)\left[\frac{(1-\varepsilon)^{-2}}{3} + (1-\varepsilon)^2\right] \qquad (48)$$

As one may verify, $\omega_n(\varepsilon)$ reduces to ω_L as given by Equation (31) if the amplitude of initial perturbation is vanishingly small, viz., $\varepsilon \to 0$. The results of the linear theory are, therefore, recovered only if the perturbation amplitude is small. This is so because the spatially nonhomogeneous base-state $H_o(x)$ reduces to the trivial solution, $H_o=h_o$ in such an instance (See Equation (40)). The time of rupture for a finite amplitude perturbation is found by maximizing the growth rate with respect to the wave number, k, and then setting $h=0$ in expansion (38a). This procedure gives

$$\tau_N = (\omega_{nm})^{-1} \ln[(1-\varepsilon)/\varepsilon)], \qquad (49)$$

where ω_{nm} is the maximum value of the growth coefficient.

For a pure free film described by the nonlinear Equation (11) and (13), the estimates (41) and (45a)-(45e) still hold. An application of the same procedure as that pursued for a film with surfactants now gives for the growth rate the expression:

$$\omega_n = \frac{h_o(1-\varepsilon)}{4\mu}\left[\frac{A}{\pi h_o^4}(1-\varepsilon)^{-4} - \sigma_o k^2\right] . \qquad (50)$$

The maximum growth coefficient for a pure free film again occurs for very large wavelengths, viz., for $k \to 0$.

$$\omega_{nm} = \omega_{L2}(1-\varepsilon)^{-3} , \qquad (51)$$

where ω_{L2} is the corresponding linear theory growth coefficient given by Equation (35b).

Unlike the case of pure film, the maximization of the growth coefficient for a film with solutes may be carried out analytically only by assuming $(D_s k^2+\omega_n)$ to be a weak function of k in Equation (48). To a first approximation then, the solution of $(\partial\omega_n/\partial k) = 0$ may be written as:

$$k_m^2 = \frac{A}{2\pi h_o^4\sigma_o}(1-\varepsilon)^{-4} - \frac{3\mu(1-\varepsilon)^{-1}}{h_o\{\mu_s+\Gamma_o|M|(D_s k_m^2+\omega_{nm})^{-1}\}} . \qquad (52)$$

This solution holds whenever k_m^2 as computed from the above expression is positive or else the dominant wave number is zero. Substitution of this dominant wave number, k_m in Equation (48) provides the maximum growth rate for a free film with insoluble surfactants:

$$\omega_{nm} = \left(\frac{A^2}{96\pi^2\mu\sigma_o h_o^5}\right)\left[(1-\varepsilon)^{-5}\{N(1-\varepsilon)^3 + 1\}^2 + 4\varepsilon(1-\varepsilon)^{-2} + 12\varepsilon(1-\varepsilon)^2\right] . \qquad (53)$$

where

$$N = \frac{6\pi\mu\sigma_o h_o^3}{A\{\mu_s+\Gamma_o|M|(D_s k_m^2+\omega_{nm})^{-1}\}} . \qquad (54)$$

The second term of Equation (53) in square braces represents the effect of amplitude on the growth coefficient derived from the linear theory. The dominant growth coefficient for the linear theory can be derived by letting the amplitude ε vanish in the set of Equations (52) to (54). That the approximation used to derive the dominant growth rate holds even for a "gaseous" monolayer type of adsorption is easily confirmed by employing the data of Ruckenstein and Jain (4). In general, even very small amounts of surfactants are sufficient to bring the parameter N close to zero or, in other words, to render the surface tangentially immobile. For example, the maximum value of the parameter N for a film 10^{-8}m (100 Å) thick and having a surface viscosity of 10^{-6} kg/s is about 10^{-1}. The presence of Gibbs elasticity, $\Gamma_0|M|$ reduces even further this value of N (Equation (54)). It is, therefore, important to consider the asymptotic case of tangentially immobile film, for which

$$N(1-\varepsilon)^3 << 1 \ . \tag{55}$$

In this case, the expression for the dominant wave number,

$$k_m^2 = (A/2\pi h_o^4 \sigma_o)(1-\varepsilon)^{-4} \tag{56}$$

is exact, and the dominant growth coefficient is given by:

$$\omega_{nm} = \left\{\frac{A^2}{96\pi^2 \mu \sigma_o h_o^5}\right\} [(1-\varepsilon)^{-5} + 4\varepsilon(1-\varepsilon)^{-2} + 12\varepsilon(1-\varepsilon)^2] \tag{57}$$

Inspection of Equations (51), (53) and (57) reveals that the growth coefficient for a finite amplitude perturbation is always larger than its linear theory counterpart. The rupture of the film is, therefore, accelerated as the disturbance amplitude is increased.

As shown earlier, the nonlinear equation describing the evolution of a tangentially immobile interface of a free film is the same as the equation that was derived by Williams and Davis (14) for a pure film in contact with a solid. It is, therefore, possible to compare the time of rupture as derived from Equations (49) and (57) to the numerical simulations of Williams and Davis (14). The dashed curve (curve labeled I) in Figure 2 represents the ratio of the times of rupture as computed by Williams and Davis and by the linear theory. Curve II of the same figure depicts the same ratio as calculated from the present theory (Equations (49) and (57)). The discrepancy between the numerical simulations and the results of present theory are within 15% for initial amplitudes as large as 0.6 of the film thickness. This appears remarkable in view of the simplicity of the present formalism. Also, William and Davis (14) report the dominant wave number for 2ε = 0.1 to be about a factor of 1.1 larger than the corresponding linear theory dominant wave number. Equation (56) also predicts this ratio to be about 1.1. Expressions (52) and (56) show that the dominant wavelength for a finite amplitude perturbation is always smaller than its linear theory counterpart. However, it is still of about the same order of magnitude and hence the long wavelength reduction procedure is justified for a finite amplitude perturbation.

The theory may be easily extended for a more general potential of the form $\phi = B/h^c$. For a tangentially immobile film, this leads to a dominant growth coefficient of the following form ($c \neq 0,2$):

$$\omega_{nm} = \omega_L[(1-\varepsilon)^{1-2c} + 4\varepsilon\{(c-2)(1-\varepsilon)^{1-c} + 3(1-\varepsilon)^2\}] \ , \tag{58}$$

where ω_L is the corresponding growth coefficient derived from the linear stability

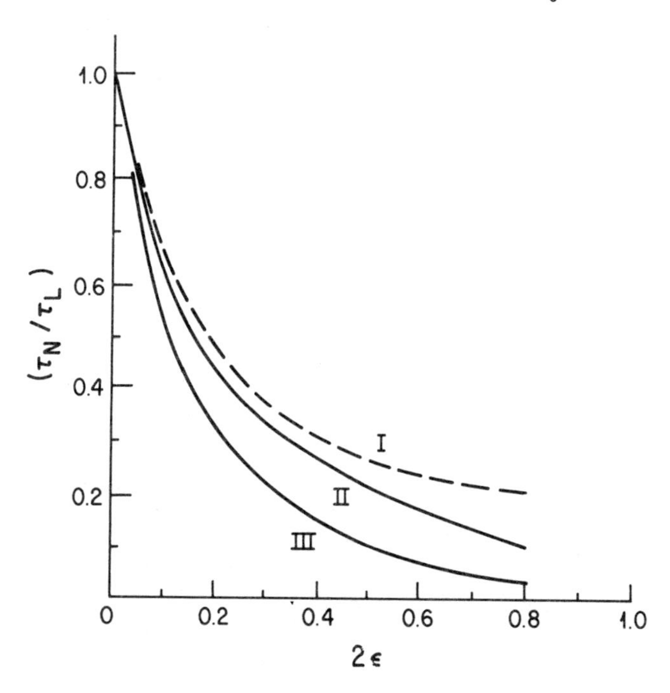

Figure 2. The ratio of the times of rupture as computed from the nonlinear and the linear theories, respectively. 2ϵ is the total nondimensional initial amplitude of the perturbation. The dashed curve (labeled I) corresponds to the numerical simulations of Williams and Davis (*14*) for a pure thin film in contact with a solid. Curve II corresponds to the present analysis for a tangentially immobile thin free film. Curve III is computed for a tangentially immobile free film with retarded dispersion interactions.

analysis. Thus for any $c>1$, the dominant growth coefficient for a finite amplitude perturbation shows a greater deviation from its linear theory counterpart as c is increased. For example, the effect of nonlinearities is even more pronounced for retarded interactions, viz., $c \simeq 4$, compared to the case of nonretarded interactions. This may be verified in Figure 2, where curve III shows the ratio between the times of rupture as derived from the nonlinear and the linear theories, respectively for the case of retarded dispersion interactions.

Finally, it is of interest to determine the conditions for neutral stability of a finite amplitude perturbation. For a film devoid of even trace amounts of solutes, this is obtained by setting the growth coefficient to zero in Equation (50), i.e.,

$$k_{cn}^2 = k_{cL}^2 (1-\varepsilon)^{-4} , \tag{59}$$

where k_{cn} and k_{cL} are the wave numbers of a neutrally stable wave as obtained from the nonlinear and linear theories, respectively. Thus, in the case of a pure free film, the neutrally stable wave has a wave number which is larger than its linear theory counterpart by a factor of $(1-\varepsilon)^{-2}$.

If surfactants are present and the film surface is tangentially immobile, the conditions for neutral stability are obtained by setting ω_n to be zero and μ_s to be infinity in Equation (48). The ratio $y = (k_{cn}^2/k_{cL}^2)$ is then determined, after some rearrangement, by the following quadratic equation:

$$y^2 - (1-\varepsilon)^{-4} y - 3\varepsilon\{(1-\varepsilon)^{-1} + \frac{(1-\varepsilon)^{-5}}{3}\} = 0. \tag{60}$$

Again, the solution of this equation indicates that $k_{cn} > k_{cL}$. For example, when the total perturbation amplitude is half of the film thickness, the dominant wavelength is about half of that derived from the linear theory. Note that according to the linear theory, all disturbances with wave numbers greater than k_{cL}, decay.

It is well known that the presence of surfactants retard the growth of perturbations and thus prolong the time of rupture significantly. The ratio of the times of rupture when the film has an excess of surfactants and when it is devoid of surfactants may be calculated by dividing Equation (51) with Equation (57). Therefore, the factor by which the time of rupture is prolonged by surfactants is given by

$$F_n = F_L \, F(\varepsilon) , \tag{61a}$$

where F_L is the stabilization factor as predicted by the linear theory, viz.,

$$F_L = 24\pi\sigma_o h_o^2/A \tag{61b}$$

and $F(\varepsilon)$ is a function of the initial amplitude which is given by

$$F(\varepsilon) = [(1-\varepsilon)^{-2} + 4\varepsilon(1-\varepsilon) + 12\varepsilon(1-\varepsilon)^5]^{-1} . \tag{61c}$$

The amplitude dependence of the stabilization factor, $F(\varepsilon)$, is plotted in Figure 3 as a function of the total initial amplitude, 2ε. It shows that $F(\varepsilon)$ is always less than one for any finite amplitude disturbance and that it is close to one only for very small amplitudes. It may thus be concluded that the stabilizing influences of the Marangoni-effect and surface viscosity are overestimated by the linear theory for a free film. This is in contrast to a thin film in contact with a solid, for

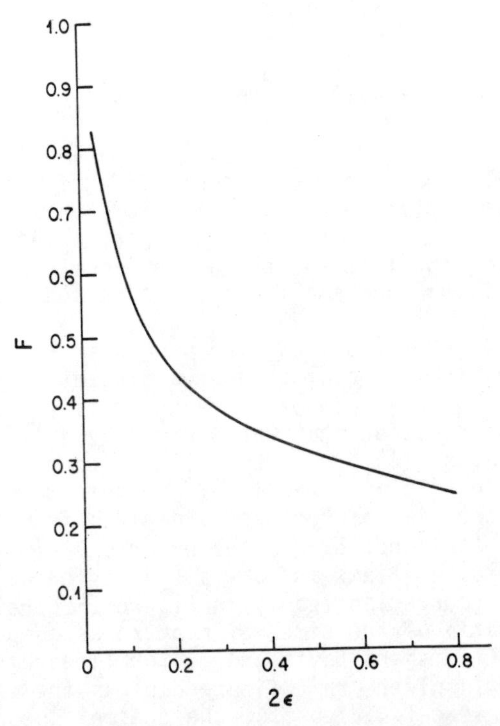

Figure 3. The maximum influence of surfactants in prolonging the time of rupture of a free film. F is the ratio of the stabilization factors predicted by the nonlinear and the linear theories, respectively. For a given amplitude, the nonlinear theory predicts a lesser influence of the Marangoni-effect and surface viscosity in prolonging the time of rupture.

which the stabilizing influence of surfactants is underestimated by the linear theory (Sharma and Ruckenstein (16)).

The surface diffusion tends to even out the surface concentration gradient and hence reduces the surface tension gradient. The role of surface diffusion is thus to undermine the stabilizing influence of the Marangoni-motion. Both the dominant wave number, k_m and the growth rate, ω_{nm} are larger for a finite amplitude perturbation as compared to their linear theory counterparts. It is in view of this that the parameter N as given by Equation (54) increases and the stabilizing influence of the Marangoni-motion decreases as the amplitude of the perturbation is increased. Therefore, the destabilizing influence of the surface diffusion is underestimated by the linear theory for a given surfactant concentration. This is also understood physically by noting that an increase in the wave number or a decrease in the wavelength results in a more rapid redistribution of surfactants because of the surface diffusion. This then undercuts the influence of the surface concentration driven Marangoni-motion.

In conclusion, dynamics of both a pure free film and a film with surfactants is analyzed when they are subjected to finite amplitude external disturbances. The expression (51) gives the dominant growth rate for a pure film and the time of rupture is computed from Equation (49). A general expression for the growth coefficient as a function of wave number is given by Equation (46) for a free film with insoluble surfactants. The approximate expressions for the dominant wave number and the dominant growth coefficient are given in this case by Equations (52) and (53), respectively. These expressions hold only if k_m^2 as calculated from Equation (52) is positive. If this is not the case, $k_m^2 \to 0$ and the film behaves like a pure free film. With this analysis, we now turn towards outlining the main results and the possible extensions and limitations of the present approach.

DISCUSSIONS

An analytical formalism is proposed to study the rupture of thin films when they are subjected to finite amplitude external perturbations. This is to be contrasted with the linear stability analysis that is expected to hold only for disturbances of vanishingly small amplitudes. The proposed analysis is only slightly more involved than the linear stability analysis and yet provides information about the nonlinear dynamics of a thin film with an ease that was offered hitherto only by the linear stability theory. The formalism is applied to a thin free film with insoluble surfactants, which is a model for foam and some emulsion films. It is shown that the lifetime of a free film is always less than that computed by the linear theory--about an order of magnitude smaller for very large perturbations. This is so because the destabilizing effect of van der Waals interactions is underestimated by the linear theory and the stabilizing influence of the surface tension is overestimated. Qualitatively, these conclusions are apparent even from the nonlinear equation of evolution (say, Equation (28)) itself. It may be observed that as the film approaches the point of rupture, viz. $h \to 0$, the term corresponding to the dispersion interactions becomes unbounded, whereas the term multiplied by the surface tension goes to zero. It is also shown that the linear theory overestimates the prolongation in the time of rupture that arises due to the Marangoni-effect (surface elasticity) and surface viscosity. For large initial perturbations, the factor by which the time of rupture is prolonged by the surfactants, may be up to an order of magnitude smaller than that given by the linear theory. Finally, it is shown that the effectiveness of surface diffusion in reducing the interfacial tension gradients, and thus in accelerating the rupture, is enhanced as the perturbation amplitude increases.

It is demonstrated that the linear theory is valid for all times, up to the point of rupture, whenever the initial amplitude of the disturbances is very small compared to the film thickness. The linear theory is, therefore, sufficient to study the dynamics of thin, isolated films in a vibrationless environment where the only interfacial disturbances are of thermal origin. Almost all of the experimental studies of thin film stability have been carried out in vibrationless environments (Radoev et al. (20), Rao et al. (21)). Thus, some of the assertions that have been made in the literature about the unsuitability of linear theories for interpreting these experiments are not valid. This contention is also supported by the numerical calculation of Williams and Davis (14) for a thin film in contact with a solid.

The generalization of the present approach to any potential of the form, $\phi = \phi(h)$ is rather straightforward. The double layer repulsion and other intermolecular interactions may thus also be included in the analysis and thus the rate of formation of extremely thin "black" films may be investigated. Here, we have reported results for a general potential of the form, $\phi = B/h^c$ (Equation (58)). It is shown that the discrepancy between the results of linear and nonlinear theories increases as the exponent, c, increases. The nonlinearities are, therefore, even more important for the retarded dispersion interactions.

The extension of the analysis to the films of other geometries is made by using the orthogonal functions appropriate for those geometries instead of trignometric functions. The response of the interface to an arbitrary initial disturbance of the form f(x) may be analyzed by Fourier decomposing the perturbation in terms of an appropriate set of orthogonal functions. The solution may then be constructed for each Fourier mode of the disturbance and owing to the linearity and homogeneity of the leading order equations, the total solution is but the sum of the individual solutions. However, one needs to exercise some caution in determining the dominant mode for such a situation. Let us denote the arbitrary Fourier mode of the initial disturbance as $\varepsilon_n f_n(x)$ and the corresponding response as $\alpha(\varepsilon_n)$ exp $(\omega_n(\varepsilon_n)t) f_n(x)$. It is apparent that the growth rate for a finite amplitude perturbation is a function of both the wave number and the amplitude of the disturbance. The dominant growth coefficient, therefore, corresponds to that Fourier mode of the disturbance for which $\omega_n(k,\varepsilon_n)$ is maximum. Further, a Fourier component of the initial disturbance may be regarded as the dominant one only if its response, $\alpha(\varepsilon_n)$ exp $[\omega_n(\varepsilon_n)t]$ exceeds greatly the response of any other Fourier mode. Since the growth rate is a function of both k and ε, it enhances the possibility that many Fourier modes of the initial disturbances have comparable responses. In such an event, the total solution is a sum of all significant individual responses.

Due to simplicity of the proposed formalism, a way is now open to include and study the nonlinear effects of many other physical phenomena that are of interest for different thin film models. The nonlinear dynamics of a thin free film with soluble surfactants was recently investigated by us (Sharma and Ruckenstein (22)). The breakup time of the tear film and its dependence on various parameters of the external eye was also investigated (Sharma and Ruckenstein (16)) based on an earlier proposed model of the tear film breakup (Sharma and Ruckenstein (23)). The formalism is thus well suited for studying the role of thin film instabilities in various biological phenomena as well.

In the present analysis, the film is assumed to be nonthinning. This assumption may be justified only if the radii of film is rather large. This is so because the velocity of film thinning is inversely proportional to the squared of the film radius. The prediction of the lifetimes and critical thickness of small radii foam and emulsion films therefore warrants the inclusion of drainage in the present analysis. The combination of the wave motion (the present analysis) and the drainage flow is possible by using one of the several methods that have been developed to combine these two motions by a quasistatic approach (Vrij and Overbeek (2), Ivanov and Dimitrov (5), Gumerman and Homsy (6), Radoev et al. (20)). The wave motion is, of course, derived from the linear theories in these approaches.

It may be noted that while the case of a tangentially immobile free film has been studied in some detail here, predictions of the critical thicknesses of the foam films should incorporate the more general expression for the growth rate, i.e., Equation (48). This is necessary for predicting the observed dependence of the critical thickness on the surfactant concentration.

Finally, while summarizing the present methodology, it was probably easy to give an impression of its infallibility. This is, however, not the case, because, for the proposed analysis to be successful in its form, it is necessary that the neutrally stable state of the system be nonoscillatory in time. This is so, because the base state considered here is nothing but the neutrally stable wave for the plane-parallel interface and this is considered to be independent of time. This condition is automatically met if the instability is purely of the dissipative type, i.e., the growth coefficient derived from the linear stability is real. It is interesting to note that a spatially nonuniform, stationary solution of the governing equation may always be constructed if the governing equations are invariant under the inversion of the sign of the lateral spatial coordinate, x. A rigorous proof of the existence of such solutions, when the equations satisfy the above condition, has been provided by V. I. Arnold in a paper of Malomed and Tribelski (24). The proposed analysis is, therefore, valid only for a purely dissipative instability of thin film, where the thin film either eventually ruptures or forms a "black" film. The study of the "saturation" of the growing perturbations and the resulting spatio-temporal oscillations (such as those displayed by ripples on a falling film) cannot be studied by the present methodology.

NOTATION

A = the Hamaker constant
a = constant of integration, Equation (14)
C = constant of integration, Equation (9)
c = exponent of film thickness for a general interaction potential, Equation (58)
D_s = surface diffusivity
$F(\varepsilon), F_L, F_n$ = parameters as defined by Equations (61c), (61b) and (61a), respectively
H = h/h_o
H_o, H_1 = zeroth and first order solutions, respectively
h = thickness of the film as a function of x and t
h_o = mean thickness of the film
h_1, h_2, h_3 = first, second and the third order harmonics, respectively, of the linear theory
k = wavenumber
M = $\partial\sigma/\partial\Gamma$ as evaluated at the equilibrium
N = parameter defined by expression (54)
p = hydrodynamic pressure
q = dimensionless wavenumber, Equation (37b)
t = time
u = x-component of the velocity
v = z-component of the velocity
x = spatial coordinate along the film surface
y = k_{cn}^2/k_{cL}^2
z = spatial coordinate normal to the film surface

Greek Letters

Γ = surface excess concentration
ε = nondimensional amplitude of the perturbation at each interface of the film
μ, μ_s = bulk and surface viscosities, respectively
μ^* = $\varepsilon/2$
ξ = rescaled x-coordinate, Equation (29)
ρ = mass density
σ = surface tension
τ = time of rupture; also rescaled time, Equation (29)
τ_N = time of rupture as determined from the present theory
ϕ = London van der Waals interaction potential per unit volume
ω = growth coefficient
ω_{L1}, ω_{L2} = linear theory growth coefficient for the tangentially immobile film and the film devoid of solutes, respectively
Ψ = parameter, Equation (7)

Subscripts

0 = zeroth order or the base case
c = critical or the neutrally stable
L = linear theory
m = dominant value of the variable
n = nonlinear theory

LITERATURE CITED

1. Scheludko, A., Adv. Colloid Interface Sci. 1, 391 (1967).
2. Vrij, A. and Overbeek, J. Th. G., J. Amer. Chem. Soc. 90, 3074 (1968).
3. Felderhof, B.U., J. Chem. Phys. 49, 44 (1968).
4. Ruckenstein, E. and Jain, R.K., Faraday Trans. 70(2), 132 (1974).
5. Ivanov, I.B. and Dimitrov, D.S., Colloid & Polymer Sci. 252, 982 (1974).
6. Gumerman, R. and Homsy, G., Chem. Eng. Commun. 2, 27 (1975).
7. Jain, R.K., Maldarelli, C. and Ruckenstein, E., A.I.Ch.E. Symp. Ser. Biorheol.74, 120 (1978).
8. Maldarelli, C., Jain, R.K., Ivanov, I.B. and Ruckenstein, E., J. Colloid Interface Sci. 78, 118 (1980)
9. Maldarelli, C. and Jain, R.K., J. Colloid Interface Sci. 90, 263 (1982).
10. Dimitrov, D.S. and Zhelev, D.V., J. Colloid Interface Sci. 99, 324 (1984).
11. Sharma, A. and Ruckenstein, E., J. Colloid Interface Sci. 106, 12 (1985).
12. Sharma, A. and Ruckenstein, E., Am. J. Optom. Physiol. Opt. 62, 246 (1985).
13. Dalle Vedove, W. and Sanfeld, A., J. Colloid Interface Sci. 84, 318 (1981).
14. Williams, M.B. and Davis, S.H., J. Colloid Interface Sci. 90, 220 (1982).

15. Desai, D. and Kumar, R., Chem. Eng. Sci. 40, 1305 (1985).

16. Sharma, A. and Ruckenstein, E., J. Colloid Interface Sci. 113, 456 (1986).

17. Benney, D.J., J. Math. Phys. 45, 150 (1966).

18. Atherton, R.W. and Homsy, G.M., Chem. Eng. Commun. 2, 57 (1976).

19. Dimitrov, D.S., Progr. Surface Sci. 14, 295 (1983).

20. Radoev, B.P., Scheludko, A.D. and Manev, E.D., J. Colloid Interface Sci. 95, 254 (1983).

21. Rao, A.A., Wasan, D.T. and Manev, E.D., Chem. Eng. Commun. 15, 63 (1982).

22. Sharma, A. and Ruckenstein, E. Langmuir, 2,480 (1986).

23. Sharma, A. and Ruckenstein, E., J. Colloid Interface Sci. 111, 8 (1986).

24. Malomed, B.A. and Tribelsky, M.I., Physica 4D, 67 (1984).

THE ROLE OF THIN LIQUID FILMS IN WETTING

P. Neogi ■ Department of Chemical Engineering, University of Missouri-Rolla, Rolla, MO 65401
C. A. Miller ■ Department of Chemical Engineering, Rice University, P.O. Box 1892, Houston, TX 77251

The kinetics of wetting a solid substrate by a thin liquid film has been reviewed here. Typically when a liquid drop spreads or retracts along a solid surface, a thin film region develops near the drop periphery. In the first part is discussed how the conventional forces in thin films are adapted to the case where the films are wedge-shaped, that is, contain a contact line—line common to the solid, the liquid, and the ambient fluid (air). In the second part the solutions to some hydrodynamic problems in wetting, especially the use of the lubrication theory approximation and the method of matched asymptotic expansions, are discussed. Particularly important features in the solution are the unbounded behavior of the stresses at the contact line and its remedy in the form of a slip boundary condition.

Teletzke, et al. (1) have defined a variety of thin films on a solid surface that one may have. In general these are sufficiently thin that their properties are affected by Hamaker forces, that is, the 'disjoining pressure' (Derjaguin (2), Blake (3)), electrical double layer forces (Sheludko (4), Felderhof (5)) and when they are curved, by the Laplace pressure difference. In the study of wetting kinetics of thin films, we also include thicker liquid layers since their movements affect the thin film regions as well. These under nonequilibrium conditions give rise to spontaneous spreading. Thin films are not known to be a characteristic feature in forced spreading. The knowledge of the wetting kinetics of thin films finds application in lubrication (Bascom, Cottington and Singleterry (6), Derjaguin and Karassev (7), Derjaguin, Karasev, Zakhavaeva and Lazarev (8)), in the study of spreading of oil slicks (Hoult (9)), in imbibition into porous media (Wooding and Morel-Seytoux (10)), in boiling heat transfer (Renk and Wayner (11), Potash and Wayner (12), Wayner, Kao and LaCroix (13), Wayner (14,15), Moosman and Homsy (16), Neogi (17), Derjaguin, Nerpin and Churaw (18)), in adhesion (Cherry and Muddarris (19)), as well as in redispersion (Ruckenstein and Chen (20,21), Ruckenstein and Chu (22), Neogi (23)) and detergency (Schwartz (24), Neogi (23)) and in biological systems (Greenspan (25)). We will confine ourselves to the spreading of a liquid on a solid surface here.

The thermodynamics of wetting is well known (Miller and Neogi (26)). A liquid drop on a solid surface is characterized by its contact angle. When the liquid is nonwetting or partially wets the solid surface the equilibrium contact angle λ is given by Young-Dupre equation

$$\cos \lambda = (\gamma_{SV} - \gamma_{SL})/\gamma_{LV} \qquad (1)$$

where γ_{SV}, γ_{SL} and γ_{LV} are the interfacial tensions at the solid (S) - vapor (V), the solid-liquid (L) and the liquid-vapor interfaces. Since according to Eq. (1), λ depends on the equilibrium properties, it is referred to as the equilibrium contact angle. A spreading coefficient

$$S = \gamma_{SV} - \gamma_{SL} - \gamma_{LV} \qquad (2)$$

is defined, which provides the difference in the surface energies before and after the wetting. Obviously if S is negative only partial wetting will take place. For a liquid on a solid surface described by Eq. (1), S can be seen to be negative. Such liquids reach equilibrium configurations where the angle subtended by the liquid-vapor interface is given by Eq. (1) and the line common to the three phases is called the contact line. In contrast, when S is positive the liquid spreads spontaneously on the solid surface and no equilibrium configuration is obtained-at least under a simple theory. During the spreading process the contact angles of these wetting liquids are seen to be zero (Bascom, Cottington and Singleterry (6), Fox and Zisman (27), Johnson and Dettre (28)). There are some experimental difficulties in obtaining reproducible values of the equilibrium contact angles. These reasons are now well-known and have been reviewed by Zisman (29), Johnson and Dettre (30) and Miller and Neogi (26)). Although many effects need to be incorporated to compare theory with the experiments, the basic features of the theory remain unaffected.

Given below are some of the experimental observations made on wetting kinetics in general and, in particular, in situations where thin film phenomena appear to be important.

Experimental

In a series of carefully conducted experiments on the spreading kinetics of wetting liquids on steel, Bascom, Cottington and Singleterry (6) gave the first description of some of the mechanisms, and distinguishing features of the wetting process by a thin film and provided quantitative results on the rates. Firstly, they found that evaporation of small amounts of volatile impurities in the liquid could give rise to small gradients of surface tension which significantly altered the nature of the profiles and the spreading rates. When the volatiles had lower surface tensions, spreading took place, but when they had surface tensions higher than the bulk liquid, the films receded. This is in keeping with the direction of the gradient of surface tension as discussed there and by Neogi (23). When the volatiles content was reasonably high, the spreading rates were found to be constant independent of time. When the liquid was stripped carefully of all volatiles the nature of the profiles changed substantially, and the spreading (and only spreading took place) rates dropped dramatically.

Secondly, it was found that even though the liquids were wetting, large contact angles were observed instead of the expected value of zero. Using ellipsometry, they found however that when film thickness dropped to about 5nm, its profile changed sharply and levelled out. Thus, a zero contact angle was subtended at the contact line in accordance with the fact that the liquids were wetting. This behavior is shown schematically in Fig. 1.

Thirdly, it was found that surface roughness played an important role and on grooved surfaces the liquid spread faster when the grooves were parallel to the direction of spreading than when they were perpendicular. Finally, they detected, by breathing against the solid surface, a thin film, probably an absorbed layer, far ahead of what could be discerned as the liquid film.

Closely related experiments were performed by Derjaguin and coworkers (Derjaguin and Karassev (7), Derjaguin, Karassev, Lavygin, Skorokhodov and Khromova (31)). On the floor of a narrow horizontal slit a thin liquid film was depositied and air was blown steadily over it. According to the basic fluid mechanics and the assumption that the contact line was pinned in a fixed position, the profiles should appear as straight lines. While this was seen to be true in general, it was found that near the contact line the profiles showed small curvatures. Later, instead of using ordinary diffraction Bascom and Singleterry (32) used ellipsometry which verified the previous findings but showed that the contact lines moved forward and against the direction of the air current. These are the film 'blow-off' experiments.

Another type of experiment has been conducted by a number of investigators. Schonhorn, Frisch and Kwei (33), Kwei, Schonhorn and Frisch (34), van Oene, Chang and Newman (35), Ogarev, Timonina, Arslanov and Trapeznikov (36) and Welygan and Burns (37) have investigated the spreading kinetics of a drop on a solid surface. Photomicrography, with $\pm$ 1μm error in the linear measurements was used. The drops were sufficiently small (<1$\mu\ell$) such that the effects of gravity could be ignored. The spreading drops of the wetting liquids had profiles of spherical caps and the dynamic contact angles were much larger than zero, the equilibrium value. In fact, from the measured values of the curvatures at the top of the drops, the drop volume V and the assumption that the profiles were spherical caps, the dynamic contact angles could be predicted with good accuracy. The spreading rates were found to be $\propto t^{-0.83}$. Later, Radigan, Ghiradella, Frisch, Schonhorn and Kwei (38), studied the spreading kinetics of a molten glass particle on a solid surface. At some stage they froze the sample suddenly and studied it using scanning electron microscopy (SEM). They found, as Bascom,

Cottington and Singleterry (1964) had before them, that the drop profiles turned sharply very close to the surface and levelled off to form films which subtended zero contact angle on the solid surface.

In summary we note that the profiles of thin spreading films of wetting liquids show sudden changes near the contact line where they turn to retain zero contact angles. This behavior appears to be quite general and relatively independent of the flow behavior of the basic system. Although this transition usually occurs at thicknesses too small to be observed by the usual optical techniques, it can occasionally be seen (Dettre and Johnson (39)). That thin films of nonwetting liquids have not been studied thus is due to the fact that such films when formed are unstable (Ruckenstein and Jain (40)) and indeed even the adsorbed amounts of such liquids are small (Johnson and Dettre (28)). However, it appears that the microscopic behavior of the contact line region of a nonwetting or partially wetting liquid has not yet been systematically investigated. One very intriguing feature of spreading of nonwetting liquids is that under the action of a gradient in surface tension. Cottington, Murphy and Singleterry (41)) showed that if the gradient of surface tension is large, even a non-wetting liquid will ''spread'' by bursting the region near the contact line. The liquid moves ahead in the form of meandering rivulets. In the remainder of this paper, such unusual forms of spreading will not be discussed. Only smooth and even spreading -- best described as wetting -- is the subject of the present discussion.

The role of thin liquid films/adsorbed layers which move ahead of the bulk liquid also appears to be interesting. Williams (42) observed instability at the junction area in the form of a scalloped contact line, Marmur and Lelah (43) observed that this film could interact with the edges of the solid surface on which the spreading was occuring with a resulting effect on the spreading rates. Thus, the rates were shown to differ somewhat when the extent and the shapes of the horizontal solid substrates were changed.

Other transport processes in such films have also been investigated. Ghiradella, Radigan and Frisch (44) found that in a spreading thin film containing an electrolyte, the film was depleted of the electrolyte. In recent experiments Damania and Bose (45) have obtained results on both equilibrium and dynamic contact angles when electrolytes and surfactants are introduced. Their effects on the kinetics of wetting are small, according to Lelah and Marmur (46). Renk and Wayner (11), in investigating the effects of vaporization in thin films found that the vaporization rates near the contact lines were very high.

Forces Near the Contact Line:

As remarked previously Hamaker forces are important near a contact line region since the thicknesses there are very small, and in general they govern the behavior of thin films. Miller and Ruckenstein (47) modelled the contact line region in the form of a wedge and obtained the excess potential at the vapor-liquid interface relative to a suitable datum potential in the form of

$$\Delta\phi = \frac{\pi}{12x^3}\left(n_L^2\beta_{LL} - n_L n_S \beta_{SL}\right)G(\alpha) - \frac{\pi}{12x^3} n_L n_S \beta_{SL}\, G(\pi-\alpha) \qquad (3)$$

where x is the distance along the vapor-liquid interface measured from the contact line, n_L and n_S are the molecular number densities of the liquid and solid, β_{SL} and β_{LL} are the solid-liquid and liquid-liquid interaction energies. Further

$$G(\alpha) = \operatorname{cosec}^3\alpha + \cot^3\alpha + \frac{3}{2}\cot\alpha \qquad (4)$$

where α is in general the non-equilibrium contact angle. From Eqs. (3) and (4) and from the requirements that the tangential force $d\Delta\phi/dx$ vanish at equilibrium and that the ratio $n_L^2\beta_{LL}/n_L n_S \beta_{SL}$ is equal to the ratio of the work of cohesion and the work of adhesion as given by Fowkes (48), they were able to show that the results were consistent with the experimental measurements on contact angles by Fox and Zisman (27). The results also agree well with Young-Dupre equation (1), (Miller and Neogi, (26)). The importance of Eq. (3) is that it encompasses the nonequilibrium situations as well. For small values of α for a wetting liquid, they obtained

$$\Delta\phi \sim -\frac{\pi\left(n_L n_S \beta_{SL} - n_L^2\beta_{LL}\right)}{16h^3} \qquad (5)$$

and for small differences in the value $(\alpha-\lambda)$ they obtained

$$\Delta\phi \sim -\frac{\pi n_L^2 \beta_{LL}}{16x^3}(\alpha-\lambda) \qquad (6)$$

Eq. (5) is the well-known 'disjoining' pressure (Deryaguin (2)), the excess potential at the vapor-liquid interface of a thin liquid film of thickness h lying on a solid surface. Ruckenstein and Lee (49) later improved the analysis by including the repulsive forces as well as including finite systems. Some indications of the effect of line tension could be obtained. It becomes obvious that for a reasonable description of the contact line region the coupling between the shape of the film in that region and the local intermolecular forces has to be maintained in some form.

Using the same geometry Neogi (50) examined the electrical double layer effects near a contact line under Debye-Huckel linear theory and found that for thicknesses much less than the Debye length κ^{-1} the electrostatic poential was given by

$$\psi \sim \frac{z}{h}(\psi_s - \psi_w) + \psi_w \qquad (7)$$

and

$$\psi \sim \psi_w - \frac{4}{\varepsilon}\sigma z \qquad (8)$$

in thin films, where constant potential ψ_w and ψ_s have been assumed at the solid - liquid and liquid - vapor interfaces in Eq. (7). The latter was replaced by a constant surface charge density σ at the liquid - vapor interface in Eq. (8). Further, h is the local film thickness, z is the coordinate in the vertical direction, ε is the dielectric constant and $\kappa^2 = 8\pi e^2 c_\infty/\varepsilon kT$, so that κ^{-1} is the usual Debye length. Here e is the charge on an electron, c_∞ is the bulk concentration of the 1:1 electrolyte, k is the Boltzmann constant and T the absolute temperature. It can be seen that both Eqs. (7) and (8) show that the charge density in the thin film is zero. This has been seen in the experiments by Ghiradella, Radigan and Frisch (44) mentioned previously. Further, it was shown by Neogi (50) that at equilibrium the film profiles were affected by the electrical forces but the contact angle λ was unchanged.

With these forces it is now possible to formulate the problem of a moving thin film. Since the film is nearly flat and thin, the lubrication theory approximation (Batchelor (51)) can be used. In this case it is assumed that the flow is sufficiently slow (in spontaneous spreading, the velocities are no more than 1 mm/min) such that quasistatic assumption can be made, and the velocity in the tangential direction is almost fully developed, that is, v_x is a function of z alone, and the normal velocity v_z can be neglected. The coordinates for the wedge-like film are shown in Fig. 2. Under these conditions the equations of motion become

$$0 = -\frac{\partial}{\partial x}(p + \phi + \phi_{el}) + \mu\frac{\partial^2 v_x}{\partial z^2} \qquad (9)$$

$$0 = -\frac{\partial}{\partial z}(p + \phi + \phi_{el}) \qquad (10)$$

Here, p is the pressure ϕ is a Hamaker type intermolecular potential and ϕ_{el} is the electrical contribution to the disjoining pressure. Also μ is the viscosity and v_x is the tangential velocity. From Eq. (10), one finds that the sum $p + \phi + \phi_{el}$ is independent of z and thus can be evaluated at the vapor-liquid interface z = h, where the normal stress balance leads to

$$p + \text{electrical stresses} = -\lambda\frac{\partial^2 h}{\partial x^2} \qquad (11)$$

where γ is the surface tension and an approximate form for the curvature has been used as appropriate under lubrication theory. Eq. (11) is the Laplace pressure condition. The electrical stresses in (11) and the electrical contribution to the disjoining pressure are yet to be calculated for the more interesting case where the film thicknesses are at least comparable to the Debye length. Finally, ϕ at z = h is given by Eq. (5) for wetting liquids and (6) for non-wetting or partially wetting liquids.

The other conditions are

$$v_x = 0 \text{ on } z = o \text{ (no slip)} \qquad (12)$$

$$-\mu\frac{\partial v_x}{\partial z} = -\frac{\partial\gamma}{\partial x} \text{ on } x = h \text{ (tangential stress balance if a surface tension gradient exists)} \qquad (13)$$

The conditions on h vary from case to case, with exception of the definition of the contact line, which is usually taken to be

$$h|_{x=x_o} = 0 \quad (14)$$

where $x_o(t)$ is the position of the contact line and dx_o/dt is the spreading rate.

These equations can be derived for axisymmetric systems under the same assumptions.

Fluid Flow Near the Contact Line:

In solving the simple viscocapillary problem Friz (52) and Ludviksson and Lightfoot (53) found that the contact line of a moving film could only be located at infinity and the shear stress and the pressure gradient there were infinite. Huh and Scriven (54) analyzed the contact line itself and found the pressure, stresses and viscous dissipation to be infinite there. Lopez, Miller and Ruckenstein (55) analyzed the case of a spreading drop under gravity. They found that although the predicted spreading rates agreed well with the experimental values, the shear stress at the contact line was infinite. They also analyzed the case of a wetting liquid moving under Hamaker forces, that is, the solution to Eqs. (5), (9)-(14) without any electrical or surface tension effects. A solution to the spreading kinetics of an axisymmetric drop could not be found. For cylindrical drops, they obtained the spreading rates as $\propto t^{-1/2}$ for the region where the film thickness was very small in good agreement with available data. A very thin and long film in the contact line region is obtained under intermolecular forces, a fact later stressed by de Gennes (56). The actual contact line, h=0, in this case is located at infinity.

Dussan and Davis (57) showed with simple experiments that the velocities at the contact line will have two values when approached along two interfaces under the most general conditions, giving rise to an infinite velocity gradient and shear stress there. The behavior of material points that reach or leave the contact line was consistent with the conventional no-slip boundary condition. Thus for a simple flow on a solid surface as shown in Fig. 3, to retain both a non-zero average velocity and the no-slip boundary condition, the velocity gradient must become unbounded at the contact line. Both Huh and Scriven (54) and Dussan and Davis (55) suggested as a remedy a slip boundary condition near the contact line.

Two kinds of slip boundary conditions have been proposed. The first of these deals with "rough" surfaces where it is seen that the scale of the surface irregularities is of the order of ~ 1μm, much greater than the 100nm distance through which the Hamaker forces are effective. Here, only the viscocapillary problem is analyzed. The slip conditions were derived by considering a flow through or over the rough surface. These slip conditions of Neogi and Miller (58) and Hocking (59) are able to eliminate the problem of infinite shear stress at the contact line. A solution to the spreading kinetics of an axisymmetric drop was found. Of further interest is the fact that machine-polished metal surfaces have roughnesses ~ 0.1 to 10 μm (Mussel and Glang (60) and thus this form of slip addresses real surfaces. The solution of Neogi and Miller (58) shows that a small droplet spreads at a rate $\propto t^{-0.9}$ which compares well with the data of van Oene, Chang and Newman (35) and with one set (where the relevant parameters have been reported) of those by Schonhorn, Frisch and Kwei (33) and Kwei, Schonhorn and Frisch (34) in the form of spreading rate $\propto t^{-0.83}$. With those data the value of the surface roughness was also found using the theory of Neogi and Miller (58) to be ~ 3μ m, which agrees well with the 1-5μm reported in the original papers. Lastly, they were also able to show that for wetting liquids in the region very close to the solid surface near the contact line the profile changed sharply to level off to maintain a zero contact angle at the contact line in accordance with the experimental observations discussed above. One important finding was that the bulk of the drop did not have a significant pressure gradient and hence had a profile of a spherical cap. The bulk of the driving force which was due to the Laplace pressure was located near the contact line in a region where liquid thickness was ~ 1μ m. This range is larger than the range of Hamaker forces, a point to which we will return later.

Whereas the above analysis proves to be both convenient and fruitful, the actual contact line is ''lost'' inside the rough surface and the problem of what happens there is not analyzed. Blake and Haynes (61) proposed a jump mechanism for the liquid molecules to move the solid surface. Ruckenstein and Dunn (62) correctly identified the forces and showed that these jumps give rise to a surface diffusion (and slip velocity) in a preferred direction. Neogi and Miller (63) obtained identical results by using a continuum formalism. Neogi and Miller (63) also solved the fluid mechanics of a drop spreading on a smooth solid surface with this slip condition. Hamaker forces were included in the analysis for a partially wetting liquid with a small but non-zero contact angle. As before, it was found that the pressure gradient for the flow was virtually absent in the bulk of the drop and was concentrated near the contact line region where the thickness was of the order of 1 μm. Consequently, the spreading kinetics were found to be independent of the Hamaker forces to the first approximation. Hence, these forces were not considered in the analysis of wetting liquids spreading on smooth surfaces. As a result, their solution do not reflect the special effects of the very thin film/adsorbed layer ahead of the drop as observed by Marmur and Lelah (43) and by Williams (42). They provide the results for the bulk of the drop alone. It is noteworthy that the wetting kinetics of the thin film alone have been analyzed by Lopez, Miller and Ruckenstein (55). The two together were not analyzed. Wayner (64) has considered the transition from a thick film to a thin film under static conditions. In addition the Hamaker interaction forces have been taken with sign opposite of what is conventionally chosen to represent wetting conditions, which is that $n_S n_L \beta_{SL} - n_L^2 \beta_{LL} > 0$. De Gennes and coworkers (56) consider a particular case where the solid surface outside the wetted region is dry and obtain a result that even wetting liquids can retain equilibrium shapes. In view of surface diffusion it is unclear as to why the dry surface would remain so at equilibrium and in view of the fact that an infinite gradient of curvature is required to balance an infinite Hamaker potential at the contact line it appears that their formulation is incomplete. Notably the Hamaker interaction potential which becomes arbitrarily large as the thickness of the drop decreases whould instead reach a finite limit (adsorption), in which case their mathemaics show that the predicted profiles may not exist in some cases.

It is worth discussing the mathematical (and physical) conditions and implications of the slip velocity. The slip velocity yields a representative length scale ℓ. This is a small or microscopic length scale, where as $\ell \to 0$, the no-slip boundary condition is recovered. The length ℓ also represents the sphere of influence of the slip. The spreading problem in general carries a macroscopic length scale L, e.g., $V^{1/3}$ for a drop, gap width in a slit, radius in a tube, etc., where $L >> \ell$. The various physical phenomena can be ranked in their importance by their length scales, and consequently in the macroscopic region conventional hydrodynamics dominates and slip does not play a role. However, near the contact line the only ''macroscopic'' dimension is the thickness $h \sim \ell$ and consequently slip becomes very important. It is also noteworthy that even though = /L is a small quantity, its effect on various hydrodynamic quantities was found by the matched asymptotic expansion method discussed below to be $\sim [\ell n|\frac{1}{\varepsilon}|]^{-1}$ (Neogi and Miller, (63)). This expression $\to 0$ as $\varepsilon \to 0$. But for = 10^{-4}, we calculate the logarithmic term as ~ 0.1, that is, the effect of a small slip in the contact line region is quite significant. In the problems studied the use of these slip conditions turned out to be adequate in eliminating undersirable features. Thus, Neogi (23) has analyzed the problem of wetting and dewetting under surface tension gradients. The problem of film 'blow-off' has also been analyzed successfully (Neogi (65)), where only the viscocapillary effects were considered and only cases where the contact angles were small but nonzero.

Because of the existence of the small quantity ε, the method of solution is based on matched asymptotic analysis. In the macroscopic region called the outer region the dimensions are of the order of 1 and an asymptotic solution is constructed on that basis. Since the outer solution is not valid near the contact line, it has as many unknowns as the number of boundary conditions in the microscopic region, which is the region in the vicinity of the contact line called the inner region. Similarly an inner solution may be constructed but now under the premise that the length scale there is comparable to ε. The inner solution also contains unknowns. By a matching process, the two solutions are made to agree with one another in the intermediate region, whence the unknowns can be obtained. The composite solution is the sum of the inner and outer solutions less the terms common to both. Often, as in the case of wetting liquids, an extra asymptotic solution is required in the intermediate region. The excellent texts by van Dyke (66) and Cole (67) may be consulted for further details on the mathematical techniques.

One practical difficulty in analyzing the spreading kinetics of wetting liquids given by the potential in Eq. (5) (but not Eq. 6) is that it predicts infinite potential at the contact line h = 0. Although h = 0 is a reasonable definition of a contact line from the continuum point of view, an adsorbed layer lies in front of it. It we consider its flux in terms of a surface diffusion, we have

$$j \text{ (moles/length/time)} = -D_s \frac{\partial \Gamma}{\partial x} \qquad (15)$$

where D_s is the surface diffusivity and is the adsorbed amount. If we define a thickness of the adsorbed layer h as $\Gamma = h\rho_L$, Eq. (15) becomes

$$j = - D_s \rho_L \frac{\partial h}{\partial x} \qquad (16)$$

The mass flow rate for a wedge under disjoining pressure has been calculated by Lopez, Miller and Ruckenstein (55) under the no-slip condition as

$$j' = - \frac{K\rho_L}{\mu} \frac{1}{h} \frac{\partial h}{\partial x} \qquad (17)$$

where $K = \frac{\pi}{6}(n_L n_S \beta_{SL} - n_L^2 \beta_{LL})$. Since j and j' in Eqs. (16) and (17) represent similar quantities a generalization can be effected by using a molecular potential for a driving force $\propto h^{-\nu}$ where ν varies with h. Such a potential may be extracted from the detailed calculations of Teletzke, et al (1) who have considered the transition from a thick film to an adsorbed layer. It is also apparent that the mobilities vary as the thickness decreases, a fact first pointed out in a different context by Derjaguin, Strakhovosky and Malysheva (68). How this may be modelled remains to be determined. The important feature to be noted is that the generalized potential may not give the trouble near the contact line that the Hamaker forces in the form of disjoining pressure presently do.

Neogi (59) in considering the effects of the electrical double layer forces, noted that it had been found previously that most of the driving force in a spreading drop was in a region of thickness ~ 1 μm. Electrical double layer forces on the other hand have a range given by the Debye length κ^{-1}, which for an aqueous solution of 1:1 electrolyte is $0.308/c_\infty^{1/2}$ nm, where c_∞ is in moles/ℓ. If c_∞ = 0.001 moles/ℓ, $\kappa^{-1} \sim 10^{-2}$ μm and the effect of electrical forces on a drop spreading would be small as seen for instance in the experiments of Lelah and Marmur (46). This argument does not hold for films which are thin throughout.

Other Processes and Unbounded Behavior at the Contact Line:

In examining Fig. 2 it is seen that in fluid flow the vapor-liquid interface is subject to the boundary condition of zero shear. On the other hand the boundary condition at the solid-liquid interface is given by no-slip, that is, zero velocity. Consequently both need to be satisfied at the contact line and it is this overspecification that contributes to unbounded behavior in various fluid mechanical quantities.

In identifying this principle it is also possible to uncover other cases where such problems might exist. In obtaining the electrostatic potentials, if the vapor-liquid interface is specified as having a potential ψ_s and the solid-liquid interface ψ_w, it becomes apparent from Fig. 2 that the potential at the contact line is again overspecified. From Eq. (7) it is seen that the electric field becomes infinite at the contact line. This problem does not arise in Eq. (8) where the vapor-liquid interface is specified by a constant charge density.

Similarly, in boiling heat transfer the vapor is a saturated vapor at the saturation temperature T_s. Then, the heat flux can be obtained in Fig. 2 as

$$q = k\,(T_w - T_s)/h \qquad (34)$$

where k is the thermal conductivity of the liquid and T_w is the constant wall temperature. Once again the temperature is overspecified at the contact line where q becomes infinite. Renk and Wayner (11) have shown that the heat fluxes there are indeed large.

In analyzing the situations in electrostatics and heat transfer it can be seen that constant ψ_w and constant T_w are idealizations for the cases where the substrate is an ideal conductor. In reality, their conductivities are finite and boundary conditions at the solid surface need to be obtained after solving the appropriate equations in the solid phase. This extension of the analyses has not been attempted, presumably because it complicates the mathematics enormously and because some reasonable results may be obtained without this sophistication. Wayner, Kao and La Croix (13) and Moosman and Homsy (16) analyzed this problem by replacing the saturation temperature at the interface T_s with its boiling point temperature T_b. The two differ because the pressures in the liquid and vapor differ by the Laplace term and the Hamaker potential. They ignored the first and showed that the flux was finite at the contact line in an evaporating thin film. Neogi (17) considered only the Laplace pressure and showed that the curvature at the contact line was equal to the curvature of the critical nucleus in boiling heat transfer. Both works assume that the ratio of the thermal conductivity of the liquid to that of the solid is very small and represents a negligible effect.

In fluid flow the use of a small but non-zero viscosity of air does not eliminate the unbounded behavior at the contact line. The use of a dynamic elastic deformation or a plastic flow in the solid phase has not been made. In any case the ratio of the viscosity of the liquid to a similar property in the solid would be very small compared to the ratio of the slip lengths to the macroscopic dimensions, and would hence constitute an effect which can probably be neglected (for an interesting account of spreading of one liquid on another as a function of viscosity ratio see Di Pietro and Cox (69)).

Conclusions

The basic transport phenomena governing wetting kinetics appear to be reasonably well established. One peculiarity of these problems is the necessity and the importance of using a slip boundary condition at the contact line. Several models have been proposed with which it has been possible to explain the key features of the wetting kinetics. In the efforts to include thin films effects and the forces peculiar to the thin film region, Hamaker forces and electrical double layer forces have also been quantified and have been used to explain some of the features seen in the experiments. A current problem with the Hamaker forces is that although the details of such forces are known where the film thins from a disjoining pressure region to an adsorbed film region, the changing mobilities with decreasing thicknesses are not known. Consequently, some of the experimental observations on the kinetics of wetting by thin films are yet to be explained theoretically.

Another interesting problem which remains is that most of the spreading kinetics experiments have utilized polymeric liquids. The special features of polymeric systems appear to have their own contributions (de Gennes, 1985). Moreover, no great attention has been paid to partially wetting liquids experimentally, where thin films are sometimes possible and their interaction with the adsorbed liquids is of great importance, for instance, in the case of autophobic liquids.

Lastly, we point out that from the extensive literature on wetting that only the material relevant to thin films has been reviewed here. Other aspects of wetting have been reviewed by Dussan (1979), Marmur (1983) and de Gennes (1985).

LITERATURE CITED

1. Teletzke, G.F., L.E. Scriven and H. Ted Davis, "Gradient Theory of Wetting Transitions", *J. Colloid Interface Sci.* 87, 550 (1982).

2. Derjaguin, B.V., ''A Theory of Capillary condensation in the Pores of Sorbents, and of other Capillary Phenomena, Taking into Account the Disjoining Action of Multimolecular Liquid Films'', Acta Physicochim. URSS 12, 181 (1940).

3. Blake, T.D., ''Investigation of Equilibrium Wetting Films of n-Alkanes on -Alumina''. JCS Faraday I 71, 192 (1975).

4. Sheludko, A., ''Thin Liquid Films'', Adv. Coll. Int. Sci. 1, 391 (1969).

5. Felderhof, B.U., ''Shape Fluctuations of Free Liquid Films with Diffuse Electric Double Layers'', J. Chem. Phys. 48, 1178 (1968).

6. Bascom, W.D., R.L. Cottington and C.R. Singleterry, ''Dynamic Surface Phenomena in the Spontaneous Spreading of Oils on Solids'', Contact Angles, Wettability and Adhesion, R.E. Gould, ed., Adv. in Chem. Series v. 43, Am. Chem. Soc., Washington, D.C., 355 (1964).

7. Derjaguin, B.V. and V.V. Karassev, ''Viscosity Studies of Liquid Boundary Layers by the Blow-Off Method'', 2nd Int. Congress of Surface Activity v.3, Butterworths, London, 531, (1957).

8. Derjaguin, B.V., V.V. Karassev, N.N. Zakhavaeva and V.P. Lazarev, ''The Mechanism of Boundary Lubrication and the Properties of the Lubricating Film'', Wear 1, 277 (1958).

9. Hoult, D.P., ''Oil Spreading on the Sea'', Ann. Rev. Fluid Mech. 4, 341 (1972).

10. Wooding, R.A. and H.J. Morel-Seytoux, ''Multiphase Flow through Porous Media'', Ann. Rev. Fluid Mech. 8, 233 (1976).

11. Renk, F.J. and P.C. Wayner, Jr. ''An Evaporating Ethanol Meniscus'', J. Heat Transfer 101, 55, 59 (1979).

12. Potash, Jr., M. and P.C. Wayner, Jr. ''Evaporation from a Two-Dimensional Meniscus'', Int. J. Heat Mass Transfer 15, 1851 (1972).

13. Wayner, Jr., P.C., Y.K. Kao and L.V. LaCroix, ''The Interline Heat Transfer Coefficient of an Evaporating Wetting Film'', Int. J. Heat Mass Transfer 19, 487 (1976).

14. Wayner, Jr., P.C., ''A Constant Heat Flux Model of the Evaporating Interline Region'', Int. J. Heat Mass Transfer 21, 362 (1978).

15. Wayner, Jr., P.C., ''Effect of Interfacial Phenomena in the Interline Region on the Rewetting of a Hot Spot'', Int. J. Heat Mass Transfer. 22, 1033 (1979).

16. Moosman, S. and G. Homsy, ''Evaporating Menisci of Wetting Fluids'', J. Colloid Interface Sci. 73, 212 (1980).

17. Neogi, P. ''Evaporation in a Microlayer'', paper presented at AIChE Nat. Mtg., Houston, 1983, paper no. 62a.

18. Derjaguin, B.V., S.V. Nerpin and N.V. Churaw, ''Effect of Film Transfer Upon Evaporation of Liquids from Capillaries'', Bull. RILEM. 29, 93 (1965).

19. Cherry, B.W. and S. Muddarris, ''Wetting Kinetics and the Strength of Adhesive Joints'', J. Adhesion 2, 42 (1970).

20. Ruckenstein, E. and J.J. Chen, ''Spreading and Surface Tension Gradient Driven Phenomena Druing Heating of Alumina-Supported Palladium Crystallites in Oxygen'', J. Catal. 70, 233 (1981).

21. Ruckenstein, E. and J.J. Chen, ''Wetting Phenomena During Alternating Heating in O_2 and H_2 of Supported Metal Crystallites'', J. Colloid Interface Sci. 86, 1 (1982).

22. Ruckenstein, E. and Y.F. Chu, ''Redispersion of Platinum Crystallities Supported on Alumninum-Role of Wetting'', J. Catal. 59, 109 (1979).

23. Neogi, P., ''Tears-of-Wine and related Phenomena'', J. Colloid Interface Sci. 105, 94 (1985).

24. Schwartz, A.M., in Colloid and Surface Science v.11, R.J. Good and R.R. Stromberg, eds., Plenum Press, N.Y., p. 305 (1979).

25. Greenspan, H.P. "On the Motion of a Small Viscous Droplet that Wets a Surface", J. Fluid Mech. 84, 125 (1978).

26. Miller, C.A. and P. Neogi, Interfacial Phenomena: Equilibrium and Dynamic Effects, Surface Science Series v. 17, M.J. Schick, ed., Marcel Dekker, N.Y. (1985).

27. Fox, H.W. and W.Z. Zisman, "The Spreading of Liquids on Low Energy Surfaces. I. Polytetrafluorethylene". J. Colloid Sci. 5, 514 (1950).

28. Johnson, Jr., R.E. and R.H. Dettre "The Wettability of Low Energy Liquid Surfaces", J. Colloid Interface Sci. 21, 610 (1966).

29. Zisman, W.A., "Relation of Equilibrium Contact Angle to Liquid and Solid Constitution", Contact Angles, Wettability and Adhesion, R.E. Gould, ed., Adv. in Chem. Series v.43, Am. Chem. Soc., Washington, D.C., 1 (1964).

30. Johnson, Jr., R.E. and R.H. Dettre, in Surface and Colloid Science v.2, E. Matijevic, ed., Wiley-Interscience, N.Y., 85 (1969).

31. Derjaguin, B.V., V.V. Karassev, I.A. Lavygin, I.I. Skorokhodov and E.N. Khromova, "Boundary Viscosity of Polydimethylsiloxane Liquids and their Binary Mixtures", Thin Liquid Films and Boundary Layers, Spec. Disc. Faraday Soc. No.1, Academic Press, London, 98 (1970).

32. Bascom, W.D. and C.R. Singleterry, "Studies with the Blowing Method of Measuring Liquid Viscosity Near a Solid Surface", J. Colloid Interface Sci. 66, 559 (1978).

33. Schonhorn, H., H.L. Frisch and T.K. Kwei, "Kinetics of Wetting of Surfaces by Polymer Melts", J. Appl. Phys. 37, 4967 (1966).

34. Kwei, T.K., H. Schonhorn and H.L. Frisch, "Kinetics of Wetting of Surfaces by Polymer Melts", J. Colloid Interface Sci. 28, 543 (1968).

35. Van Oene, H., Y.F. Chang and S. Newman, "The Rheology of Wetting by Polymer Melts", J. Adhesion 1, 54 (1969).

36. Ogarev, V.V., T.N. Timonina, V.V. Arslanov and A.A. Trapeznikov, "Spreading of Polydimethylsiloxane Drops on Solid Horizontal Surfaces", J. Adhesion 6, 337 (1974).

37. Welygan, D.G. and C.M. Burns, "Spreading of Viscous, Well-Wetting Liquids on Plane Surfaces", J. Adhesion 10, 123 (1979).

38. Radigan, W., H. Ghiradella, H.L. Frisch, H. Schonhorn and T.K. Kwei, "Kinetics of Spreading of Glass on Fernico Metal", J. Colloid Interface Sci. 49, 241 (1974).

39. Dettre, R.H. and R.E. Johnson, Jr., "The Spreading of Molten Polymers", J. Adhesion 2, 61 (1970).

40. Ruckenstein, E. and R.K. Jain, "Spontaneous Rupture of Thin Liquid Films", JCS Faraday II 70, 132 (1974).

41. Cottington, R.L., C.M. Murphy and C.R. Singleterry, "Effect of Polar-Nonpolar Additives on Spreading of Oils, with Application to Non-spreading Oils", Contact Angles, Wettability and Adhesion, R.E. Gould, ed., Adv. in Chem. Series v 43, Am. Chem. Soc., Washington, D.C., 341 (1964).

42. Williams, R. "The Advancing Front of a Spreading Liquid", Nature (London) 266, 153 (1977).

43. Marmur, A. and M.D. Lelah, "The Dependence of Drop Spreading on the Size of the Solid Surface", J. Colloid Interface Sci. 78, 262 (1980).

44. Ghiradella, H., W. Radigan and H.L. Frisch, "Electrical Resistivity Changes in Spreading Thin Films", J. Colloid Interface Sci. 51, 522 (1975).

45. Damania, B. and A. Bose, ''Dynamic Surfactant Effects on the Spreading of Liquids on Solid Surfaces'', paper presented at AIChE Nat. Mtg., Houston, 1985, paper no. 32a.

46. Lelah, M.D. and A. Marmur, ''Spreading Kinetics of Drops on Glass'', J. Colloid Interface Sci. 482 518 (1981).

47. Miller, C.A. and E. Ruckenstein, ''The Origin of Flow During the Wetting of Solids'', J. Colloid Interface Sci. 48, 368 (1974).

48. Fowkes, F.M., ''Dispersion Forces Contributions to Surface and Interfacial Tensions, Contact Angles and Heats of Immersion'', Contact Angles, Wettability and Adhesion, R.E. Gould, ed., Adv. in Chem. Series v.43, Am. Chem. Soc., Washington, D.C., 99 (1964).

49. Ruckenstein, E. and P.S. Lee, ''The Wetting Angle of Very Small and Large Drops'', Surface Sci. 52, 298 (1975).

50. Neogi, P., ''Electrical Double-Layer Effects Near a Contact Line'', J. Colloid Interface Sci. 98, 425 (1984).

51. Batchelor, G.K. An Introduction to Fluid Dynamics, Cambridge University Press, Cambirdge, p. 219 (1980).

52. Friz, G., ''Überden Dynamischen Randwinkle in fall der Vollsfandigen Benetzung'', Z. Angew. Phys. 19, 374 (1965).

53. Ludviksson, V. and E.N. Lightfoot, ''Deformation of Advancing Menisci'', AIChEJ 14, 675 (1968).

54. Huh, C. and L.E. Scriven, ''Hydrodynamic Model of Steady Movement of a Solid/Liquid/Fluid Contact Line'', J. Colloid Interface Sci. 35, 85 (1971).

55. Lopez, J., C.A. Miller and E. Ruckenstein, ''Spreading Kinetics of Liquid Drops on Solids'', J. Colloid Interface Sci. 56, 460 (1976).

56. De Gennes, P.G., ''Wetting Statics and Dynamics'', Rev. Mod. Phys. 57, 827 (1985).

57. Dussan, V, E.B. and S.H. Davis, ''On the Motion of a Fluid-Fluid Interface Along a Solid Surface'', J. Fluid Mech. 65, 71 (1974).

58. Neogi, P. and C.A. Miller, ''Spreading Kinetics of a Drop of a Rough Solid Surface'', J. Colloid Interface Sci. 92, 338 (1983).

59. Hocking, L.M., ''A Moving Fluid Interface on a Rough Surface, Parts I and II'', J. Fluid Mech 76, 801 (1976); 79, 209 (1977).

60. Mussel, L.I. and R. Glang, eds., Handbook of Thin Film Technology, McGraw-Hill Book Co., N.Y., 1970.

61. Blake, T.D. and J.M. Haynes, ''Kinetics of Liquid-Liquid Displacement'', J. Colloid Interface Sci. 30, 421 (1969).

62. Ruckenstein, E. and C.S. Dunn, ''Slip Velocity During Wetting of Solids'', J. Colloid Interface Sci. 59, 135 (1977).

63. Neogi, P. and C.A. Miller, ''Spreading Kinetics of a Drop on a Smooth Solid Surface'', J. Colloid Interface Sci. 86, 525 (1982).

64. Wayner, Jr., P.C., ''Interfacial Profile in the Contact Line Region of a Finite Contact Angle System'', J. Colloid Interface Sci. 77, 495 (1980); 88, 294 (1982).

65. Neogi, P., ''The Film 'Blow-Off' Experiments'', J. Colloid Interface Sci. 89, 358 (1982); 90, 554 (1982).

66. Van Dyke, M. Perturbation Methods in Fluid Mechanics, Annotated Ed., Parabolic Press, 1975.

67. Cole, J.D. Pertubation Methods in Applied Mathematics, Blaisdell Pub. Co., 1968.

68. Derjaguin, B.V., G. Strakhovosky and D. Malysheva, ''Measurement of the Viscosity of the Wall-Adjacent Boundary Layers by Film Blow-Off Method'', Acta Physicochimica URSS 19, 541 (1944).

69. DiPietro, N.D. and R.G. Cox, ''The Spreading of a very Viscous Liquid on a Quiescent Water Surface'', Q. Jl. Mech. Appl. Math. 32, 355 (1979).

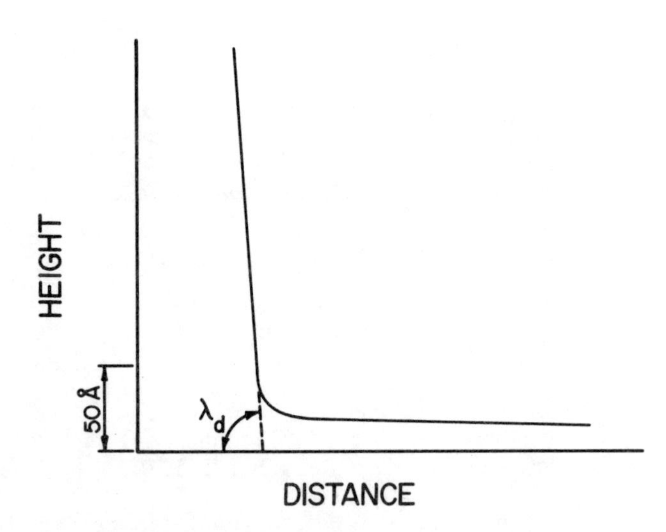

Figure 1. The observations of Bascom, Cottington and Singleterry (1964) have been shown schematically. The dynamic contact angle λ_d has been indicated, and the equilibrium value is zero.

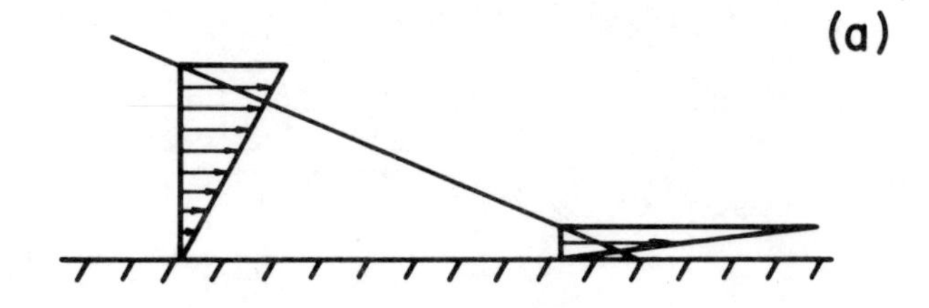

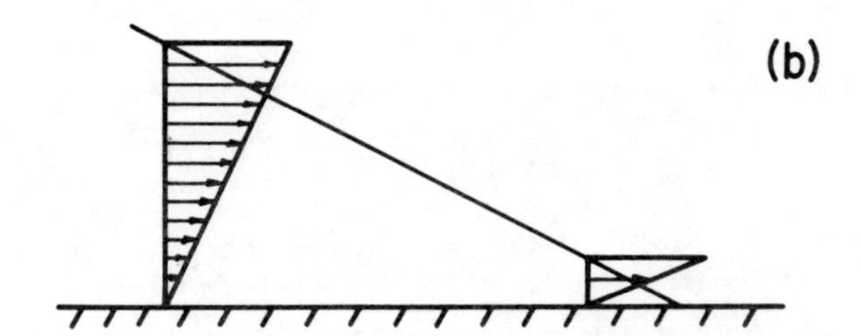

Figure 3. Shows the velocity profile in the thin liquid as the contact line is approached (a) without slip and (b) with slip.

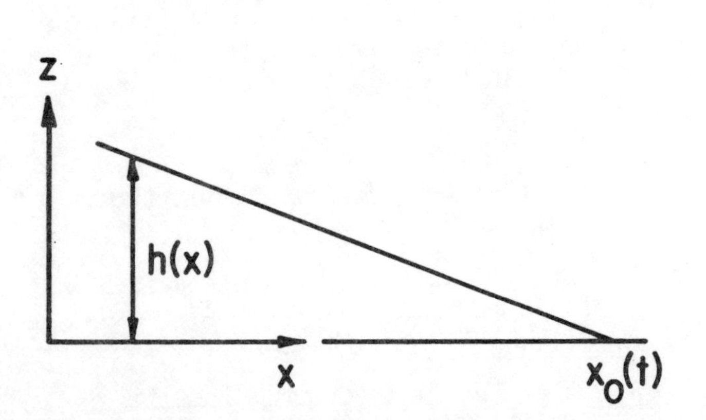

Figure 2. Shows the basic shapes of a moving wedge-like liquid film. The coordinates used in Eqs. (9) onwards are also shown. The position $x = x_0(t)$ denotes the contact line where h=0.

FLUID FLOW AND EVAPORATION IN AN ULTRA-THIN FILM OF A BINARY MIXTURE

B. G. Volintine ■ U.S. Department of Energy, Washington, D.C. 20585
P. C. Wayner, Jr. ■ Department of Chemical Engineering
Rensselaer Polytechnic Institute, Troy, NY 12180-3590

The physicochemical phenomena associated with fluid flow (due to surface shear and the London-Van der Waals dispersion force) in an evaporating ultra-thin film of an ideal binary liquid mixture in the contact line region are discussed. The dependence of the thin film vapor pressure on the film thickness ("disjoining pressure" effect), temperature, and composition is obtained from an extension of the Kelvin equation and Raoult's law. The kinetic theory is applied to obtain the liquid evaporation rate from the film vapor pressure.

Interline shear stresses arising from composition gradients in an evaporating film of a dilute mixture are estimated, and on this basis, the interline thickness profiles and evaporation rates of dilute mixtures are compared with those of "pure" fluids. The volumetric flow rates arising from surface shear and the London-Van der Waals potential gradient are compared.

INTRODUCTION

In this study, the physicochemical characteristics of an evaporating "spreading" liquid mixture in the region near the stationary triple interline (Figure 1) of a non-evaporating adsorbed film of null curvature, an evaporating meniscus, and saturated vapor are examined. Previously, Wayner, Kao, and LaCroix (1) gave an effective method for estimating the interline heat sink capacity of an adsorbed evaporating film of a pure liquid; the vapor pressure of the film was assumed to be a function of its thickness but not of its curvature. Moosman and Homsy (2) modeled the transport phenomena that occur in an evaporating meniscus of a pure component affected by both capillarity and multilayer adsorption. These analyses are extended in this paper to dilute ideal mixtures to estimate the interline stresses that arise from a composition change due to evaporation, and to explain experimentally observed differences in behavior of single and two-component systems.

The Marangoni effect due to a composition gradient near the interline is emphasized. It is noted, however, that the liquid mixtures under consideration are ideal, and that the composition gradient in the film is assumed to arise from distillation only, and not to differential adsorption upon the solid phase. Furthermore, the liquid-vapor interfacial temperature is assumed to be constant.

There is extensive literature on Marangoni phenomena. Much of the historical literature (pre-1960) on the subject was reviewed by Scriven and Sternling (3); another review was given by Kenning (4). Apparently the first correct qualitative interpretation of surface tension driven flows was given by J. Thomson (5). The first quantitative analyses of the fluid dynamical instability associated with a surface tension gradient were made by Sternling and Scriven (6) and (independently) by Pearson (7).

Marangoni instability in the contact region of a spreading film of a pure liquid was studied by Neogi and Berryman (8). The film was assumed to be thin enough so that its vapor pressure was a function of its thickness but not of its curvature. Neogi and Berryman concluded that Marangoni instability due to a temperature gradient does not ordinarily occur in the film; in their case, however, the fluid motion due to the surface tension gradient is assumed to be away from the interline of Figure 1, and in this analysis the direction of the liquid motion caused by distillation is assumed to be toward the triple interline.

The well-known phenomenon of the "tears of a strong wine" and other phenomena whose cause is related to a surface tension gradient were analyzed by Neogi (9). Profiles of the leading front of liquid films in the film region where the potential for fluid flow is related to the curvature of the

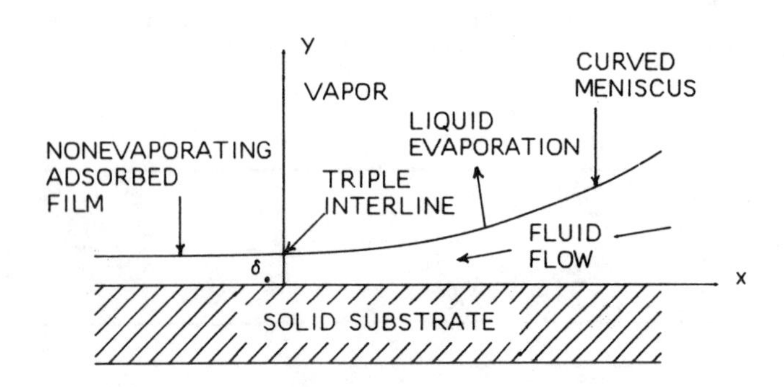

Figure 1. Triple interline region of a spreading thin liquid film.

interface were obtained, for both "positive" and "adverse" surface tension gradients.

Of particular significance to the present work are the pioneering experimental studies of Bascom, et al (10) on the spontaneous spreading of oils on solids due to the unequal evaporative depletion of one of the components of a mixture. Bascom's work was extended by Ludviksson and Lightfoot (12); in the latter experimental studies, the plate was nonisothermal and the warm end of the plaste was partly immersed. The present study was inspired by recent experiments that demonstrated the stable nature of fluid flowing towards a heat source while undergoing high rates of heat transfer [e.g., Tung, et al (13) and Wayner, et al (14)]. In this study the volumetric flow rates of liquid towards the heat source arising from the surface tension gradient and the London-van der Waals potential are compared.

ANALYSIS

This analysis is restricted to dilute binary spreading liquid mixtures that are ideal in the sense that the reduction of the vapor pressure of the major component in bulk is strictly proportional to composition. Furthermore, the vapor pressure of the minor component is assumed to be much lower than that of the major component at all (subcritical) temperatures. Thus to a first approximation, only the "lighter" liquid component evaporates. In addition, the standard "thin film" (or lubrication) approximations are invoked, and these approximations justify the neglect of bulk convection of momentum, heat, and mass in the thin film. Experiments demonstrate that the characteristic thickness of the interline region (δ_0) is much smaller than its axial extent.

The disjoining pressure in thin films of mixtures has been studied by Derjaguin and Churaev (15, 16). In general the disjoining pressure in films of mixtures is a complicated function of composition, but in the case of mixtures of similar components (such as two alkanes), the dielectric and optical properties of the components may be sufficiently close such that the molecular component of the disjoining pressure can be assumed independent of the composition

$$\Pi_m = \frac{\bar{A}}{\delta^3} \tag{1}$$

where $A_{12} = -6\pi\bar{A}$ is the Hamaker constant for a thin adsorbed film.

The coefficient $\bar{A}$ in Equation (1) is a measure of the relative effects of liquid-liquid, liquid-solid interactions; the vapor phase is assumed to have the dielectric properties of vacuum. A disjoining isotherm in inverse third power of film thickness has been recommended by Churaev (17) for thin films ($\delta < 150$(Å)). It is assumed that there are no composition gradients in the direction normal to the substrate.

The net mass flux crossing a liquid-vapor interface can be estimated with a result from the kinetic theory of gases:

$$\dot{m} = \sigma_e \left(\frac{M_1}{2\pi R}\right)^{1/2} \left\{\frac{P_{\ell v}(T_{\ell v}, x_1, \delta)}{\sqrt{T_{\ell v}}} - \frac{P_e(T_d, x_{1\infty})}{\sqrt{T_d}}\right\} \tag{2}$$

where σ_e is an evaporation coefficient, T_d is the dew-point temperature of the vapor, and M_1 is the molecular weight of the component "1". If $T_{\ell v}$ is close to T_d, then Equation (2) can be approximated as

$$\dot{m}_1 = \sigma_e \left(\frac{M_1}{2\pi R\bar{T}}\right)^{1/2} \{P_{\ell v}(T_{\ell v}, x_1, \delta) - P_e(T_d, x_{1\infty})\} \tag{3}$$

where $\bar{T} = \sqrt{(T_{\ell v} T_d)}$.

The equilibrium pressure in the vapor phase is given by Raoult's law

$$P_e = x_{1\infty} P_1^{sat}(T_d) + (1-x_{1\infty}) P_2^{sat}(T_d)$$

$$\simeq x_{1\infty} P_1^{sat}(T_d) \quad (4)$$

where $x_{1\infty}$ is the composition present in a liquid pool located at $x = +\infty$ from the interline region. The vapor pressure of component "2" is assumed to be much lower than that of component "1".

The vapor pressure is reduced in the film relative to that of pure component "1" in bulk by dilution and the "disjoining pressure" effect. Alternatively, it is increased by an increase in temperature; $T_{\ell v} > T_d$.

Assuming liquid phase ideality,

$x_1 P_1^{sat}(T_{\ell v})$ represents the effect of composition on the vapor pressure. The reduction of vapor pressure due to the liquid-solid interactions can be estimated from a modified form of the Kelvin relation

$$P_{\ell v}(T_{\ell v}) = P_1^{sat}(T_{\ell v}) \exp\left(\frac{-\Pi_m V_\ell}{RT_{\ell v}}\right) \quad (5)$$

The effect of curvature on the vapor pressure has been neglected.

The vapor pressure at $T_{\ell v}$ can be computed from the Clapeyron equation

$$P_1^{sat}(T_{\ell v}) = P_1^{sat}(T_d) \exp\left(\frac{\beta H_{\ell v}}{RT_{\ell v}}\right) \quad (6)$$

where the liquid superheating parameter

$$\beta = (T_{\ell v} - T_d)/T_d \quad (7)$$

and $H_{\ell v}$ is the enthalpy of vaporization of component "1". Mixing heats are neglected in the liquid phase.

It is assumed that the effects of composition, temperature, and "pressure" on the mass flux distribution can be combined in Equation (3)

$$\dot{m}_1 = \sigma_e \left(\frac{M_1}{2\pi R\bar{T}}\right)^{1/2} P_1^{sat}(T_d)$$

$$*\left\{x_1 \exp\left(\frac{-\Pi_m V_1}{RT_{\ell v}}\right) \exp\left(\frac{\beta H_{\ell v}}{RT_{\ell v}}\right) - x_{1\infty}\right\} \quad (8)$$

Expansion of exponentials in Equation (8) to first order (and retaining first order terms only) yields the relation

$$\dot{m}_1 = \sigma_e \left(\frac{M_1}{2\pi R\bar{T}}\right)^{1/2} P_1^{sat}(T_d)$$

$$*\left\{(x_1 - x_{1\infty}) - \frac{x_1 \bar{A} V_\ell}{RT_{\ell v}\delta^3} + \frac{x_1 \beta H_{\ell v}}{RT_{\ell v}}\right\} \quad (9)$$

It is assumed that the film is so thin that resistance to heat conduction in the liquid pahse can be neglected, i.e., the temperature of the liquid-vapor interface is a constant $T_{\ell v}$.

The interline thickness (thickness for zero evaporation)[1] is obtained from Equation (9):

$$\delta_o^3 = \left[\frac{\bar{A} V_\ell}{\beta H_{\ell v} - RT_{\ell v}(x_{1\infty}/x_{1o} - 1)}\right] \quad (10)$$

[1]This is not the thermodynamic equilibrium film thickness for a binary mixture, because it is based on the assumption that the second (heavier) component in the mixture does not evaporate. Thus for a fixed liquid superheat there is a small evaporation rate of component "2" at the film thickness given by Equation (10), and a residual evaporation rate of component "1" at that film thickness as well. It is noted that expansion of exponentials and square root terms appearing in Equation (2) are not necessary approximations, and that a sharper estimate of the interline thickness (under the assumption that the second component does not evaporate) is available at once from Equation (2).

where x_{1o} is the composition in the non-evaporating portion of the film. Since $x_{1\infty} > x_{1o}$, the required effect of the disjoining pressure decreases as a result of the decrease in vapor pressure of the liquid from composition change. Thus at the same liquid superheating, an evaporating film of a mixture does not thin as much as an evaporating film of a pure component, even though the molecular forces of attraction between the liquid and the solid are approximately the same for the two systems. Based on preliminary calculations made in this study, it is postulated that a shear stress arising from a composition gradient could support fluid flow.

Evidently Equation (10) has meaning only for

$$1 + \frac{\beta H_{\ell v}}{RT_{\ell v}} > (x_{1\infty}/x_{1o}) \qquad (11)$$

and furthermore for $x_{1o} > 0$; i.e., there is component "1" left to evaporate. In the case that $x_{1o} \to 0$, Equation (8) no longer applies, and the evaporation rate of component "2" must be considered.

The total variation in the liquid-vapor interfacial tension is

$$\frac{d\sigma}{dx} = \frac{\partial\sigma}{\partial x_1}\frac{dx_1}{dx} \qquad (12)$$

It is assumed that

$$-\tau_o(x) = \frac{d\sigma}{dx} \qquad (13)$$

where $\tau_o(x) > 0$ is to be determined. The coefficient of the composition gradient can be assumed constant, and it is negative if the lower boiling fraction has a lower surface tension at all (subcritical) temperatures. The composition distribution in the film is then

$$x_1 = x_{1o} + (1/|\partial\sigma/\partial x_1|)\int_0^x \tau_o(s)ds \qquad (14)$$

for negligible axial dispersion and convection of mass by bulk flow.

First the fluid flow due only to surface shear is evaluated. Assuming that the liquid density is approximately constant, the x-component of the momentum equation is

$$\frac{dp_1}{dx} = \mu\frac{d^2u(y)}{dy^2} = 0 \qquad (15)$$

neglecting the curvature and disjoining gradients. Thus the assumed flow profile is plane

$$u(y) = \frac{-\tau_o(x)y}{\mu} \qquad (16)$$

The total volumetric flow rate (per unit of cross-sectional width) can be computed from Equation (16):

$$Q = \int_0^\delta u(y)dy = \frac{-\tau_o(x)\delta^2}{2\mu} \qquad (17)$$

The evaporation rate of component "1" is obtained by material balance

$$\dot{m}_1 \simeq \dot{m} = \frac{1}{2\nu}\frac{d(\tau_o\delta^2)}{dx} \qquad (18)$$

Sufficiently close to the interline, the assumption

$$\tau_o(x) = \tau_o\text{, a positive constant} \qquad (19)$$

could hold; but not uniformly for all x because the composition must remain bounded. From (14) and (19), the assumed composition dependence is linear near the interline

$$x_1(\xi) = x_{1o} + \xi \qquad (20)$$

where the dimensionless position variable

$$\xi = x\tau_o / \left| \partial\sigma/\partial x_1 \right| \qquad (21)$$

The composition variable is subject to the constraint

$$0 < x_{1o} \leq x_1(\xi) \leq x_{1\infty} < 1 \qquad (22)$$

Let

$$\eta = \delta/\delta_o \qquad (23)$$

and define the dimensionless evaporative heat transfer rate

$$\Lambda_1 = (\dot{m}_1/L_1) \frac{RT_{\ell v}}{\beta H_{\ell v}} \qquad (24)$$

where

$$L_1 = \sigma_e \left(\frac{M_1}{2\pi R\overline{T}}\right)^{1/2} P_1^{sat}(T_d) \qquad (25)$$

Equations (9) and (18) are equated, and rearranged to yield

$$\Lambda_1 = \{\frac{RT_{\ell v}}{\beta H_{\ell v}} [(1-x_{1\infty}/x_1) - (1-x_{1\infty}/x_{1o})/\eta^3]$$

$$+ 1 - 1/\eta^3\} \, x_1 = \alpha \frac{d\eta^2}{d\xi} \qquad (26)$$

where the dimensionless ratio

$$\alpha = \frac{\tau_o^2 \delta_o^2}{2\nu \left| \partial\sigma/\partial x_1 \right| L_1} \frac{RT_{\ell v}}{\beta H_{\ell v}} \qquad (27)$$

It is noted that the effect of "disjoining pressure" on the vapor pressure is retained in Equation (26) but the effect of disjoining pressure on the flow is not retained.

RESULTS AND DISCUSSION

Equations (20) and (26) govern the evaporative heat transfer in the thin film, but only the functional dependence of the composition distribution is given.

If a composition change is specified for a given range of film thickness, then Equation (26) can be integrated to obtain the shear stress required to accomplish this change. Here an initial value problem governed by Equation (26) is obtained by specifying a liquid-phase superheat and values of the system parameters.

The particular liquid-solid system selected for investigation is a 2% (weight) Tetradecane(2)/Decane(1) liquid mixture on steel.

The following property values are assumed for the liquid mixture (18) and liquid-solid interactions; the property values were evaluated at an assumed dew-point temperature T_d = 300(K); all but the liquid-vapor interfacial tension are assumed to be independent of temperature. All of the fluid properties except vapor pressure are assumed to be independent of pressure (and hence of film thickness).

$x_{1\infty}$ = 0.986; $H_{\ell v}$ = 5.0E+04(J)/(mol);

V_ℓ = 2.0E-04(m)^3/(mol);

ν = 1.2E-06(m)^2/(s);

$P_1^{sat}(300)$ = 200.(N)/(m)^2;

$P_2^{sat}(300)$ = 2.5(N)/(m)^2;

σ_e = 1; $\overline{A}$ = 1.0E-20(J) [Churaev, (17)]

and

$$\sigma(x_1,T) = (50.8 - 0.0920T)x_1$$

$$+(51.8 - 0.0869T)(1-x_1) \text{ (dyn)/(cm)}$$

where T is in Kelvins.

The interline boundary conditions (10) fixes the interline thickness for values of the superheat β and the composition x_{1o}. An estimate of the composition at δ_o can be obtained.

If all of the flow were caused by capillarity and disjoining pressure, then

$$\sigma(T_{\ell v}, x_{1o}) = \sigma(T_d, x_{1\infty}) \qquad (28)$$

assuming that the liquid pool is at temperature T_d. If on the other hand there is Marangoni flow then

$$\sigma(T_{\ell v}, x_{1o}) > \sigma(T_d, x_{1\infty}) \qquad (29)$$

so that the estimate of x_{1o} given by Equation (28) is an upper bound of the composition in the nonevaporating portion of the film. To obtain a more precise estimate of x_{1o}, a more extensive problem must be considered. Compositions x_{1o} computed from Equation (28) are graphed in Figure 2; the diagram shows that there is a range of liquid superheating for which the inequality (11) holds, for the particular values of the system parameters.

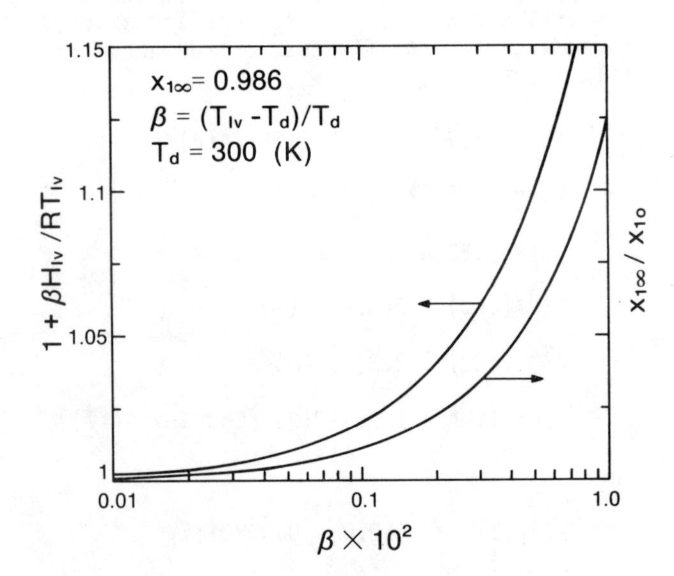

Figure 2. Interline compositions (x_{1o}) computed on the basis of null interfacial shear stress in an evaporating decane/tetradecane film.

For a small change in liquid composition, the coefficients x_1 in Equation (26) can be approximated as constants (and equal to their initial values). This approximation gives the constant-coefficent d.e.

$$C_1 \, [1 - 1/\eta^3] = \alpha \frac{\delta \eta^2}{d\xi} \qquad (30)$$

where

$$C_1 = x_{1o} \left\{ \frac{RT_{\ell v}}{\beta H_{\ell v}} (1 - x_{1\infty}/x_{1o}) + 1 \right\} \qquad (31)$$

This approximation overestimates the effect of composition on the vapor pressure. On the other hand, this approximation partially offsets the previously neglected diffusional effect on the vapor pressure in the one-dimensional model.

The differential Equation (30) has an analytic integral:

$$C_1 \xi = \alpha \left\{ \eta^2 + \frac{2}{3} \ln (1 - 1/\eta) - \frac{1}{3} \ln (1/\eta^2 + 1/\eta + 1) - \frac{2}{\sqrt{3}} \arctan \left(\frac{2/\eta + 1}{\sqrt{3}} \right) + C_2 \right. \qquad (32)$$

The integration constant C_2 is obtained from the initial condition imposed on Equation (30):

$$\eta(0) = 1 + \varepsilon \qquad (33)$$

where ε is a small positive parameter. For computational purposes, the parameter ε was assigned the same value as the degree of superheat β.

For a particular liquid superheating parameter $\beta > 0$ a composition $x_1 > x_{1o}$ can be specified at $\eta = 3$, and Equations (20), (27), (32) and (33) can be combined to evaluate the interfacial shear stress τ_o. The analysis is restricted to three initial film thicknesses; it is known from experiment that this range of film thickness is much thinner than that portion of a spreading film where fluid flow is controlled by capillarity.

As a specific computational example, for $\beta = 0.005$, the composition in the non-evaporating portion of the film stipulated by Equation (28) is x_{1o} = 0.930. Then the constants C_1 = 0.369, C_2 = 4.10 (for ε = 0.005); and from Equation (10), δ_o = 2.7E-09(m). For x_1 = 0.98 specified at η = 3, Equation (20) gives ξ = 0.05, and from Equation (32), α = 1.56E-03. Then from Equation (27), τ_o = 2.6E+02(N)/(m)^2.

Interfacial stresses τ_o evaluated in this manner are graphed as a function of film thickness in Figure 3. The stresses are made dimensionless with respect to the magnitude of the disjoining pressure at that film thickness. The stresses are small compared with the disjoining pressure for this range of film thickness and liquid conditions, but the effects of the stresses are large.

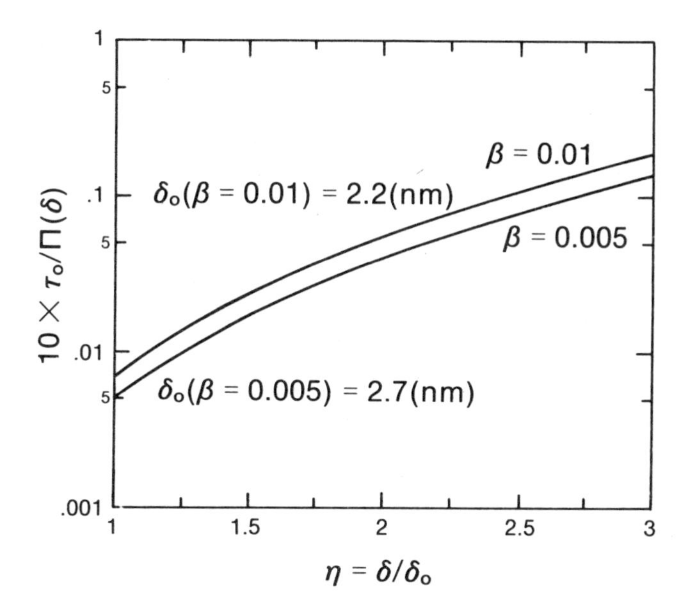

Figure 3. Dimensionless interfacial stresses evaluated to give a specified composition change over three initial film thicknesses in an evaporating decane/tetradecane film.

In the case of pure component "1" (decane), it is assumed that the origin of the flow is a thickness gradient; the pressure gradient for fluid flow is given by (Wayner, et al (1); Neogi and Berryman (8))

$$P_1' = -3\bar{A}\delta'/\delta^4 \tag{34}$$

where the prime denotes differentiation with respect to position. The volumetric flow due to the pressure gradient (34) can be computed from the implied parabolic velocity distribution

$$Q_d = \frac{-\bar{A}}{\mu}(\ell n(\delta))' \tag{35}$$

The mass flux rate is related to the gradient of volumetric flow

$$\dot{m}_1 = \frac{\bar{A}}{\nu}(\ell n(\delta))'' \tag{36}$$

which, in turn, is related to the evaporation rate of pure component "1", obtained from Equation (9) for $x_{1o} = x_{1\infty} = 1$. The resulting d.e.

$$\frac{\bar{A}}{\nu}(\ell n(\eta))'' = L_1[1 - 1/\eta^3]\frac{\beta H_{\ell n}}{RT_{\ell v}} \tag{37}$$

together with the initial conditions

$$\eta(0) = 1 + \varepsilon \tag{38}$$

$$\eta'(0) = 0 \tag{39}$$

for β fixed was integrated numerically (by orthogonal collocation). The parameter ε was given the same value[2] as it was for the initial condition specified for a film of mixture (Equation (33)). In Equation (37), $\eta = \delta/\delta_o$ with δ_o given by Equation (10) for x_1 = 1. The ratio of volumetric flow rates arising from an assumed surface tension gradient in an evaporating film of a dilute mixture (Equation (17)) to the volumetric flow in a "pure" liquid film arising from a thickness gradient (Equation (35)) is plotted as a function of film thickness in Figure 4, for a fixed liquid superheat.

The figure shows that the flow available from the estimated shear stress is a significant fraction of the volumetric flow due to the thickness gradient in the film of the pure liquid at the same liquid superheat. It is emphasized that the computed shear stress depends critically upon the composition at δ_o; this composition has been estimated but is unknown. As $\delta \to \infty$, Q becomes more important than Q_d as the disjoining pressure becomes small.

---

[2] The parameter ε is needed in Equations (33) and (38) because Equations (30) and (37) have a logarithmic singularity at η = 1. The significance of the singularity in relation to the dynamics of a spreading film is discussed in Ref. (8).

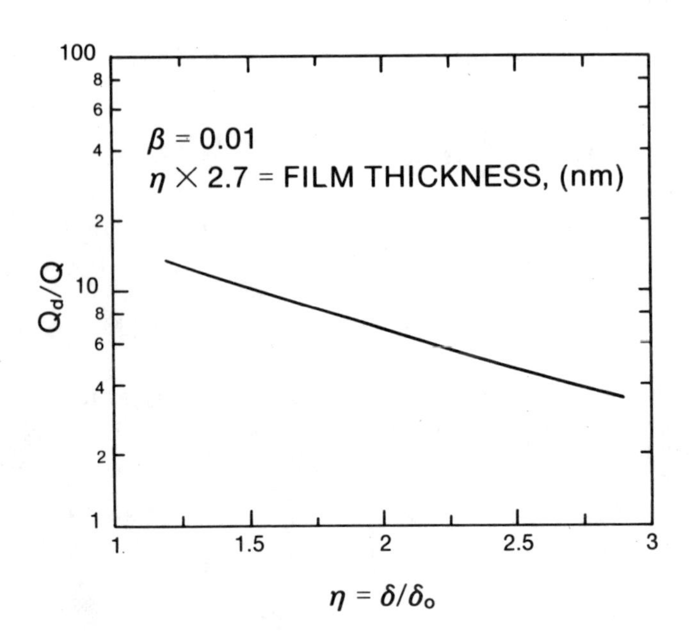

Figure 4. A comparison of liquid volumetric flow rates as a function of film thickness in evaporating decane/tetradecane films and films of pure decane, at the same liquid superheating.

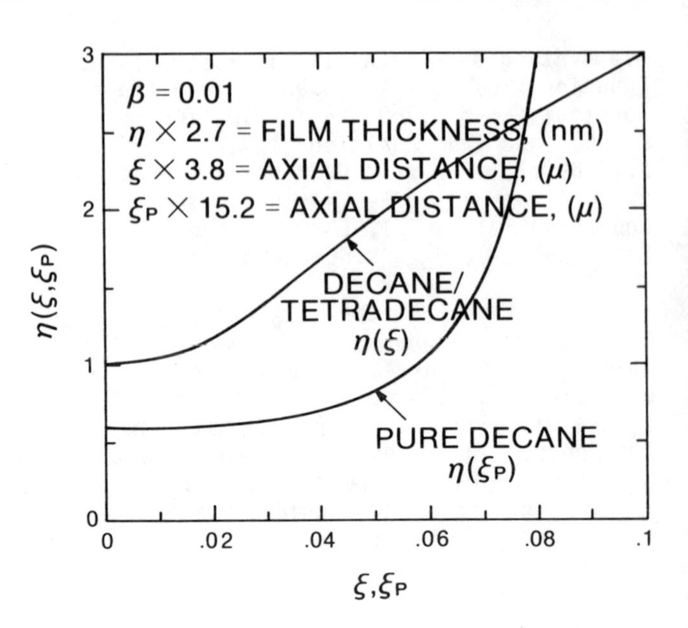

Figure 5. Profiles of evaporating films of decane/tetradecane and decane, based on fluid flow arising from shear stress and film thickness gradient, respectively.

Further information is obtained from the qualitative features of the film profiles which are plotted as a function of distance from the interline in Figure 5. It is noted that the profile of the pure component (decane) is compressed (by a factor of four) onto the ξ- coordinate scale, and that the profile of the pure component is plotted relative to δ_0 for a mixture. The figure shows that the slopes of the mixture profile are significantly decreased toward the liquid pool, compared with the slopes of the film profile of the pure component.

It is suggested that if capillarity had been included in the equations of transfer, then the presence of the additional mechanism of transport would cause the magnitude of the curvature gradient to decrease. Also, a reduction in the film slope toward the thicker portion of the film (relative to that of a pure component) would indicate that the maximum in the film curvature would occur farther from the interline in a mixture than for a pure component.

Although approximate solutions of a steady-state dynamical problem have been obtained, the analysis is actually "pseudo-steady" because at very long times the interline region of a spreading film may be depleted of volatile components. In this case the results obtained in this study are not valid. The results obtained in this study apply to the initial stages of spreading, for which the tendency to spread is known to be large [Cottington, et al (11)]. Film profiles for the "long-time" have been predicted by Neogi (9), which agree with film profiles determined experimentally by Bascom, et al (10).

The results of the analysis given in this study are contingent on the assumption that $\tau_0(x)$ could be considered constant over a range of film thickness. It would be reasonable to assume that $x_1 \to x_{10}$ asymptotically, so that $\tau_0 \to 0$ near the interline. Such an asymptotic decay of τ_0 could occur within the range of initial values $\eta \leq 1+\varepsilon$. The vanishing of τ_0 leaves a significant potential gradient for diffusion, of course, but diffusion has been neglected entirely.

CONCLUSIONS

The potential for fluid flow arising from a surface tension gradient in a very thin liquid film of a mixture is large, for even a modest composition change. Fluid flow that results from the surface tension gradient in a dilute mixture was shown to be large towards the thicker portion of the film, and it is suggested that some distance from the contact line the surface tension gradient replaces the "disjoining pressure" gradient in importance. It has been demonstrated experimentally that the curvature gradient (capillarity) is important for the control of fluid flow in both films of mixtures and films of pure liquids in the thicker portion of the film.

The results suggest that there may be conditions for which the "disjoining pressure" gradient can be neglected in studies of fluid transport in evaporating films of mixtures, depending on the magnitudes of the system parameters.

ACKNOWLEDGEMENT

One of the authors (B.G. Volintine) gratefully acknowledges the financial support of a Union Carbide Corporation Fellowship. In addition, this material is based upon work supported by the National Science Foundation under Grant No. MEA82-13690. Any opinions, findings and conclusions or recommendations expressed in this publication are those of the authors and do not necessarily reflect the views of the National Science Foundation or the Union Carbide Corporation.

NOTATION

$\bar{A}$ = Molar dispersion constant, (J)

C_1, C_2 = Integration constants [Equations (30), (32)]

H = Molar enthalpy, (J)/(mol)

L_1 = Coefficient of equation from kinetic theory, (Kg)/(m)^2(s) [Equation (25)]

M_1 = Molecular weight of component "1", (g)/(mol)

P = Pressure, (N)/(m)^2

Q = Volumetric flow rate per unit of crosssectional width in a film of a mixture, (m)^3/(s)(m)

Q_d = Volumetric flow rate per unit of cross-sectional width arising from a disjoining gradient in a film of a "pure" fluid, (m)^3/(s)(m)

R = Universal gas constant, (J)/(mol)(K)

T = Absolute temperature, (K)

V = Molar volume, (m)^3/(mol)

$\dot{m}_1$ = Evaporation rate of pure component "1", (Kg)/(M)^2(s)

u = x-component of velocity vector, (m)/(s)

x = Dimensional axial coordinate along solid substrate, (m)

x_1 = Mole fraction of component "1" in liquid mixture

y = Dimensional coordinate orthogonal to solid substrate, (m)

Greek

Λ_1 = Dimensionless evaporative flux rate of component 1", [Equation (24)]

Π_m = Molecular component of disjoining pressure, (N)/(m)^2

α = Dimensionless stress ratio, [Equation (27)]

β = Superheating parameter, [Equation (7)]

δ = Dimensional film thickness, (m)

δ_0 = Interline thickness, (m), [Equation (10)]

ε = Small positive parameter

η = Dimensionless film thickness, [Equation (23)]

μ = Absolute liquid viscosity, (kg)/(m)(s)

ν = Kinematic liquid viscosity, (m)^2/(m)(s)

ξ = Dimensionless axial coordiate, [Equation (21)]

σ = Liquid-vapor interfacial tension, (N)/(m)

σ_e = Evaporation coefficient, [Equation (2)]

τ_0 = Interfacial shear stress, (N)/(m)^2

Subscripts

d = Dew point; disjoining

e = Equilibrium

ℓ = Liquid phase

ℓv = Liquid-vapor interfacial

o = Interline

p = Calculation for $x_1 = 1$

∞ = Liquid pool (bulk) condition

Superscripts

overstrike ($\bar{\ }$) Geometric mean film

sat = Saturation (equilibrium) pressure

LITERATURE CITED

1. Wayner, P.C., Jr., Y.K. Kao, and L.V. LaCroix, Int. J. Heat Mass Transfer, 19, 487-492 (1976).

2. Moosman, S. and G.M. Homsy, J. Coll. Interf. Sci., 73, 212-223 (1980).

3. Scriven, L.E. and C.V. Sternling, Nature, 187, 186-188 (1960).

4. Kenning, D.B.R., Appl. Mech. Rev., 21, 1101-1111 (1968).

5. Thomson, J., "On certain curious motions observable on the surfaces of wine and other alcoholic liquors," Phil. Mag. ser., 4, 10, 330 (1855). Reprinted in Thomson, W., Popular Lectures and Addresses, 1, 56-59, MacMillian and Co., London (1891).

6. Sternling, C.V. and L.E. Scriven, AIChE J., 73, 514-523 (1959).

7. Pearson, J.R.A., J. Fluid Mech., 4, 489-500 (1958).

8. Neogi, P. and J.B. Berryman, J. Coll. Interf. Sci., 88, 100-110 (1982).

9. Neogi, P., J. Coll. Interf. Sci., 105, 94-101 (1985).

10. Bascom, W.D., R.L. Cottington, and C.R. Singleterry, "Dynamic surface phenomena in the spontaneous spreading of oils on solids," in Contact Angle, Wettability, and Adhesion, Advances in Chemistry Series, 43, (R.F. Gould, ed.), pp. 355-379, American Chemical Society, Washington, D.C. (1964).

11. Cottington, R.L., C.M. Murphy, and C.R. Singleterry, "Effect of polar-nonpolar additives on oil spreading on solids, with applications to nonspreding oils," in Contact Angle, Wettability, and Adhesion, Advances in Chemistry Series, 43, ibid., pp. 341-354 (1964).

12. Ludviksson, V. and E.N. Lightfoot, AIChE J., 17, 1166-1173 (1971).

13. Tung, C.-Y., M. Tirumala, and P.C. Wayner, Jr., "Experimental study of evaporation in the contact line region of a mixture of decane and 2% tetradecane," Proc. 7th Int. Heat Tr. Conf., vol. IV, pp. 101-106, Sept. 6-10, München FRG (1982).

14. Wayner, P.C., Jr., C.-Y. Tung, M. Tirumala, and J.H. Yang, ASME J. Heat Transfer, 107, 182-189 (1985).

15. Derjaguin, B.V. and N.V. Churaev, Kolloidnyi Zhur. (English translation) 37, 970-976 (1976).

16. Derjaguin, B.V. and N.V. Churaev, J. Coll. Interf. Sci., 62, 369-380 (1977).

17. Churaev, N.V., Kolloidnyi Zhur. (English translation) 36, 287-290 (1974).

18. Reid, R.C., J.M. Prasunitz, and T.K. Sherwood, The Properties of Gases and Liquids, 3rd ed., McGraw-Hill, NY (1977).

APPLICATION OF THE GRADIENT-DRIVEN INSTABILITY MECHANISM TO POLYMERIC MEMBRANE MORPHOLOGY

Bruce A. Nerad, Kevin J. Gleason, and
William B. Krantz ■ Department of Chemical Engineering, University of Colorado
Campus Box 424, Boulder, CO 80309
Roderick J. Ray ■ Bend Research, Inc., 64550 Research Road, Bend, Oregon 97701-8599

Solvent-cast asymmetric polymeric membranes are produced by rapid removal of a volatile or water-soluble solvent from a polymer/solvent solution. This procedure results in a relatively dense, thin interfacial skin which serves as the permselective layer, underlaid by a considerably less dense highly permeable supporting substructure. The resulting solidified membrane is rich in spatial structure ranging from large micron-scale "fingers" down to submicroscopic nodules which significantly influence its permselectivity properties. This hierarchy of spatial structures is thought to arise from gradient-driven interfacial instabilities which are generated by the very large concentration gradients created during the solvent-casting process. This paper applies the gradient-driven instability mechanism in order to describe several features of asymmetric membrane morphology.

Ever since the energy crisis of the 1970's, there has been a renewed interest in the search for energy-efficient separations processes. Polymeric membranes may satisfy this need since they avoid the energy-consuming, entropy-intensive, phase-transition step inherent in many conventional separation processes, such as distillation. However, this potential was not realisable until a significant advancement in membrane technology was made in 1962 when Loeb and Sourirajan (1) developed the first asymmetric membrane. An asymmetric membrane is characterized by a very thin (< 100 nm) relatively dense skin supported by a thicker (~ 100 μm) porous substructural layer. The skin is responsible for the solute rejection properties, whereas the substrate provides the mechanical strength to withstand the high pressures employed in reverse osmosis.

Besides the overall asymmetry, there are other extensive spatial structures in these membranes. These appear as surface structures or ''nodules'' (Figure 1) on the skin and internal macrovoids or ''fingers'' (Figure 2) within the bulk of the membrane. It is believed by Kesting (2), Borens et al. (3), and others that the solute permeation in membrane separations is predominantly through the less dense interstitial regions between adjacent nodules. If so, this suggests that membrane permeation might be increased by decreasing the nodule diameter. The macrovoids or fingers on the other hand are responsible for the high porosity of the highly permeable substructural layer which gives the membrane its mechanical strength. This finger structures also can be used as reservoirs for liquid membranes or immobilized enzyme reactors. Clearly one would like to control finger formation to optimize the substructural porosity of a membrane for a variety of applications.

Asymmetric membranes found their first commercial application in the desalination of seawater. However, as a variety of new applications for membranes has developed, the focus of membrane research has been redirected towards developing task-specific membranes. This change in focus necessitates a better understanding of the membrane formation process than is currently available. Until now, the ''recipe'' for casting asymmetric membranes has been developed by empirical trial-and-error methods because of

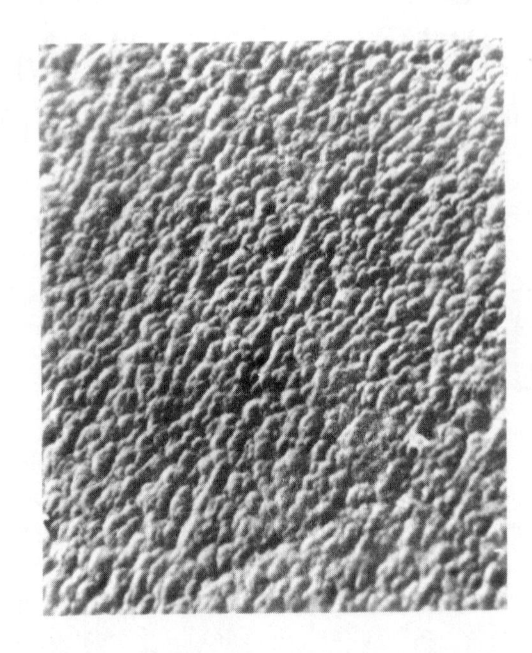

Figure 1. Scanning electron micrograph (1000X) of the outside surface of a Kynar hollow fiber showing 5 μm nodular structures.

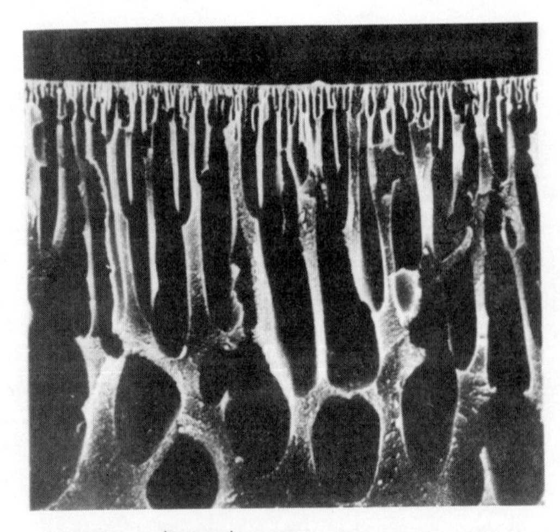

Figure 2. Scanning electron micrograph (375X) of an asymmetric membrane made from 20 percent polyamide in DMF and 5 percent LiCl precipitated in water showing macrovoid or finger structures (courtesy of H. K. Lonsdale, Bend Research, Inc.).

the lack of predictive models which relate the membrane skin properties to relevant formation process parameters. Likewise, it is not known what mechanism is responsible for the morphological features of asymmetric membranes.

better understanding of the mechanism of asymmetric membrane formation by investigating the cause of membrane morphology. The investigation involves the development of a mathematical model based on the premise that the spatial structures arise from steep concentration gradients present during the membrane casting process. These steep concentration gradients, caused by the rapid transfer of solvent out of the precipitating polymer solution, correspond to a gradient in the excess intermolecular potential energy. This excess intermolecular potential energy gradient arising from the long range van der Waals dispersion force, induces spontaneous free convection and interfacial corrugations in the polymer/solvent layer. These instabilities are subsequently frozen into place when the membrane solidifies; thus, a fossilized structure in the form of fingers and nodules resulting from the instabilities is left behind.

First, this paper will describe the casting procedure for manufacturing asymmetric membranes in order to provide a physical basis upon which the model can be built. Next, a brief overview of the theoretical development will be given. The model predictions then will be presented and interpreted physically. These predictions then will be compared with experimental data on membrane morphology available in the literature. Finally, the conclusions will be summarized and recommendations for future work will be suggested.

PHYSICS OF MEMBRANE CASTING

Prior to ascertaining the mechanism of the instabilities thought to be responsible for the morphology of asymmetric membranes, it is necessary to have an understanding of the procedure used to cast these membranes. The Loeb-Sourirajan procedure for preparing solvent-cast asymmetric membranes is shown schematically in Figure 3 and consists of the following four steps:

1. A solution of polymer dissolved in a suitable solvent (or solvent mixture) is

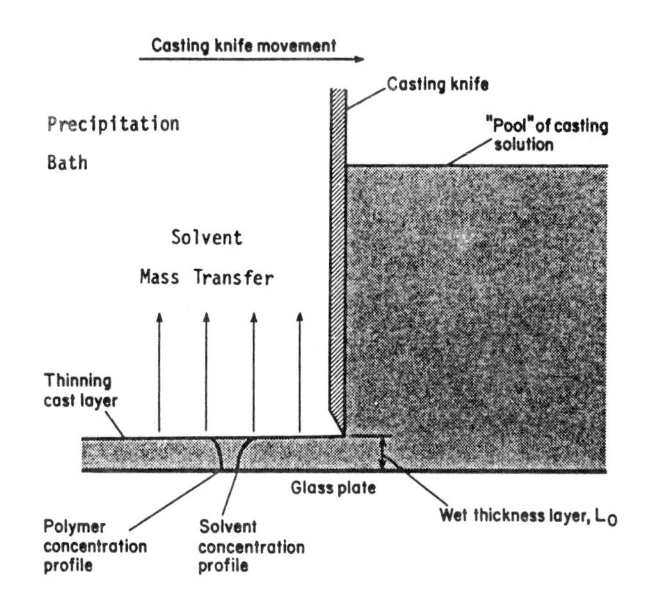

Figure 3. Schematic of the asymmetric membrane casting process.

cast as a thin film (~ 100 to 1000 μm) on a smooth solid surface.

2. The cast film may then receive an optional exposure to an ambient gas phase for a prescribed period of time.

3. Immediately after casting, or after the air evaporation step, the cast film is immersed in a precipitation bath containing a liquid in which only the solvent is miscible, but a nonsolvent for the polymer. The film remains in the precipitation bath until the polymer completely solidifies.

4. The solidified film is annealed in hot water which densifies the membrane, thereby decreasing its porosity.

During steps 2 and 3, solvent transfers out of the cast phase and concentration gradients develop. Near the upper surface of the film, the concentration of solvent decreases and the concentration of polymer increases; hence, the viscosity increases locally. Likewise, the solvent diffusivity, which is highly concentration dependent, decreases and larger gradients are required to supply the interface with solvent. Also, the interface translates downward due to local film shrinkage caused by solvent loss as well as the excess volume of mixing effect. There also may be some diffusion of the precipitation bath liquid into the cast phase, however, this effect will be neglected in the present analysis.

For purposes of modeling the casting process, the system is assumed to be a horizontally infinite liquid layer of finite depth. The fluid depth is on the order of 100 μm, which is much too thick for a ''thin film'' instability. The instability believed responsible for the membrane morphology arises because of the concentration gradient established when the solvent leaves the initially uniform concentration cast film. This solvent loss is characterized by a lumped parameter approach via a mass transfer coefficient. The concentration dependence of the density, diffusivity and viscosity are taken into account. Thus, the nonuniform shrinkage of the cast layer due to solvent loss and excess volume of mixing is included in the model. This idealized conception of the membrane casting process is translated into mathematics in the following section. The manner in which this can be done is aided by the analogy between the thin film and concentration gradient-induced instability mechanisms which has been discussed elsewhere (4).

THEORETICAL DEVELOPMENT

Solution Methodology

A strong similarity exists between the mechanisms leading to thin film break-up and that which is thought to cause asymmetric membrane morphology. This similarity will only be summarized here; for more details, the interested reader is referred to Ray et al. (5) and Nerad and Krantz (4). A thin film may break-up or rupture if the thickness of the film is less than the region of influence of the long range intermolecular forces. In this case, the molecules in the gas/liquid interfacial region are able to ''feel'' the presence of another gas/liquid interface or a solid/liquid interface. Hence, any perturbations in the interface will alter the film thickness and change the molecular environment. If this change lowers the system free energy, (this is true if the attractive forces between molecules in the liquid are stronger than their attractions with molecules in surrounding phases), then the interfacial perturbation will grow and eventually rupture the film. This problem has been analyzed by Ruckenstein and Jain (6), Maldarelli (7), and others.

The similarity between the mechanism operative in the thin film problem and that occurring in the membrane casting problem is that the intermolecular attractive forces can also cause an interfacial corrugation in membrane casting solution to grow. In the casting of membranes, there is a very steep concentration gradient which develops because of rapid mass transfer of solvent out of the cast film. Because of the unsteady-state nature of the solvent loss, much of this concentration change occurs within the region of influence of the long range intermolecular forces. Thus, a molecule at the interface of the casting solution is able to ''sense'' the different molecular environment in the bulk, in the same way a molecule at one interface of the thin film ''feels'' the other interface. Again, any perturbation in the interface will either grow or decay depending on whether the surface corrugation increases or decreases the magnitude of the total attractive forces between molecules. This instability problem has been analyzed by Ray et al. (5) and Nerad and Krantz (4).

In these aforementioned papers on thin films and membrane casting, the intermolecular potential energy gradient, which is introduced as a body force in the Navier-Stokes equation, induces spontaneous free convection in the liquid layer. This fluid motion or instability was modeled in the following manner. First, the unsteady-state, one dimensional equations describing the system without any instability were solved to yield the basic-state solution. Next, infinitesimal perturbations of the basic-state interface and velocities were impressed upon the system and a linear stability analysis was performed to determine if these perturbations grew or decayed.

It should be noted that finger formation in asymmetric membranes consists of two steps: initiation and propagation (11). The instability mechanism modeled here will provide an explanation for the initiation as well as for their regular spacing. However, the mechanism responsible for the subsequent propagation of the fingers into the substrate of the cast layer will not be discussed here.

Basic-State Solution

The basic state corresponds to the diffusion of solvent out of a binary liquid solution of polymer and solvent, and is described by the unsteady-state, one-dimensional, convective-diffusion equation. The convective transfer of solvent is because of the film shrinkage arising from the solvent mass loss and the excess volume of mixing effect. Solution of the convective-diffusion equation is further complicated by the exponential dependence of the diffusivity on concentration. A numerical solution of this problem for solvent evaporation into air has been obtained by Ray et al. (8). Their solution can be applied to membranes fabricated by immediate immersion into a precipitation bath provided that the precipitant does not counterdiffuse into the polymer/solvent film. Since this will be assumed here, the basic-state results of Ray et al. (8) summarized in Figure 4 will apply. This figure shows the diffusing solvent mass concentration as a function of distance from the solid surface for five successive times after casting up to the time at which the interface solidified

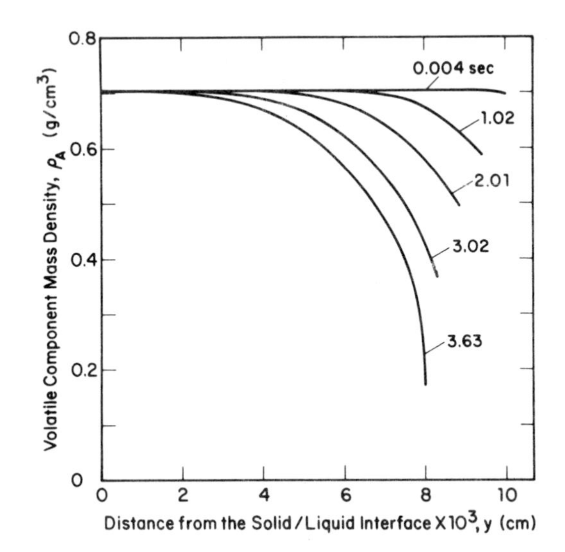

Figure 4. Diffusing solvent mass concentration as a function of the distance from the solid surface for five successive times after casting up to the time at which the interface solidified (3.63 s) for the evaporative casting of a 100 μm film of a cellulose acetate/acetone solution.

(3.63 s) for the evaporative casting of a 100 μm film of a cellulose acetate/acetone solution. The concentration gradient, or correspondingly the intermolecular potential energy gradient, near the free surface is seen to be very steep immediately prior to interfacial skinning or solidification. This steep gradient is thought to be the driving force responsible for the membrane morphology. Supporting evidence for this contention will emerge from the linear stability analysis discussed in the next section.

Linear Stability Analysis

As mentioned above, the gradient of excess intermolecular potential energy is believed to be the driving force for the formation of nodules and fingers in asymmetric polymeric membranes. This gradient in intermolecular potential energy is represented as a body force in the governing equations of motion used in the stability analysis. This term is derived by choosing a molecule and summing its pairwise interactions with all the other molecules in the system, then subtracting off the intermolecular potential energy corresponding to the constant concentration associated with the reference system. A detailed derivation of the excess intermolecular energy gradient can be found in Nerad and Krantz (4). The result is given by

$$\left.\frac{\partial W_0}{\partial z}\right|_{z=0} = 2\pi A_{ij}\left[\frac{b_i\rho_j^s}{3\delta_{ij}^3} - \frac{b_i b_j}{4\delta_{ij}^2} + \frac{\rho_i^s b_j}{3\delta_{ij}^3}\right] - 2\pi A_{ik}\frac{b_i\rho_k^s}{3\delta_{ij}^3} \quad (1)$$

where A_{ij} is the magnitude of the intermolecular potential energy between two molecules i and j, b_i is the slope of the concentration profile of i at the interface, ρ_i^s is the surface concentration of i, and, δ_{ij} is the distance of closest approach between molecules i and j. The indices i and j refer to molecules in the polymer/solvent phase, and the index k denotes the precipitant. In the derivation of the above result, the precipitant concentration profile was assumed to be constant and the solvent concentration gradient in the precipitant bath was assumed to be negligible. These assumptions are reasonable when one considers that the diffusion coefficient in the polymer solution near the interface is orders of magnitude smaller than the diffusion coefficient in the precipitation bath. This implies steep solvent concentration gradients in the polymer solution relative to the precipitation bath and relatively little penetration of the precipitant into the casting solution.

Note that the slope of the concentration profile, b_i, is time dependent and decreases as the mass transfer proceeds. Hence, the gradient of the excess intermolecular potential energy also decreases in time and eventually damps out entirely. Therefore, the influence of this instability on the membrane morphology only occurs because the polymer solidifies prior the disappearance of the concentration gradients.

Since the time scale for the growth of the instabilities is far shorter than that characterizing the time evolution of the basic state, a frozen time analysis is justified. That is, in solving the linear stability problem, the basic state will be assumed to be time independent. Furthermore, the direct coupling between the equation of motion and the convective-diffusive equation will be ignored. Thus, coupling enters in the equation of motion only through the concentration dependence of the viscosity and of the excess intermolecular potential energy which appears as a body force in the equation of motion. Only two-dimensional disturbances will be considered in this linear stability analysis, since Squires' Transformation (9) may permit obtaining the results for three-dimensional disturbances from the two-dimensional analysis.

In view of the above considerations, the appropriate form of the x-component of the linearized equation of motion is given by

$$\rho\frac{\partial u}{\partial t} = -\frac{\partial P}{\partial x} - \frac{\partial W}{\partial x} + \mu(z)\frac{\partial^2 u}{\partial x^2} + \frac{\partial}{\partial z}\left(\mu(z)\frac{\partial u}{\partial z}\right) \quad (2)$$

where ρ is the overall mass density of the polymer solution, u is the horizontal velocity component, w is the vertical velocity component, P is the pressure, W is the excess intermolecular potential, and $\mu(z)$ is the viscosity, whose value in this frozen time analysis is assumed to be an exponential function of distance below the polymer solution/precipitation bath interface at any given time and is given by

$$\mu(z) = \mu_0 + (\mu_s - \mu_0)e^{-Az} \quad (3)$$

where μ_s and μ_0 are the viscosities at the upper interface and at the lower solution/solid interface, respectively. The z-component of the linearized equation of motion is

$$\rho\frac{\partial w}{\partial t} = -\frac{\partial P}{\partial z} - \frac{\partial W}{\partial z} + \mu(z)\frac{\partial^2 w}{\partial x^2} + \frac{\partial}{\partial z}\left(\mu(z)\frac{\partial w}{\partial z}\right) \quad (4)$$

Finally, the continuity equation is given by

$$\frac{\partial u}{\partial x} + \frac{\partial w}{\partial z} = 0 \quad (5)$$

Two of the four boundary conditions required to solve the above equation must be applied at the free surface. Hence, a kinematic condition is also required to relate the amplitude of the surface disturbance to the velocity components at the interface. The boundary conditions at the lower solid/solution interface are that the interface is impermeable

$$w = 0 \quad \text{at } z = L \quad (6)$$

and the no-slip condition

$$u = 0 \quad \text{at } z = L \quad (7)$$

The linearized normal stress balance at the upper interface is given by

$$P - P_r - 2\mu(z)\frac{\partial w}{\partial z} - \sigma\frac{\partial^2 \xi}{\partial x^2} = 0 \quad \text{at } z = \xi \quad (8)$$

where P_r is the pressure of the gas phase and σ is the surface tension. The variable, ξ, describes the local position of the perturbed interface about its unperturbed basic-state location. The linearized tangential-stress balance for a constant surface tension fluid is

$$\frac{\partial u}{\partial z} + \frac{\partial w}{\partial x} = 0 \quad \text{at } z = \xi \quad (9)$$

Finally, the linearized kinematic surface condition is given by

$$w = \frac{\partial \xi}{\partial t} \quad \text{at } z = \xi \quad (10)$$

Note that Equations (2) and (4) contain the excess intermolecular potential energy terms, $\partial W/\partial x$ and $\partial W/\partial z$. However, only an expression for $\partial W/\partial z$ is needed, because the excess intermolecular potential energy does not appear in the combined form of Equations (2) and (4), but only in the normal stress balance in the form of the z-derivative. Therefore, the expression for the gradient of the excess intermolecular potential energy in the vertical direction, Equation (1), is all that is necessary to describe the effect of the steep concentration profiles.

In order to determine the behavior of this system in response to interfacial corrugations, Equations (2) through (10) were nondimensionalized. Nondimensionalization

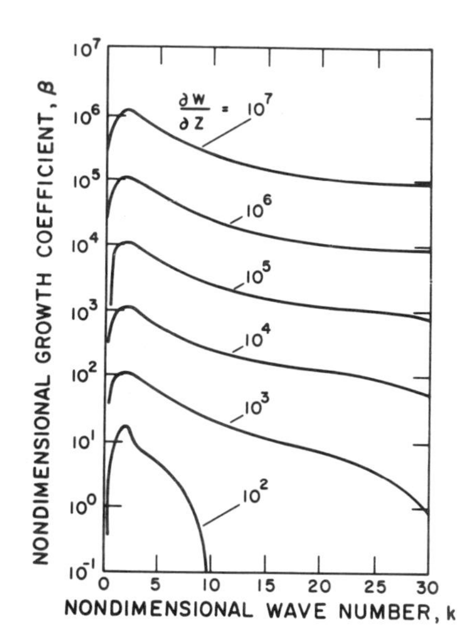

Figure 5. Nondimensional growth coefficient as a function of the nondimensional disturbance wave number for several values of the nondimensional intermolecular potential energy gradient W_z.

allows the results to be presented in a generalized form accounting for any size or type of system. Next, the resulting equations were combined into a single fourth-order partial differential equation in the perturbed vertical velocity component. Finally, a normal-mode, frozen-time analysis was carried out on the system of equations to determine the growth rate of surface perturbations of various wave numbers. The details of this analysis can be obtained from either Ray (10) or Nerad and Krantz (4).

Unlike the approximate solution of Nerad and Krantz (4), this system of equations is impossible to solve analytically due to the nonconstant coefficients resulting form the highly variable viscosity. A solution thus was obtained numerically to obtain the generalized results shown in Figure 5. Figure 5 is a plot of the nondimensional growth coefficient versus the nondimensional wave number for various magnitudes of $\partial W_0/\partial z$. The growth coefficient, wavenumber, and excess intermolecular potential energy gradient, have been nondimensionalized by $\sigma/\mu_0 L$, $1/L$, and σ/L, respectively, where L is the solution depth. One striking result is that regardless of the magnitude of $\partial W_0/\partial z$, the most highly amplified disturbance has a nondimensional wave number of k=2.11. However, as the magnitude of the driving force is increased, both the growth coefficient and the range of unstable wavenumbers also increase. This implies that a broader range of more highly developed structures might be seen for increased driving forces. As can be seen from Equation (1), some parameters that would increase the magnitude of $\partial W_0/\partial z$ are a steeper concentration profile and stronger molecular interactions. These and other effects will be discussed further in the next section.

DISCUSSION OF RESULTS

Mass Transfer Effects

According to the model developed here, anything that increases the concentration gradients (and hence, the dimensionless intermolecular potential energy gradient) near the liquid/liquid interface also increases the growth coefficient or growth rate of the spatial structures. The concentration gradient is increased by either increasing the mass transfer coefficient in the precipitation bath or by lowering the diffusivity in the polymer solution. Increasing the concentration of volatile component in the precipitation bath, on the other hand, will decrease the rate of transfer of solvent out of the cast layer, thus, decreasing the concentration gradient. The experiments of Strathmann et al. (11) confirm that indeed less prominent spatial structures result when the cast polymer/solvent film is precipitated in a water bath containing increasing amounts of the solvent.

Another way to affect the solvent mass transfer is to use a precipitation bath with varying affinities for the solvent. A precipitation bath with a higher affinity for the solvent will result in a higher mass transfer rate as well as a significant temperature rise on mixing with the solvent. Furthermore, a higher affinity corresponds to a larger A_{ik}. For example, since dimethylacetamide has a high temperature rise on mixing with water (hence a high affinity), but a relatively low temperature rise on mixing with glycerine (thus a low affinity), membranes precipitated into higher water content baths should have a more highly developed structure. Strathmann et al. (11) show several scanning electron micrographs of cross-sectional views of Nomex/dimethylacetamide membranes precipitated into various mixtures of glycerine and water. As expected, as the concentration of water (and hence, the affinity between solvent and precipitation bath) increases, the prominence and complexity of the spatial structure also increases.

Further confirming evidence for this model comes from the rate of precipitation of the cast layer. When the solvent mass transfer rate out of the cast phase is high, the rate of precipitation is high (i.e., the time to precipitation is short) since the interfacial concentration of solvent decreases very quickly. When the solvent mass transfer rate is slow, the rate of precipitation is slow (i.e., time to precipitate is long) because the bulk can adequately supply the interface with solvent. Matz (12) has calculated an ''experimental average rate of precipitation'' for membranes cast from 20 percent cellulose acetate in six different solvents precipitated into a water bath. Matz' order of increasing rate of precipitation corresponds to an increasing prominence and complexity of the spatial structure as well as correlating with the order of increasing solvent/water solubility, or higher affinity, as determined by So et al. (13).

Viscosity Effects

Along with surface tension, viscous dissipation retards the formation of structure. Hence, as the viscosity of the polymer/solvent casting solution is increased, the growth rate of the spatial structures is severely dampened. Both Cabasso et al.

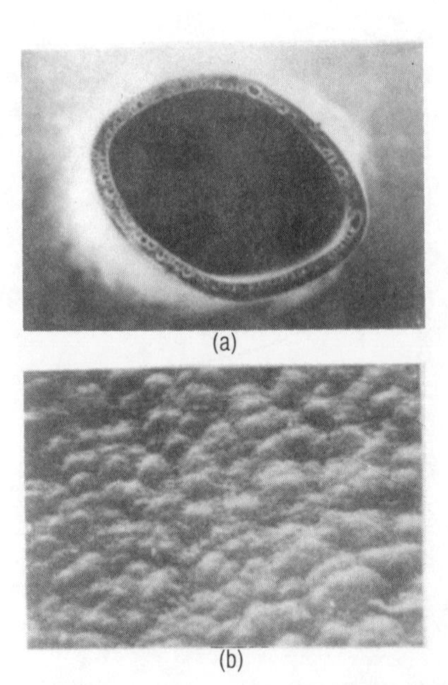

Figure 6. Scanning electron micrographs of the cross section (a) and the outside surface (b) of a Kynar/DMAc hollow fiber. Prepared from 12 percent Kynar/DMAc. Magnifications are: (a) 100X and (b) 10,000X.

(14) and Strathmann et al. (11) have demonstrated experimentally that by increasing the concentration of the polymer (thereby increasing the viscosity), they were able to virtually eliminate all prominent spatial structures. This effect of viscosity is shown dramatically in Figures 6 and 7. The upper panel in each of these figures shows a scanning electron micrograph at 100X) of a cross-section of a freeze-fractured Kynar hollow fiber membrane which was prepared from dimethylacetamide (DMAc) solution cast into a water bath. The lower panel shows a scanning electron micrograph (at 10,000X) of the outer skin of this hollow fiber membrane. Figure 6 corresponds to a 12 weight percent solution of Kynar in DMAc, whereas Figure 7 corresponds to 15 percent. The 15 percent solution is more viscous than the 12 percent solution and one sees that the finger structures in the upper panels as well as the nodular structures in the lower panel are far less prominent for the 15 percent solution. This strongly suggests that both fingers and nodules are due to the same mechanism as is implied by the model. It also suggests a method whereby

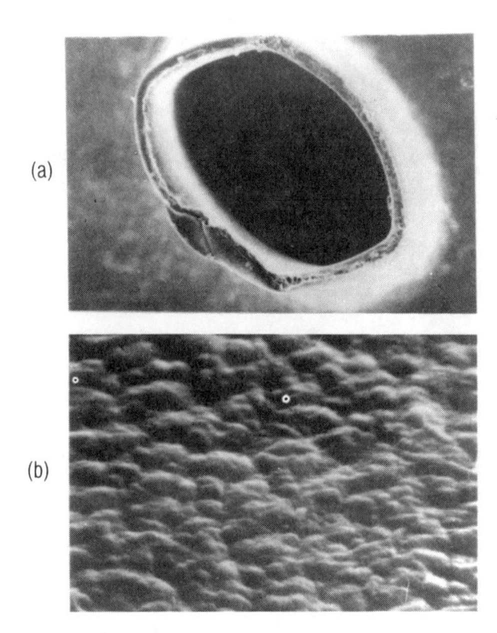

Figure 7. Scanning electron micrographs of the cross section (a) and the outside surface (b) of a Kynar/DMAc hollow fiber. Prepared from 15 percent Kynar/DMAc. Magnifications are: (a) 100X and (b) 10,000X.

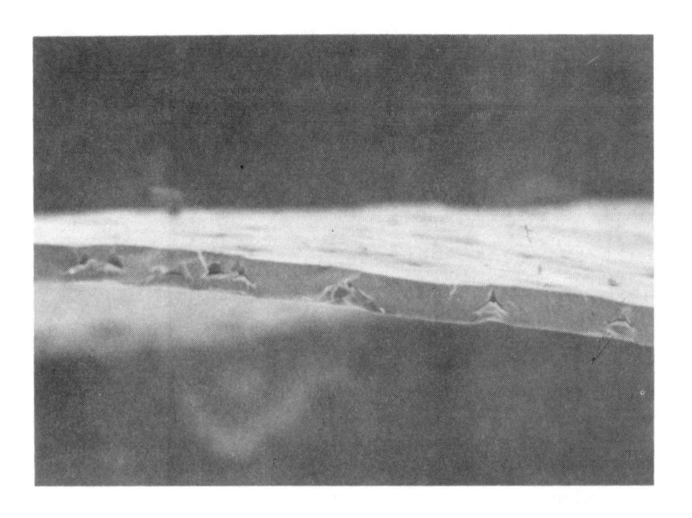

Figure 8. Wide-angle scanning electron micrograph of the cross-section of an asymmetric membrane made from 25 percent cellulose acetate (CA-389-10) in dimethylformamide, 100X.

the smaller scale less highly amplified features can be eliminated, thereby permitting one to test the theoretical prediction for the most highly amplified wave.

Recall that the model predicts that the most highly amplified wave will have a nondimensional wave number of k=2.11. This corresponds to a nondimensional wave length of λ/L=4.19. As suggested above, the viscosity of the casting solution can be increased such that only the most highly amplified wave is observed (see Figure 8). Figure 8 is a scanning electron micrograph of a freeze-fractured cross-sectional view of a membrane showing fingers corresponding to a single lateral spacing wave number. Note that the finger spacing is approximately four times the thickness of the membrane. This procedure was implemented on various polymer solvent systems. The resulting data are summarized in Ray et al. (5) and are plotted in Figure 9. Although there is considerable scatter (standard deviation of 0.59), the mean for these data, λ/L = 4.13, is remarkably close to the predicted value of λ/L = 4.19. Much of the scatter in these data result from a parallax

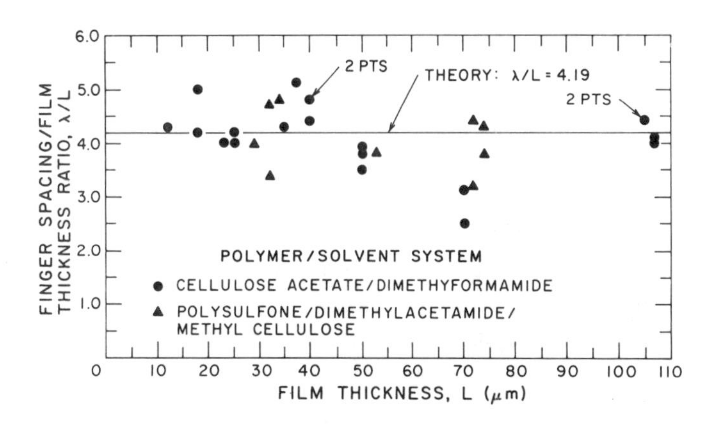

Figure 9. Ratio of finger spacing width to film thickness as a function of film thickness for cellulose acetate membranes cast from dimethylformamide (circles), and polysulfone membranes cast from dimethylacetamide (triangles).

effect associated with the manner in which these fingers are observed in a freeze fractured cross-section of the membrane; That is, not all the fingers seen in a cross-section view lie along a straight line.

Surface Tension Effects

It has been suggested that the spatial structures in asymmetric membranes might be due to the Marangoni instability, a convective instability caused by lateral gradients in the interfacial tension. However, the growth rates predicted by a Marangoni analysis for the system of interest are much too slow to account for membrane morphology which is frozen in on a time scale of a few seconds at most (Reference (10), p. 366). Nevertheless, experiments were performed to test this contention. The details of the experiments can be found in Ray (10), (page 375), and will be summarized here. It is possible to observe the growth of the ''fingers'' under an optical microscope by placing a sample of the casting solution between two microscope slides and initiating contact with the precipitation bath. A corrugated interface rapidly develops upon contact between the casting solution and the precipitation bath. The finger structures appear to grow because of precipitant penetration at the less dense interstitial regions between adjacent corrugations. The rate at which these fingers grow in length should be related to the rate at which the interfacial corrugations are established. The upper panels in Figure 10 show optical micrographs (approximately 100X) of a time sequence for the growth of fingers resulting from the contact of a cellulose acetate/dimethylformamide casting solution with a water precipitation bath. The lower panel shows a scanning electron micrograph (11,000X) of the resulting casting solution/water bath solidified interface. These microscope slides experiments were used by Ray (10) to study the effect of adding surfactants to the water bath in the finger initiation and growth process. Ray's

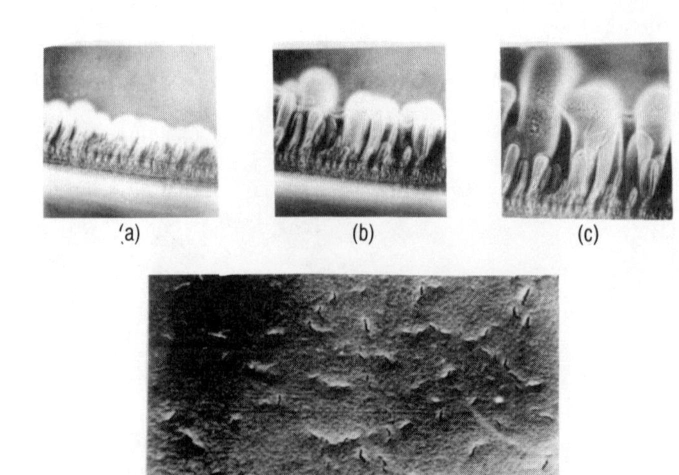

Figure 10. Observed micrographs (100X) showing a time sequence for finger growth at the cellulose acetate/dimethylformamide water precipitation bath interface at (a) 20 s (b) 45 s and (c) 75 s; scanning electron micrograph (11,000X) of the solidified interface (d).

(10) results are shown in Figure 11 for four different precipitation baths, two of which contained a small amount of Triton X-100 surfactant. This plot of finger length (in arbitrary units) versus time clearly demonstrates that surfactants enhance the initiation and growth of fingers in these casting experiments. Whereas surfactants stabilize Marangoni instabilities via surface elasticity effects appearing in the tangential stress boundary condition, they can destabilize the concentration-gradient-driven instability mechanism proposed here via a reduction of the surface tension appearing in the normal stress boundary condition. Thus, these microscope slide experiments strongly suggest that the finger structures observed in solvent-cast membranes do not arise because of the Marangoni instability.

CONCLUSIONS AND RECOMMENDATIONS

In conclusion, a model has been developed to explain the morphology of solvent-cast asymmetric polymeric membranes. The model suggests that the finger and nodular structures in these membranes are the result of an interfacial instability arising from the

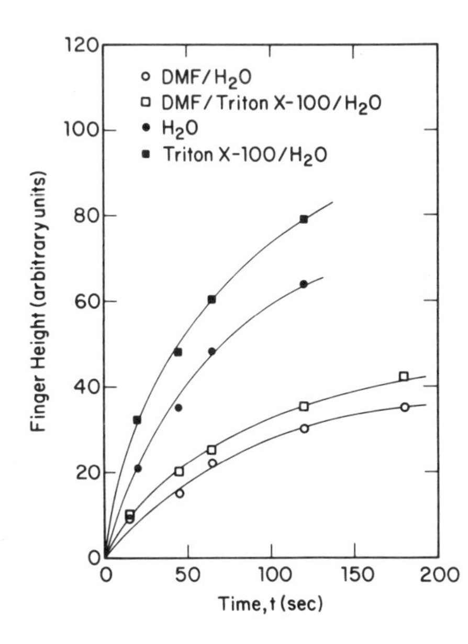

Figure 11. Finger height as a function of time for cellulose acetate/dimethylformamide solution cast in precipitation baths with and without Triton X-100 surfactant.

steep concentration gradient inherent during the precipitation step in the membrane formation process; that is, these structures are a fossil of the surface corrugations present when the polymer/solvent cast film solidified. The predictions of this model have been confirmed by experimental studies on the effects of precipitation bath composition, viscosity and surface tension on the prominence of these structures. Furthermore, it has been shown that the most highly amplified wave length predicted by this model corresponds to the largest scale most prominent fingers observed in these membranes.

The model developed here can only explain the structures present in membranes formed during a precipitation step. However, membranes processed via an air evaporation step, such as cellulose acetate, also possess similar structures. It is possible that these structures are due to an analogous steep temperature gradient-driven instability mechanism as has been suggested by Nerad and Krantz (4). This is an area of current research.

ACKNOWLEDGMENTS

The authors acknowledge the following sources of financial support of this work: the NSF Graduate Fellowship Program, the NSF Grant No. CPE 8121841, the National Bureau of Standards Cooperative Agreement No. NB84RAH45130, and the National Center for Atmospheric Research Advanced Study Program.

NOMENCLATURE

Roman

A_{ij} = Constant proportional to the strength of the intermolecular attractions

b = Slope of the density profile at the surface

k = Wave number

L = Thickness of the cast film

P = Pressure

t = Time

u = x-component of the velocity

w = z-component of the velocity

W = Excess intermolecular potential energy

x = Lateral coordinate

z = Coordinate perpendicular to the planar surface

Greek

δ_{ij} = Distance of closest approach between molecules i and j

μ = Viscosity

ξ = Coordinate of surface disturbance

ρ = Density

σ = Surface tension

Subscripts

i = Molecular index

j = Molecular index

k = Molecular index

t = Partial derivative with respect to time

x = Partial derivative with respect to x

z = Partial derivative with respect to z

0 = Basic state quantity

Superscripts

s = Value at the surface

LITERATURE CITED

1. Loeb, S., and S. Sourirajan, Adv. Chem. Ser., 38, (1962), 117.

2. Kesting, R. E., Pure and Appl. Chem., 50, (1978), 633.

3. Broens, L., F. M. Altena, and C. A. Smolders, Desalination, 32, (1980), 33.

4. Nerad, B. A., and W. B. Krantz, A.I.Ch.E. Symposium Series (this volume).

5. Ray, R. J., W. B. Krantz, and R. L. Sani, J. Mem. Sci., 23, (1985), 155.

6. Ruckenstein, E., and R. Jain, J. Chem. Soc., Faraday Trans. II, 70, (1974), 132.

7. Maldarelli, C., ''The Linear Hydrodymanic Stability of of Interfacially Perturbed Thin Films with Appllications to the Onset of Small Scale Biological Cell Membrane Motions,'' Ph.D. Thesis, Columbia University, New York, N. Y., 1981.

8. Ray, R. J., W. B. Krantz, R. L. Sani, and K. J. Gleason, accepted for publication in J. Mem. Sci.

9. Squire, H. B., Proc. Roy. Soc., A142, (1933), 621.

10. Ray, R. J., ''Interfacial Instabilities Arising from Excess Intermolecular Potential Gradients: Application to Asymmetric Membrane Morphology,'' Ph.D. Thesis, University of Colorado, Boulder, 1983.

11. Strathmann, H., K. Kock, P. Amar and R. W. Baker, Desalination, 16, (1975), 179.

12. Matz, R., Desalination, 10, (1972), 1.

13. So, M. T., F. R. Eirich, H. Strathmann, and R. W. Baker, Polymer Letters, 11, (1973), 201.

14. Cabasso, I., E. Klein and J. K. Smith, J. Appl. Polym. Sci., 21, (1977), 165.

A MECHANICAL FORMULATION OF CHEMICAL INDUCED DEFORMATIONS OF THE RED BLOOD CELL PLASMA MEMBRANE

Charles Maldarelli ■ Institute of Applied Chemical Physics, Department of Chemical Engineering
City College of New York
Zeev Dagan ■ Department of Mechanical Engineering, City College of New York

A mechanical model is detailed, and a shell theory is formulated, to describe shape changes of the plasma membrane of the red blood cell which are induced by chemical effects. The model is based on the concept that bending deformations of the lipid bilayer structure of the red cell membrane originate in differences between the (in-plane) stresses on the two halves of the bilayer. Chemical effects can cause this stress difference by either altering the interaction energy between the membrane lipids and the aqueous intra- and extra-cellular fluids, or by causing relative area expansions between the two faces of the membrane. Chemical induced relative area expansions occur via several mechanisms, as, for example, direct intercalation of large macromolecules preferentially onto one membrane face, or a chemical reaction which promotes network formation or conformational changes in membrane constituents.

A variational principle is used to derive shell equations which describe, in a unified way, all these chemo-mechanical routes. In these equations, the chemo-mechanical coupling is formulated through surface tensors which contribute to the bending moment of the membrane. It is shown how the shell equations, together with species conservation equations, can be used to determine membrane shapes.

Surrounding the periphery of the human erythrocyte (red blood cell or RBC) is a plasma membrane. Chemically the plasma membrane consists of phospholipids, cholesterol and integral and peripheral proteins and structurally it is arranged as a fluid mosaic with the lipids forming a bilayer and the cholesterol and integral proteins interspersed throughout the lipid matrix. The most important peripheral component is the protein spectrin; it is located as a fibrous network on the cytoplasmic side of the plasma membrane. (For reviews of the biochemistry and ultrastructure of the red cell plasma membrane see Chien (1).) When suspended in a quiescent medium of physiological saline, the configuration of the RBC plasma membrane is that of a biconcave disk. Although it has not been conclusively established that the discocyte form is the red cell membrane's natural unstressed configuration, mechanical modeling studies lend supporting evidence to this view (Fischer, et al. (2)). This resting shape is, however, strongly influenced by chemical effects which can give rise to bending deformations of the discocyte form. Direct evidence of such transformations are the configurational alterations induced by the so-called "shape changing compounds." These are amphiphilic molecules which physically adsorb onto the plasma membrane. Detailed and numerous studies of their effects (Deuticke (3), Ponder (4), Seemen (5) and Bessis (6)) indicate that these compounds cause one of two different types of deformations: A shape change in which the membrane becomes crenated (an echinocytotic shape change), and a deformation in which the cell becomes cupped (a stomatocytotic change).

Other evidence of chemically induced shape transformations of red cells involve the chemical reaction of membrane constituents rather than physical adsorption. Suspension of erythrocytes in a medium containing an antibody specific to a RBC membrane-bound antigen causes a change in shape from the discocyte form (van Oss and Mohn (7)). The chemical reaction involved in this example is the complexing of antigen and antibody; however, the connection between the complexing and the deformation is as yet not understood.

A second example of chemical reaction induced shape changes is the echinocytotic formations which result when RBCs are suspended in an isotonic medium unsupplemented with ATP or precursors required for the cell synthesis of ATP (Makato et al. (8), Weed et al. (9), Sheetz and Singer (10), and Birchmeier and Singer (11)). The chemo-mechanical mechanism involved in this example is also not resolved. Most studies have attributed the crenations to deviations, from steadily maintained values, of the concentrations of "shape determining" membrane constituents. The prescribed values are sustained by chemical reactions in which ATP participates, and the reduction in the ATP concentration effects the deviations by altering the regulating kinetics.

Despite the extensive documentation, and notwithstanding the significant progress which has been made in elucidating the biochemistry, ultrastructure and mechanical properties of the red cell membrane, chemo-mechanical models which can describe mathematically the chemical origin of red cell membrane morphological change (i.e. which can describe how a "shape changing" chemical species acts) have only been partially developed, and no unifying models have been suggested in the literature. Thus mechanisms for transformations induced by physical adsorption have been proposed, and separately models for reaction induced transformations has been posited. In addition, the models which have been proposed have, for the most part, not been formulated within the framework of a rapidly evolving shell theory which has been developed to describe the mechanics of the red cell. In particular, this theory has constructed area streching, shearing and bending elastic strain energy functions for the RBC membrane, and has determined numerical values for the parameters in these functions from measurements obtained in carefully controlled membrane deformations, particularly one in which the red cell is partially aspirated into a pipette. (cf. Skalak et al. (12), Evans (13, 14), and Evans et al. (15) for studies connected to the red cell membrane 's elastic response to shear extensions and area dilitation, Evans and Hochmuth (16), Chien et al. (17), Hochmuth et al. (18), and Tozeren et al. (19) for studies of viscoelastic effects and Evans (20) for measurements of elastic bending moduli. Many of these results are summarized in the monograph by Evans and Skalak (21).)

The aims of this article are twofold. The first is to critically review the literature of chemo-mechanical mechanisms, and to construct, from this survey, a model for chemically induced deformations of red cells which draws upon and unifies concepts developed by several research groups. The second is to develop the mechanics of the model within the already developed framework of red cell membrane mechanics. The result will be a well posed set of shell equations which can be used as the starting point for the description and understanding of chemically induced morpological changes in RBCs.

CHEMO-MECHANICAL MODELS OF CHEMICALLY INDUCED SHAPE CHANGES OF PLASMA MEMBRANES

The earliest efforts at chemo-mechanical model development were undertaken by A. Sanfeld and his research group, and were directed at understanding how chemical reactions can, in general, cause plasma membrane deformations (Steinchen and Sanfeld (22), Sanfeld and Steinchen (23), Sorensen et al. (24, 25), Dalle Vedove and Sanfeld (26, 27); see also the work on dissipative surface structures caused by a chemical reaction by Ibanez and Velarde (28) and Joos and van Bockstaele (29)). In the Sanfeld studies, the plasma membrane is modeled as a Newtonian interface under an isotropic tension σ. Two bulk layers representing the intra- and extra-cellular fluids adjoin the interface. On the interface are assumed to be reactable, bulk insoluble chemical species which, upon compression, exert a surface pressure (Π) which lowers the membrane tension ($\sigma = \sigma_0 - \Pi$). Inasmuch as the surface pressure is a function of the surface concentrations of the reactable species, spatial inhomogeneities in these concentrations (established by the chemical reaction network) give rise to interfacial tension gradients. The gradients exert tangential tractions on the fluid sublayers adjoining the interface causing fluid motion in these layers. This chemo-mechanical mechanism is termed the Gibbs-Marangoni effect, and it is schematically described in Fig. 1 (see below).

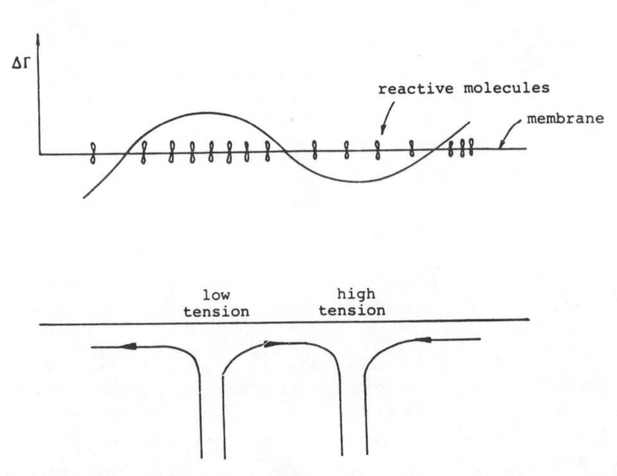

Figure 1. *The Sanfeld Model:* A chemical reaction scheme creates a surface concentration pattern ($\Delta\Gamma$ is the deviation of the surface concentration from a value for which the rate of reaction is zero). The pattern creates tangential surface tractions which give rise to convection cells. Note the essence of the model is the generation of fluid streaming, not bending deformations.

(It should also be mentioned that one direction of the Sanfeld studies focused on a surface tension which was an increasing function of the reactable species, the intention being to model the action of contractile microfilaments in cytokinetic processes. (Cf. Sorensen and Castillo (30) and Sorensen (31) and some of the related work of Greenspan (32 and 33). Contractile elements are usually located in a viscous cytoplasmic layer adjacent to the endoface. However, no evidence exists which indicates that the red cell membrane has such structures, and any processes which resemble microfilament action (especially spectrin conformational changes or network formations as discussed below in connection with Singer's explanations of ATP depletion echinocytosis) can better be understood within the context of bilayer moments as described below.)

A key concept in the Sanfeld model which is in agreement with the other models to be discussed is the idea that the tension within the membrane is a function of the number density of chemical species, and that increasing this density reduces the tension by creating compression forces due to molecular crowding. The model is, however, not a model for membrane bending deformations. As is evident from the illustration, the interfacial tractions entrain flows in the surrounding fluids. Interfacial deflections play a minor role.

The tension gradients of the Sanfeld model cannot induce bending because the membrane structure was modeled as a Newtonian interface with no capacity for developing bending moments. Later work by Evans (34) and Sheetz and Singer (35, 36) examined more carefully plasma membrane ultrastructures, and showed how the chemically induced compressions can yield bending moments.

Evans ((34); see also Evans and Waugh (37)) examined the chemo-mechanics of a planar, amphiphilic bilayer vesicle exposed to an aqueous environment. In his study, stress resultants in each of the two halves of the bilayer are distinguished. In each half of the structure, the stress is assumed to arise from two effects: a tension arising from the interaction energy between the water molecules (free to exchange with the membrane surface) and the amphiphilic molecules and (ii) compression forces derived from molecular crowding (intermolecular repulsion) among the amphiphilic molecules which do not exchange with the environment. In the unstressed planar state, these stresses balance one another on either face of the membrane. In addition, the bilayer halves are assumed to be attached, and hence extensions in the transverse direction of the lamella are disallowed since these would involve energetically unfavorable elongations of the amphiphilic molecules. In a bilayer whose mechanics is so formulated, bending moments arise when different stress resultants develop between the two halves. This difference cannot be accomadated soley by extension of one area with respect to the other since sliding is not permitted, and so a bending couple develops in order to simultaneously satisfy the stress difference and the transverse constraint.

Evans hypothesized that a chemically induced bending moment arises when the water-amphiphilic tension on one side of the bilayer is altered. This can occur when, for example, molecules binding to one face of the membrane lowers the interaction energy between amphiphiles and the aqueous environment. Although Evans only considered chemically induced stresses derived from changes in the interaction energy between the aqueous phase and the membrane amphiphiles, his results can be extended when considering more involved bilayer structures. In the case of the red cell membrane, molecules other than the amphiphiles (for example, the shape changing compounds or proteins) can exert their own surface pressure. The surface pressure is a function of the surface concentration of the molecules; variations in the surface pressure on one face of the membrane owing, for example, to adsorption or reaction of the pressure exerting species would then give rise to a bending moment.

Evans suggested that his formulation is suitable to the description of red cell deformations excepting that, in the case of the red cell membrane, local sliding between the lipid layers is possible. This conclusion is based on the X-ray diffraction data of Mateu *et al*. (38) and Tardieu *et al*. (39) which indicates that the nonpolar lipid tails are not connected. However, Chien (1) notes that in the red cell membrane exists the glycoprotein glycophorin which runs perpendicular to the plane of the membrane and protrudes out of each face. This glycoprotein may also be joined to the spectrin cytoskeleton. In any case, the presence of a protein spanning the entire cross section will certainly stiffen the membrane to transverse extensions, and may, as well, reduce sliding between the layers because it fortifies the membrane matrix. Note also that in the RBC membrane sliding of bilayers will in addition be limited by the fact that the geometry of the membrane is closed.

While the suggestion that bilayer stress differences are the underlying mechanism for RBC membrane bending seems likely, Evans' identification of the change in the aqueous-amphiphathic energy (or surface pressure changes) as the primary cause of these differences appears too restrictive in light of other studies which indicate that more dominant effects may be at the root of the imbalance. Sheetz and Singer's (35 and 36) experimental results on a series of shape changing compounds led them to conclude that the action of these compounds is to intercalate themselves preferentially onto one of the two sides of the bilayer. Compounds which intercalate themselves onto the extracellular half of the bilayer (the exoface) increase the area of this half, and thereby create compressive stresses on the exoface. A bending moment is then created since the intracellular (endoface) half of the membrane is unaffected. This moment causes the membrane to crenate in order to accomadate the increase in exoface area. Alternatively, compounds binding to the endoface, create couples which force the cell to cup in order to accommadate the swelling of the intracellular face area. The intercalation explanation is illustrated in Fig. 2 (below).

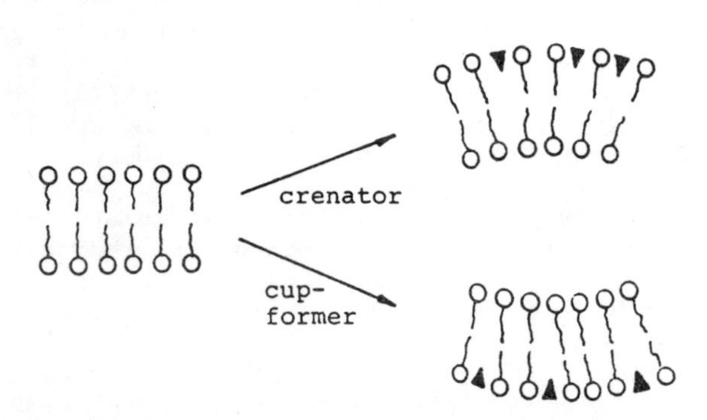

Figure 2. *The Bilayer Couple Hypothesis:* Compounds which intercalate themselves onto the exoface cause the cell to crenate in order to accommodate the increase in area while molecules binding to the endoface cause the cell to invaginate.

The Sheetz and Singer concept identifies relative area expansions by chemical species intercalation as the origin of the chemical moment. The explanation has been used successfully in understanding the action of many shape changing compounds. The concept can be generalized since chemical species intercalation need not be the only mechanism for creating relative area changes between the endoface and exoface. Some examples of chemical reaction phenomena which may give rise to area differences are described below and schematically detailed in Fig. 3.

(i) A transconformational reaction of a membrane integral protein which is located on only one half of the bilayer (Fig. 3a).

(ii) Complexing of antigens bound to the RBC exoface by multifunctional antibodies. In this example, area contraction occurs if the antigen-antibody complex does not adsorb onto the membrane surface, but the distance between antigens in the complex is less then the distance in the uncomplexed state (Fig. 3b; note that for this case the antigen must be bound to the membrane lipid matrix)). Expansion occurs if the complex itself is large and intercalates itself directly into the membrane with a significant area change.

(iii) ATP-mediated chemical reaction network formations in the spectrin complex (Fig. 3.c). Singer's group have used this example to explain ATP depletion induced echinocytosis (Sheetz and Singer (10) and Birchmeier and Singer (11); see also Pinder et al. (40)). In their view, phosphorylation of one subunit of the spectrin protein promotes polymerization or network formation. The subunit is phosporylated by ATP through the competing effects of endogeneous kinase and phospatase activity. They hypothesize that dephosporylation has the reverse effect. In the case of depletion induced echinocytosis, the reduction in the extent of phosphorylation of the spectrin subunit decreases the area of the endoface and causes the cell to crenate. The matter, however, is far from settled and related studies have offered either different biochemical routes by which spectrin conformational changes may cause the deformations or identified other ATP-mediated interconversion cycles of shape controlling bilayer components as the cause of the change (see, for example, Patel and Fairbanks (41)).

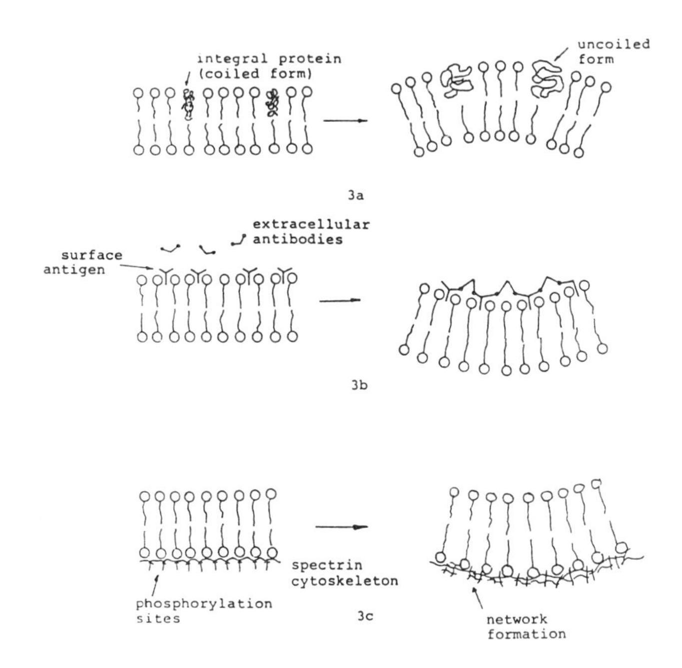

Figure 3. *Bending Induced by Chemical Reaction:* (a) A transconformational reaction of a globular integral protein expands the exoface. (b) Antigen-antibody complexing induces invagination. (c) Phosphorylation of spectrin promotes network formation and cupping by expansion of endoface.

The mathematics of mechanical deformations induced by area expansions was considered by Landman and Greenspan (42) in a study concerned with shape changes induced by adsorption of molecules onto a membrane surface. This study modeled the cell as a spherical fluid droplet (of fixed volume) encapsulated by an interface under isotropic tension, and studied the stability of a base state of uniform adsorption. The adsorbed species is assumed to exert a surface pressure; in the base state adsorption drives increasing surface concentrations and decreasing interfacial tensions. A spherical cell configuration is maintained since the adsorption increases only the surface concentration with no concomitant change in surface area. The results of a linear stability analysis on the base state indicate that when this state is perturbed, crenation of the spherical form occur immediately after the disturbance because the surface behaves incompressibly during this time period, and therefore additional deposition of drug increases the surface area. The shape change is simply due to the fact that the area is being increased with the volume held constant. (Recall that the sphere represents the smallest surface area to volume ratio.). Landman showed that the ruffles eventually disappear as the surface concentration increases to accommodate the deposited material. The reason the surface deformations ultimately disappear is that the surface was compressible, and the adsorbing species did not intercalate to effect a permanent increase in area. Note also that because the analysis does not account for the bilayer structure of the membrane, it cannot distinguish between the differing actions of crenators and cup-formers; in the Landman study all adsorbing species yield the same type of surface deformations.

From the above review of chemo-mechanical models, it is clear that the bending moment induced by differing tensions between the two halves of the lipid bilayer is a feasible mechanism for describing chemically induced deformations. The more difficult question is whether the tension difference arises from surface pressure changes or amphiphilic-aqueous tension changes as suggested by Evans, or direct exoface or endoface area changes as posited by Sheetz and Singer. Probably both are viable alternatives, and the determination as to which is relevant is dependent upon the particular chemical event. Most likely, if the event involves molecular adsorption of small molecules (as, for example, Ca^{++} binding to the endoface of the red cell) then intercalation area changes are not as significant and a tension reduction owing to a change in the amphiphilic-aqueous energy is the dominant effect. Alternatively, when adsorption of macromolecules is involved, or chemical reactions which alter large membrane constituents (e.g. integral proteins) or the cytoskeletal structure are occurring, direct area expansion is the proper framework.

Evans constructed a simplified and restricted set of shell equations to describe curvature changes induced by energy variations on one face of the membrane, and Sheetz and Singer detailed no mechanical analysis of their direct area expansion concept. Detailed below is a complete mechanical formulation of the chemical induced bending moment which is derived in a sufficiently general way so as to incorporate both ideas.

MECHANICAL ANALYSIS

The starting point for the formulation of the mechanics of chemical induced deformations is the identification of the principal components of the membrane fluid mosaic - i.e. the lipid and cholesterol molecules, and the spectrin cytoskeleton - as the material particles of the membrane continuum. To describe the relative displacements of the material particles the Kirchhoff hypothesis is utilized. In the framework of this hypothesis, stretching, shearing and bending of the membrane is described through the deformation of the midplane (regarded as a material surface), and it is assumed that (i) material planes initially aligned parallel to the midsurface remain, upon deformation, parallel to that surface and at the same (i.e. initial) distance from it and (ii) transverse material lines initially normal to a family of material planes which are parallel to the midsurface remain normal to these planes (the deformation tensors of the shell under this hypothesis are calculated explicitly in Naghdi and Nordgren (43) and the hypothesis is discussed in detail in the review article of Naghdi (44)). Formulation of the membrane deformation through the Kirchhoff hypothesis is restrictive in that it does not allow sliding of the bilayer chains in the direction perpendicular to the membrane nor sliding of the two bilayers in the plane of the membrane. In unclosed bilayer vesicles without any constraining structures (e.g. protein networks), the hypothesis may not be as useful for describing the lamella mechanics, and other liquid crystal formulations are under scrutiny (see, for example, the recent article by Ljunggren and Eriksson (45)).Nevertheless, as mentioned in the discussion above, for the case of the red cell membrane, the geometry is closed and this fact together with the rigidity imposed by the the glycophorin and spectrin structures limits sliding motions. Thus the Kirchhoff hypothesis provides a good starting point.

Equations of motion for the membrane consistent with the Kirchhoff hypothesis and which include a chemo-mechanical coupling can be derived by formulating a variational principle: Consider the isothermal, equilibrium states of a segment of the membrane. Under isothermal conditions, the change in the Helmholtz free energy of the segment between equilibrium states is equal to the sum of the work done by tractions exerted on the segment periphery and the amount of energy entering the segment by mass exchange at the boundary.

The Helmholtz free energy of the membrane segment can be conceptually divided into two terms. The first is the strain energy of the continuum, and represents the resistance of the membrane to stretching, shearing and bending deformations. The second are the interfacial energies of the membrane faces; these account for the interaction energies between the water molecules and the amphiphiles on either face of the membrane which are exposed to the aqueous phase. To introduce chemo-mechanics, it is assumed that on one face of the membrane (the upper face for illustration) a chemical species A is present. This species alters the interaction energy between the membrane amphiphiles and the aqueous phase, as suggested by Evans. In addition, it is assumed that A is area controlling in the manner suggested by Sheetz and Singer, i.e., A intercalates itself within the upper surface.

An expression for the change in the strain energy when the segment is deformed can be derived as follows. Locate a convected spatial coordinate system (ξ^i: ξ^1,ξ^2,ξ^3) such that the surfaces of constant ξ^3 are the material planes parallel to the midsurface (ξ^3=0 denotes this surface), and locate a surface convected system for the midplane (ν^Δ: ν^1, ν^2) by the assignments $\xi^1=\nu^1$, $\xi^2=\nu^2$. (In the equations which follow, Latin indices are understood to vary from 1 to 3 and Greek indices from 1 to 2; the summation convention is employed, and partial differentiation is denoted by a comma.) As a consequence of the Kirchhoff hypothesis, the location of material particles of the membrane at any time t during the variation may be expressed by the equation:

$$\underline{R}(\xi^i,t) = \underline{r}(\nu^1(=\xi^1),\ \nu^2(=\xi^2),\ t) + \xi^3\underline{n}(\nu^1(=\xi^1),\ \nu^2(=\xi^2),\ t) \quad (1)$$

where $\underline{R}(\xi^i,t)$ and $\underline{r}(\nu^1(=\xi^1),\ \nu^2(=\xi^2),\ t)$ denote position vectors to, respectively, a material point in space and on the midsurface and $\underline{n}$ denotes the local (unit) normal to the midsurface pointing in the direction of increasing ξ^3. Material strains within the membrane can be expressed from Eq. (1) in terms of the metric ($a_{\Delta\Omega}$) and curvature ($b_{\Delta\Omega}$) surface tensors of the midplane; thus $\epsilon_{\Delta 3} = \epsilon_{33} = 0$, and

$$\epsilon_{\Delta\Omega} = {}_{(o)}e_{\Delta\Omega} + {}_{(1)}e_{\Delta\Omega}\xi^3 + {}_{(2)}e_{\Delta\Omega}(\xi^3)^2 \quad (2)$$

where ϵ_{ij} denotes the components of the spatial strain tensor (in the convected system) and the surface tensors $_{(o)}e_{\Delta\Omega}$, $_{(1)}e_{\Delta\Omega}$ and $_{(2)}e_{\Delta\Omega}$ are measures of the membrane strain and are defined below.

$$2_{(o)}e_{\Delta\Omega} = a_{\Delta\Omega}(\nu^{\Delta}, t) - a_{\Delta\Omega}(\nu^{\Delta}, t_o)$$

$$_{(1)}e_{\Delta\Omega} = -(b_{\Delta\Omega}(\nu^{\Delta}, t) - b_{\Delta\Omega}(\nu^{\Delta}, t_o))$$

$$2_{(2)}e_{\Delta\Omega} = b_{\Delta\Sigma}(\nu^{\Delta}, t)b^{\Sigma}_{\Omega}(\nu^{\Delta}, t) - b_{\Delta\Sigma}(\nu^{\Delta}, t_o)b^{\Sigma}_{\Omega}(\nu^{\Delta}, t_o) \quad (3)$$

In the above definitions, t_o denotes the time in which the membrane is in its reference configuration.

The strain energy of the segment can be represented in terms of a strain energy density function Σ (per unit initial area of the midsurface) which is dependent upon the local values of the strain measures $_{(o)}e_{\Delta\Omega}$, $_{(1)}e_{\Delta\Omega}$ and $_{(2)}e_{\Delta\Omega}$ and represents the elastic resistance of the membrane to stretching, shearing and bending. Expressions for this dependence have been developed, and are discussed in the monograph of Evans and Skalak (21). Consider a displacement $\delta\underline{R}(= \frac{\partial\underline{R}}{\partial t}\delta t)$ of material particles of the segment; by requiring that this deformation be consistent with the Kirchhoff hypothesis $\delta\underline{R}$ may be expressed, from Eq.(1), in terms of $\delta\underline{r}$ (the displacement of material particles of the midsurface) and $\delta\underline{n}$, the material derivative of the unit normal. Since $\delta\underline{n}$ can be evaluated as a function of $\delta\underline{r}$, the membrane deformation is completely specified by the latter vector. The variation in the strain energy of the segment is given by the following relation:

$$\delta(\iint_s \Sigma\sqrt{(a(t_o)/a(t))}ds) = \iint_s (\tau^{\alpha\beta}\delta_{(o)}e_{\alpha\beta} + \gamma^{\alpha\beta}\delta_{(1)}e_{\alpha\beta}\gamma^{\alpha\beta} + \Omega^{\alpha\beta}\delta_{(2)}e_{\alpha\beta})ds \quad (5)$$

In the above, a denotes the determinant of the metric tensor of the deformed midsurface. Further, s indicates that the integration is over the midsurface area of the segment ($ds = \sqrt{a}d\nu^1 d\nu^2$) and $\tau^{\alpha\beta}$, $\gamma^{\alpha\beta}$ and $\Omega^{\alpha\beta}$ are defined below.

$$\tau^{\alpha\beta} = \frac{\partial\Sigma}{\partial_{(o)}e_{\alpha\beta}}(a(t_o)/a)^{1/2} \quad (6)$$

$$\gamma^{\alpha\beta} = \frac{\partial\Sigma}{\partial_{(1)}e_{\alpha\beta}}(a(t_o)/a)^{1/2} \quad (7)$$

$$\Omega^{\alpha\beta} = \frac{\partial\Sigma}{\partial_{(2)}e_{\alpha\beta}}(a(t_o)/a)^{1/2} \quad (8)$$

Note that the variations $\delta_{(o)}e_{\alpha\beta}$, $\delta_{(1)}e_{\alpha\beta}$, and $\delta_{(2)}e_{\alpha\beta}$ can be expressed in terms of $\delta\underline{r}$.

The free energy of interaction (per unit deformed area) of the upper surface amphiphiles with the aqueous environment can be represented by a free energy function $\tilde{f}$ (defined per unit deformed area of the membrane face); to incorporate the change in this energy with the amount of A present on the surface, $\tilde{f}$ can be regarded as a function of the surface concentration of A. (Note also that Evans and Skalak (21) remark that $\tilde{f}$ will be a function of the area strain of the exoface since as this surface is stretched more of the hydrophobic core is exposed to interaction with water molecules. This effect can easily be incorporated into the expression for the strain energy of the membrane, and so is not considered separately here.) Thus the variation in the aqueous - amphiphile interaction energy is simply given by:

$$\delta(\iint_{\tilde{s}} \tilde{f}\, d\tilde{s}) = \iint_{\tilde{s}}(\mu_A\delta\Gamma_A + \delta\tilde{a}/2\tilde{a})d\tilde{s} \quad (9)$$

where $\tilde{s}$ denotes integration about the deformed area of the upper surface of the membrane segment, and μ_A denotes the surface chemical potential of species A and is defined below.

$$\mu_A = (\frac{\partial\tilde{f}}{\partial\Gamma_A}) \quad (10)$$

A similar formulation can be undertaken for the opposite face with the dependence on Γ_A excluded. Thus:

$$\delta(\iint_{s^*} f^*\, ds^*) = \iint_{s^*}\delta a^*/2a^* ds^* \quad (11)$$

where the "*" denotes the lower face.

Under isothermal conditions, the differential change in the Helmholtz free energy between equilibrium states of the membrane segment is equal to the sum of the work done on the segment at its edge (δW) and the flux of energy into the segment owing to mass exchange (δU_M). Formulations of these contributions are given below (cf. Naghdi and Nordgreen (43) for an explanation of the work term δW).

$$\delta W = \int_{\ell} \nu_\alpha (q^\alpha \delta w + T^{\alpha\gamma} a_{\gamma\nu} \delta u^\nu - M^{\alpha\gamma}(\delta w)_{,\gamma} - M^{\alpha\gamma} b_{\nu\gamma} \delta u^\nu) d\ell \qquad (12)$$

$$\delta U_M = \int_{\tilde{\ell}} (\lambda\ \delta n) d\tilde{\ell} \qquad (13)$$

In the above two equations, q^α, $T^{\alpha\nu}$, and $M^{\alpha\gamma}$ denote, respectively, the transverse stress resultant, the (in plane) membrane stress tensor and the bending moment tensor. The variables ℓ and $\tilde{\ell}$ denote, respectively, line integration on the curves bounding the midplane and face of the segment, and ν_α denotes the local outward normal to midplane bounding curve (ℓ) in the plane of the midsurface. Finally, δn denotes the differential molar flux (per unit length) of A, δw and δu^ν are the normal and tangential components of the displacement vector ($\delta \underline{r}$) of the midsurface particles and λ is an as yet undetermined coefficient.

The variations in Equations (9), (11), (12) and (13) are constrained by the requirements of species A conservation and, more significantly, the relation between the variation in the surface concentration of A (Γ_A, per unit deformed area of the upper face) and the local area strain of the upper membrane face. The constraint of conservation of species A is straightforwardly formulated as:

$$\delta\left(\iint_{\tilde{s}} \Gamma_A d\tilde{s}\right) = \int_{\tilde{\ell}} \delta n\ d\tilde{\ell} \qquad (14)$$

To formulate the area constraint, note that the change in the number of molecules of A in the control segment multiplied by the area the upper face is expanded per molecule of A intercalated (this area is denoted by Θ) is equal to the area change of the segment's upper face. Thus:

$$\iint_{\tilde{s}} (\sqrt{\tilde{a}} - \sqrt{\tilde{a}}(t_o)) d\nu^1 d\nu^2 = \Theta \iint_{\tilde{s}} (\Gamma_A \sqrt{\tilde{a}} - \Gamma_A(t_o)\sqrt{\tilde{a}}(t_o)) d\nu^1 d\nu^2 \qquad (15)$$

From relation (15), the variation in $\tilde{a}$ with respect to $\delta\Gamma_A$ can be computed

$$\delta\sqrt{\tilde{a}} = (\Theta\sqrt{\tilde{a}}/(1-\Theta\Gamma_A))\delta\Gamma_A \qquad (16)$$

Upon adjoining the constraints to the variational equation

$$\delta\left(\iint_{\tilde{s}} \Sigma\sqrt{(a(t_o)/a(t))} ds\right) + \delta\left(\iint_{\tilde{s}} \tilde{f} d\tilde{s}\right) + \delta\left(\iint_{s^*} f^* ds^*\right) = \delta U_M + \delta W \qquad (17)$$

via Lagrange multipliers and rewriting all deformation variations in terms of δu^ν and δw, the following field equations and relations for the stress variables are obtained:

$$T^{\alpha\beta} b_{\alpha\beta} + M^{\beta\alpha}{}_{|\beta\alpha} = 0 \qquad (18)$$

$$T^{\alpha\beta}{}_{|\alpha} - M^{\nu\alpha}{}_{|\nu} b^\beta_\alpha = 0 \qquad (19)$$

$$T^{\alpha\beta} = \tau^{\alpha\beta} + \theta_1^{\alpha\beta} + \theta_2^{\alpha\beta} - b^\beta_\nu(\gamma^{\nu\alpha} + h\theta_1^{\nu\alpha} - h\theta_2^{\nu\alpha}) \qquad (20)$$

$$M^{\alpha\beta} = \gamma^{\alpha\beta} + h\theta_1^{\alpha\beta} - h\theta_2^{\alpha\beta} - b^\beta_\nu(\Omega^{\nu\alpha} + h^2\theta_1^{\nu\alpha} + h^2\theta_2^{\nu\alpha}) \qquad (21)$$

$$\theta_1^{\alpha\beta} = \tilde{\phi}(\sqrt{(a/\tilde{a})})(a^{\alpha\beta} - 2h\epsilon^{\beta\sigma}\epsilon^{\alpha\nu} b_{\nu\sigma} + h^2\epsilon^{\alpha\nu}\epsilon^{\beta\sigma} b_{\nu\phi} b^\phi_\sigma) \qquad (22)$$

$$\theta_2^{\alpha\beta} = \phi^*(\sqrt{(a/a^*)})(a^{\alpha\beta} + 2h\epsilon^{\beta\sigma}\epsilon^{\alpha\nu} b_{\nu\sigma} + h^2\epsilon^{\alpha\nu}\epsilon^{\beta\sigma} b_{\nu\phi} b^\phi_\sigma)$$

where

$$\tilde{\phi} = \tilde{\sigma} - \beta/(1-\Theta\Gamma_A);\ \phi^* = f^* \qquad (23)$$

and

$$\tilde{\sigma} = \tilde{f} - \mu_A\Gamma_A;\ \sigma^* = f^* \qquad (24)$$

Two final results are that $\lambda = \mu_A$, and μ_A is uniform along the surface. In the above set of equations, β is the force resultant arising from the area constraint, $\epsilon^{\alpha\nu}$ denotes the permutation tensor of the midsurface, h is equal to one-half the film thickness and the slash indicates covariant differentiation with respect to the midsurface metric tensor. Equations (18)-(24) comprise the complete set of shell equations; the characteristics of these equations are discussed below.

CHARACTERISTICS OF THE DERIVED EQUATIONS

The terms in the set of shell equations which account for the chemo-mechanical coupling are the surface tensors $\theta_1^{\alpha\beta}$ and $\theta_2^{\alpha\beta}$; these contribute both to the in-plane stress tensor ($T^{\alpha\beta}$) and the bending moment tensor ($M^{\alpha\beta}$) (note Equations (20) and (21)). In the absence of any chemical effects, and neglecting the energy of interaction between the membrane amphiphiles and the water molecules, $\tilde{f}$, f^*, $\theta_1^{\alpha\beta}$ and $\theta_2^{\alpha\beta}$ are all equal to zero. The remaining equations are simply the mechanical equilibrium equations for a shell (Naghdi (44); Eq.(18) is the normal stress balance and Eq.(19) the tangential balance), and the expressions for the in-plane stress tensor and bending moment in terms of a strain energy function (Eqs.(20) and (21), respectively). Thus the derived equations reduce correctly in the absence of chemical effects.

The chemical surface tensors θ_1 and θ_2 consist of two terms. The first, the $\tilde{\sigma}$ (or σ^*) term, accounts for stresses derived from the interaction energy between the membrane amphiphiles and the water molecules. The second (the β term) incorporates the force resultants arising from chemically induced area changes. The variable β is undefined, and is computed in the solution to the mechanical equations and area constraint (Eq. (15)) in the same manner as the pressure is calculated in the solution of an incompressible flow problem. Note that in the expression for θ_2 this latter term is missing since the area controlling species (A) is only located on one face of the membrane.

Consider first the case in which A is not area controlling, i.e. $\beta = 0$. To illustrate simply how the chemical tensors cause membrane bending, assume the membrane is very thin so that $\theta_1^{\alpha\beta} \approx \tilde{\sigma}(\sqrt{(a/\tilde{a})})a^{\alpha\beta}$ and $\theta_2^{\alpha\beta} \approx \sigma^*(\sqrt{(a/a^*)})a^{\alpha\beta}$ (i.e. the h and h^2 terms in relation (22) are negligible). Surface gradients in Γ_A induce gradients in $\theta_1^{\alpha\beta}$ and finally variations in $M^{\alpha\beta}$. These variations give rise to a bending couple since $M^{\alpha\beta}{}_{|\alpha}$ is nonzero.

When the chemical species is area controlling, it is clear that since β is undefined the variable $\tilde{\phi}$ is also undefined and thus $\tilde{\sigma}$ in the definition of $\tilde{\phi}$ does not have to be specified. Thus, importantly, when a species is area controlling the action of modulating the membrane area dominates any bending couples derived from the amphiphile-water interaction energy.

Although the equations which have been developed describe equilibrium states of membrane deformation, they can be applied as well to dynamic deformations involving dissipative processes such as chemical reaction or adsorption. In the dynamic case, if membrane inertia and viscoelastic effects are neglected, Equations (18) and (19) have only to be supplemented by terms representing normal and tangential stresses exerted by the surrounding fluids on the membrane. In addition, the distribution of the chemical species on the membrane face (Γ_A) is computed, in the dynamic case, by the solution of a surface reaction-diffusion equation (this replaces the result that the chemical potential of A is uniform along the surface). The reaction diffusion equation is of the form:

$$\frac{D}{Dt}\Gamma_A + (\Gamma_A/(2\tilde{a}))\frac{D}{Dt}\tilde{a} = -(J_A{}^{\alpha}{}_{|\alpha}) + R_A + H_A \qquad (25)$$

where $\frac{D}{Dt}$ denotes the substantial derivative, $J_A{}^{\alpha}$ denotes the diffusive flux of A, and H_A and R_A are, respectively, the rates of reaction and adsorption of A. (In the above, the covariant derivative is with respect to the upper face metric tensor; it can, however, be rewritten in terms of the midsurface metric.) If species A is area controlling, then in addition to (25) the area constraint relation (15) is necessary; the dynamic form of this relation is:

$$\frac{D}{Dt}(\sqrt{\tilde{a}}) = \Theta(\frac{D}{Dt}(\Gamma_A\sqrt{\tilde{a}})) \qquad (26)$$

Note that if A is area controlling, its diffusive flux ($J_A{}^{\alpha}$) is most likely negligible since the molecular size is large and A may, in fact, be attached to the membrane matrix. Combining (25) and (26) and neglecting $J_A{}^{\alpha}$ yields the following relationship between reaction and adsorption of A and area expansion:

$$(1/\Theta)(\frac{D}{Dt}(\sqrt{\tilde{a}})) = R_A + H_A \qquad (27)$$

It should be noted that in the case of chemical reaction ($H_A = 0$), the rate of reaction of the area controlling species may be controlled by smaller diffusing molecules (e.g. ATP). For this case, species conservation equations would then have to be formulated for these diffusing species.

In any case, the shell equations which have been formulated, together with reaction-diffusion equations and area controlling relations, represent a rational starting point for the computation of stationary configurations and dynamic motions of the red cell membrane. They also provide a framework for investigating how the nature of a reaction network is expressed by the morphological shape of the cell (cf. Eq.(27)). This latter type of information is essential in relating pathological cell shapes to alterations in reaction networks.

ACKNOWLEDGEMENT

This work was supported in part by a grant from the National Science Foundation (Thermodynamics and Transport Phenomena Program; CBT-8420098) and a grant from the Department of Energy (DE-AC02-80ER10599).

LITERATURE CITED

1. Chien, C.S., "Red cell membrane and hemolysis," in Cardiovascular Flow Dynamics and Measurements, Hwang, N. and N. Normann, (Eds.), University Park Press, Baltimore (1977).

2. Fischer, T., C. Haest, M. Stohr-Liesen, H. Schmid-Schonbein and R. Skalak, Biophysical Journal, 34, 409 (1981).

3. Deuticke, B., Biochim. Biophys. Acta., 163, 494 (1968).

4. Ponder, E., Hemolysis and Related Phenomena, Grune and Stratton, New York (1971).

5. Seeman, P., Pharmacol. Rev., 24, 583 (1972).

6. Bessis, M., "Red cell shape," in Red Cell Shape, Physiology, Pathology, and Ultrastructure, Bessis M., R.I. Weed and P.L. Leblond, (Eds.), Springer Verlag, New York (1973).

7. Van Oss, C.J. and J.F. Mohn, Vox. Sang., 19, 432 (1970).

8. Makato, N., T. Nakao, and S. Yamazoe, Nature, 187, 945 (1960).

9. Weed, R.I., P.L. LaCelle and E.W. Merrill, J. Clin. Invest., 48, 795 (1969).

10. Sheetz, M.P. and S.J. Singer, J. Cell Biology, 73, 638 (1977).

11. Birchmeier, W. and S.J. Singer, J. Cell Biology, 73, 647 (1977).

12. Skalak, R., A. Tozeren, R.P. Zarda, and S. Chien, Biophysical J., 13, 245 (1973).

13. Evans, E, Biophysical J,. 13, 926 (1973).

14. Evans, E, Biophysical J,. 13, 941 (1973).

15. Evans, E., R. Waugh, and L. Melnik, Biophysical J., 16, 585 (1976).

16. Evans, E. and R.M. Hochmuth, Biophysical J., 16, 1 (1976).

17. Chien, S., K.P. Sung, R. Skalak, S. Usami, and A. Tozeren, Biophysical Journal, 24, 463 (1978).

18. Hochmuth, R.M., Worthy and E. Evans, Biophysical J., 26, 101 (1979).

19. Tozeren, A., R. Skalak, P.S. Kuo-Li and S. Chien, Biophysical J., 39, 23 (1982).

20. Evans, E., Biophysical J., 43, 27 (1983).

21. Evans, E. and R. Skalak, Mechanics and thermodynamics of biomembranes, CRC Press, Boca Raton, Florida (1980).

22. Steinchen-Sanfeld, A. and A. Sanfeld, Chemical Physics, 1, 156 (1973).

23. Sanfeld, A. and A. Steinchen, Biophysical Chemistry, 3, 99 (1975).

24. Sorensen, T.S., M. Hennenberg, A. Steinchen-Sanfeld, and A. Sanfeld, Progr. Colloid and Polymer Sci., 61, 64 (1976).

25. Sorensen, T.S., M. Hennenberg, A. Steinchen, and A. Sanfeld, J. Colloid and Interface Sci., 56, 191 (1976).

26. Dalle Vedove, W. and A. Sanfeld, J. Colloid and Interface Sci., 84, 318 (1981).

27. Dalle Vedove, W. and A. Sanfeld, J. Colloid and Interface Sci., 84, 328 (1981).

28. Ibanez, J.L. and M.G. Velarde, Journal de Physique, 12, 1479 (1977).

29. Van Bockstaele, M. and P. Joos, An. Quim., 71, 889 (1975).

30. Sorensen, T.S. and J. Castillo, J. Colloid and Interface Sci., 76, 399 (1980).

31. Sorensen, T.S., J.C.S. Faraday Trans. II, 76, 1170 (1980).

32. Greenspan, H.P., J. Theor. Biology, 65, 79 (1977).

33. Greenspan, H.P., J. Theor. Biology, 70, 125 (1978).

34. Evans, E. Biophysical J., 14, 923 (1974).

35. Sheetz, M.P. and S.J. Singer, Proc. Nat. Acad. Sci. USA, 71, 4457 (1974).

36. Sheetz, M.P. and S.J. Singer, Journal of Cell Biology, 70, 247 (1976).

37. Evans, E.A. and R. Waugh, J. Colloid and Interface Sci., 60, 286 (1977).

38. Mateu, L., V. Luzzati, Y. London, R.M. Gould, F.G.A. Vosseberg and J. Olive, J. Mol. Biol., 75, 697 (1973).

39. Tardieu, A., V. Luzzati and F.C. Reman, J. Mol. Biol., 75, 711 (1973).

40. Pinder, J., E. Ungewickell, D. Bray and W.B. Gratzer, J. Supramolecular Structure, 8, 439 (1978).

41. Patel, V. and G. Fairbanks, J. Cell Biology, 88, 430 (1981).

42. Landman, K.A. and H.P. Greenspan, Studies in Applied Mathematics, 66, 189 (1982).

43. Naghdi, P.M. and R.P. Nordgren, Quart. Appl. Math, 21, 49 (1963).

44. Naghdi, P.M., "The Theory of Shells and Plates," in Handbuch der Physik, Flugge, S. (Ed.), Springer-Verlag, N.Y. (1972).

45. Ljunggren, S. and J.C. Ericksson, J. Colloid and Interface Sci., 107, 138 (1985).

EFFECT OF GLUCOSE, GALACTOSE AND TRANSMEMBRANE POTENTIAL ON SHEAR ELASTIC MODULUS OF ERYTHROCYTE MEMBRANE

T. T. Traykov and R. K. Jain ■ Department of Chemical Engineering
Carnegie Mellon University, Pittsburgh, PA 15213

The deformability of erythrocytes suspended in buffered and non-buffered isotonic glucose-saline and galactose-saline solutions was determined using micropipette aspiration technique. In this method, a negative pressure was applied to the erythrocyte membrane via a micropipette and the resulting deformation was analyzed using a Kelvin model to yield a membrane elastic modulus E. When glucose concentration was increased from 0 to 0.3 M/lit in the extracellular media, E increased seven-fold with most increase occurring in the 0-0.1 M/lit range. In contrast, galactose had no significant effect on E up to a concentration of 0.28 M/lit. However, in NaCl-free, non-buffered galactose solution (0.308 M/lit), E decreased by a factor of three. This decrease was not observed in buffered glactose solution. Increase in E in the presence of glucose may be due to binding of glucose to the membrane and intracellular proteins. Decrease in E in the presence of galactose in NaCl-free non-buffered solution may be due to the transmembrane potential generated by chloride ion. Presence of buffer presumably suppressed the generation of transmembrane potential. The increase in E due to glucose in less than 1 hour may cause reduction in the tumor blood flow rate during hyperglycemia.

Several investigators have shown that glucose can decrease blood flow rate of tumors without significantly affecting the blood flow rate of several normal tissues (1). Furthermore, galactose, a sugar not metabolized by tumors, can reduce blood flow rate of both normal and tumor tissues, although significantly more of tumors than of normal tissues (1). Glucose can also lower the pH of tumors without significantly changing the pH of most normal tissues (2). Since it is easier to heat a tissue which has a reduced blood supply, and since low pH makes cells more vulnerable to heat, it seems reasonable to combine glucose with heat in the treatment of cancer (3). Before such a combination can be used optimally, it is important to understand the cause of blood flow reduction due to glucose and galactose.

Blood flow in a tissue is governed by three key parameters: blood viscosity, vascular geometry and pressure drop across the vasculature. Blood viscosity in turn is governed primarily by plasma viscosity and erythrocyte concentration, aggregation and deformability. Crandall et al (4) have shown that low pH can increase the rigidity of erythrocytes. However this mechanism of increased rigidity requires very low pH ($\leq$6.5) and long exposure times ($\geq$ 1 hour). In contrast, reduction in tumor blood flow due to hyperglycemia is a faster process (<1 hour) (1,2). We, therefore, wanted to find out if glucose increased the rigidity of erythrocytes directly and in a shorter time.

The objective of our work was, therefore, to measure changes in the membrane elastic modulus of rat erythrocytes suspended in glucose-saline and galactose-saline solutions using the micropipette aspiration technique (5,6,7). In these experiments, glucose and galactose concentrations were increased up to 0.308 M/lit and NaCl concentration was decreased accordingly to maintain isotonicity (iso-osmolarity) in the suspending media. The experiments were carried out both in the presence and absence of a buffer (8,9). In the absence of a buffer, the decrease of the chloride concentration in the continuous phase generates a transmembrane potential (10,11). It has been shown that under some conditions large transmembrane potentials can be generated in erythrocytes and other cells (12,13). Therefore the results of this investigation should be useful in understanding the effect of glucose and galactose as well as the effect of transmembrane potential on erythrocyte rigidity.

EXPERIMENTAL APPROACH AND DATA ANALYSIS

The determination of the shear elasticity modulus of erythrocyte membrane was performed following the procedure described in (7): a cell was aspirated at several step negative pressures ΔP from -1.0 to -25.0 mm water using pipettes of radius, R_p, between 0.3 to 0.7 μm. After application of each step negative pressure, the cell was allowed to equilibrate for 40-50 sec. The maximum projection length of the cell membrane D_{pm} was measured using a video micrometer. The elastic modulus, E, was obtained by fitting the following equation (6,7):

$$D_{pm}/R_p = 0.246 + (0.408/E)(\Delta P . R_p) \quad (1)$$

The details of our experimental set-up are described elsewhere (8,9).

RESULTS

For convenience we will discuss the results obtained in a buffered solution, and then those obtained in a buffer-free solution. The latter results delineate the role of transmembrane potential on the viscoelastic properties of erythrocyte membrane.

In the presence of a buffer, the elastic modulus, E, increased about seven-fold as the concentration of glucose was increased from 0 to 0.3 M/lit (carbohydrate/saline solution ≤ 100) with most increase occurring in the 0-0.1 M/lit range. Further, the time constant of this increase was ∿15 minutes (8). In contrast, E did not change in the presence of galactose in the same concentration range (8) (Table 1).

The results for non-buffered solutions of glucose and galactose were similar to those for buffered solutions up to concentrations of 0.28 M/lit (Table 1) (9). In salt-free pure glucose solution, the value of E decreased by less than 10% ($2.65 \pm 0.26 \times 10^{-2}$ dyn/cm at 0.300 M/lit to $2.44 \pm 0.22 \times 10^{-2}$ dyn/cm at 0.308 M/lit). In contrast, the decrease in E in the salt-free galactose solution was about three-fold ($4.43 \pm 0.51 \times 10^{-3}$ dyn/cm at 0.277 M/lit to $1.31 \pm 0.29 \times 10^{-3}$ dyn/cm at 0.308 M/lit) (9).

DISCUSSION

The objective of our investigations was to measure changes in the membrane elastic modulus, E, of erythrocytes due to glucose and galactose. Our results clearly show that glucose increases E even at moderate concentrations (Table 1). Galactose, on the other hand, has no effect on E except in salt-free, non-buffered solutions of galactose where membrane elasticity decreases by a factor of three. In what follows, we will attempt to explain these results in terms of the data available in the literature (8).

TABLE 1

EFFECT OF CLUCOSE ON ERYTHROCYTE MEMBRANE ELASTICITY MODULUS*

Concentration (M/lit)	Glucose ($\times 10^2$ dyn/cm) Buffered	Nonbuffered
0	.372 ± .026	.385 ± 0.94
0.3	.813 ± .036	----
0.1	1.51 ± 0.17	1.28 ± 0.24
0.15	1.81 ± 0.16	2.04 ± 0.32
0.175	1.83 ± 0.13	----
0.20	2.42 ± 0.27	2.38 ± 0.18
0.27	2.72 ± 0.23	2.62 ± 0.23
0.30	2.71 ± 0.14	2.65 ± 0.26

*In galactose buffered solution (0-0.3 M/lit)E = $0.38 \pm 0.011 \times 10^{-2}$ dyn/cm

and in nonbuffered solution (0-0.28 M/lit)E = $0.399 \pm 0.031 \times 10^{-2}$ dyn/cm.

The three-fold reduction in E in salt-free, non-buffered galactose solution is presumably due to the electrical breakdown of the membrane caused by large transmembrane potential. Recall that the suspending media were prepared by mixing isotonic non-buffered solutions of NaCl and galactose. Hence, as galactose concentration increased, NaCl concentration in the solution decreased. Note that the erythrocyte membrane is highly permeable to Cl^- (10). Therefore, as Cl^- concentration in the external non-buffered medium is progressively diluted (with increasing galactose concentration), a transmembrane potential is generated. This potential practically coincides with the theoretical Nernst potential:

$$V_m = (RT/F) \ln ([Cl^-]_e/[Cl^-]_i) \quad (2)$$

(10,11). A plot of V_m against $\log[Cl^-]_e$ would yield a straight line with a slope of 60 mV. As a result, the value of V_m is zero for erythrocytes suspended in isotonic saline and increases with about 60 mV at every ten-fold dilution of the external Cl^-. In the case of erythrocytes suspended in pure galactose solution, its value is very large and may cause an electrical breakdown of the membrane and lower the value of E.

The effect of dilution of Cl^- in the external phase is better illustrated in Fig. 1. This semilogarythmic plot shows E as a function of $C_{Cl^-}=[Cl^-]_e$ instead of galactose concentration. Note that this curve can be conveniently divided into three parts. The first part shows the independence of E on galactose. In the second part of curve ($10^{-4}<[Cl^-]_e<10^{-3}$; $130<V_m<180$ mV), E falls by a factor of approximately 3. With the Cl^- concentration less than 10^{-4} M/lit, in the third part, E remains practically constant up to pure galactose. As a whole, this curve suggests a phase transition in the erythrocyte membrane in the region ($130<V_m<180$ mV). Putvinskii et al. (13) have experimentally established critical values of V_m around 150 mV which cause the electrical breakdown of the lipid core of a membrane (see also (11)). These values are in support of our data (Fig. 1).

To our knowledge, no other theoretical or experimental studies have related E with V_m (14). Our results on the effect of galactose have two interesting implications: (i) Under normal physiological conditions, the effect of transmembrane potential on erythrocyte membrane elasticity can be neglected since in such systems $[Cl^-]_e$ can never fall to such low values ($<10^{-3}$ M/lit); (ii) Transmembrane potential can be generated by increasing permeability of some nonuniformly distributed cations (e.g. K^{++}, Ca^{2+}, H^+) by adding appropriate ion carrier (15,16,17,18). Such situations are possible not only in vitro but also in vivo in some pathological cases (12).

It has been shown that the spectrin-actin network lining of the endoface of the erythrocyte membrane plays a significant role in determining mechanical properties of the membrane (19). Changes in the molecular organization of spectrin, e.g. a change in the dimer/tetramer ratio, can increase the rigidity of the membrane. Since erythrocyte membrane is permeable to glucose, it is possible for glucose to bind to the membrane and intracellular proteins in such a way as to make the cell less

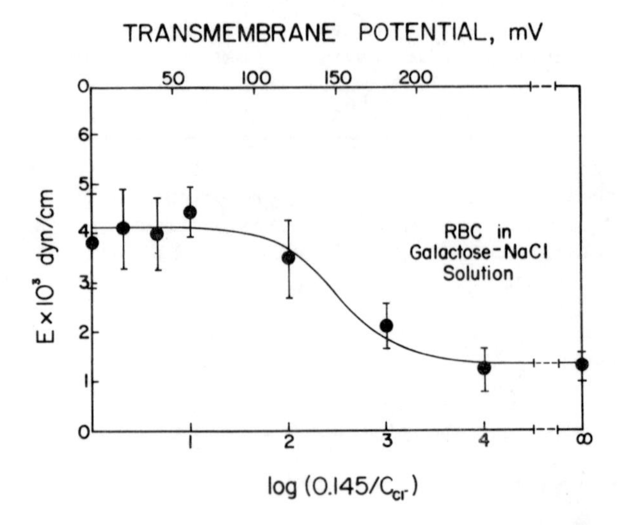

Figure 1. Effect of transmembrane potential on the elastic modulus of erythrocytes suspended in NaCl-galactose isotonic solutions. Transmembrane potential was generated as a Nernst chloride diffusion potential by dilution of saline with isotonic non-buffered galactose solution [from (9), with permission].

deformable. Further, the products of glucose metabolism may also change the intracellular osmolarity of erythrocytes making them more rigid. It is possible that the increased rigidity of the spectrin-actin network caused by glucose masks the effect of transmembrane potential on E. This masking is absent in the presence of galactose, a compound which is physicochemically similar to glucose. SDS-polyacrylamide gel electrophoresis (PAGE) of cells exposed to glucose and galactose may verify this hypothesis.

As stated in the introduction, the current hypothesis of pH induced rigidity can not explain the rapid reduction in tumor blood flow (in less than 30 minutes) caused by hyperglycemia. We suggest that the decrease in erythrocyte deformability caused by glucose itself in the same time frame may play an important role in vascular stasis. It is possible that the increase in rigidity is even more rapid in tumors in vivo, because of the presence of specific endogenous substances, e.g. insulin, lower pH and higher temperature where the rate of biochemical reactions can be greater than in our experimental conditions. By injecting rigid erythrocytes into circulation, Simchon et al. (20) have shown that blood flow reduction in various organs is proportional to the tortuosity of their blood vessels. Therefore, rapid increase in rigidity of erythrocytes coupled with sluggish blood flow through highly tortuous vasculature in tumors may contribute to the flow reduction in less than 30 minutes post glucose injection (1). As shown in (8), E does not go down to the baseline value after glucose is removed from the extracellular medium. These results suggest that the tumor blood flow would not return to its baseline value as soon as glucose is cleared from plasma. This is what we found in our in vivo experiments (2). Since it is not always possible to extrapolate in vitro results to in vivo situations, it would be worthwhile to measure erythrocyte rigidity and local blood viscosity in tumor vessels in vivo to confirm our hypothesis.

ACKNOWLEDGEMENTS

This work was supported by grants from the National Science Foundation (INT-82-09490 and CPE-84-13191) and the National Institutes of Health (CA-36358). We thank Dr. S.A. Shah, Ms. V.A. Rebar, Dr. D.S. Dimitrov and Dr. K.-L.P. Sung for their help in setting up the micropipette aspiration system. This article is based on the material published in references (8) and (9).

During the course of this work, T.T. Traykov was on a leave for one year from the Department of Biophysics, Medico-Biological Institute, Medical Academy, 1431 Sofia, Bulgaria.

NOTATION

$[Cl^-]_e$ = chloride ion concentration in external medium

$[Cl^-]_i$ = chloride ion concentration inside cell

D_{pm} = maximum projection length of cell

E = elastic modulus

F = Faraday's constant

ΔP = negative pressure applied to cell

R = gas constant

R_p = radius of pipette

T = temperature

V_m = Nernst potential

LITERATURE CITED

1. Ward-Hartley, K. and R.K. Jain, "Effect of Glucose and Galactose on Microcirculatory Flow in Normal and Neoplastic Tissue in Rabbits," Cancer Res. In press (1986).

2. Jain, R.K., S. Shah and P. Finney, "Continuous Non-Invasive Monitoring of pH and Temperature in Rat Walker 256 Carcinoma During Normo- and Hyperglycemia," J. Nat. Cancer Inst. 73, 429-436 (1984).

3. Jain, R.K. and K. Ward-Hartley, "Tumor Blood Flow-Characterization, Modification, and Role in Hyperthermia," IEEE Trans. Sonics and Ultrasonics SU31, 504-526 (1984).

4. Crandall, E., A. Critz, A. Osher, D. Keljo, and R. Forster, "Influence of pH on Elastic Deformability of the Human Erythrocyte Membrane," Am. J. Physol. 235, C269-C278 (1978).

5. Rand, R.P. and A.C. Burton, "Mechanical Properties of the Red Cell Membrane," Biophys. J. 4, 115-135 (1964).

6. Evans, E.A., "New Membrane Concept Applied to the Analysis of Fluid Shear and Micropipette-Deformed Red Blood Cells," Biophys. J. 20, 303-313 (1973).

7. Chien, S., K.-L.P. Sung, R. Skalak, S. Usami and A. Tozeren, "Theoretical and Experimental Studies on Viscoelastic Properties of Erythrocyte Membrane," Biophys. J. 24, 463-487 (1978).

8. Traykov, T.T. and R.K. Jain, "Effect of Glucose and Galactose on Red Blood Cell Membrane Deformability," Int. J. Microcirculation: Clinical and Experimental. In press (1986).

9. Traykov, T.T. and R.K. Jain, "Effect of Transmembrane Potential on the Deformability of RBC's Suspended in Carbohydrate-Saline Solution," J. Colloid Interface Sci. In press (1986).

10. Macey, R.L., J.S. Adorante and F.W. Orme, "Erythrocyte Membrane Potentials Determined by Hydrogen Ion Distribution," Biochim. Biophys. Acta 512, 284-295 (1978).

11. Gavelya, I.V., T.V. Puchkova, O.M. Parnev, and Yu.A. Vladimirov, "Electrical Breakdown of Erythrocyte Membranes under Ultraviolet Radiation," Biophysics 28, 721-723 (1983).

12. Putvinskii, A., A. Potapenko, E. Puchkov, D. Roshchupkin, and Yu. Vladimirov, "Biological Effect of UV-Irradiation on the Membranes," Studia Biophys. 64, 17-22 (1977).

13. Putvinskii, A., S.A. Popov, T.V. Puchkova, Yu.A. Danilov, and Yu. Vladimirov, "Electrical Breakdown of Erythrocyte Membranes through the Diffusion Potential Difference," Biophysics 28, 543-544 (1983).

14. Pethica, B.A. and D.G. Hall, "Electric Field Effects on Membranes," J. Coll. Interface Sci. 85, 41-51 (1982).

15. Gunn, R.B. and D.C. Tosteson, "The Effect of 2,4,6-trinitro-m-cresol on the Cation and Anion Transport in Sheep RBC," J. Gen. Physiol. 57, 593-607 (1971).

16. Gardos, G., "Ion Transport Across the Erythrocyte Membrane," Haematologia 6, 237-245 (1973).

17. Reed, P.W., "Effect of the Divalent Cation Ionophore A23187 on Potassium Permeability of Rat Erythrocytes," J. Biol. Chem. 251, 3489-3493 (1976).

18. Sarkadi, B., I. Sarsz and G. Gardos, "The Use of Ionophores for Rapid Loading of Human Red Cells with Radioactive Cations for Cation-Pump Studies," J. Membrane Biol. 26, 357-370 (1976).

19. Chabanel, A., K.-L.P. Sung, J.T. Pachal, J. Paleck, and S. Chien, "Contributions of Lipids and Proteins to Viscoelasticity of Red Cell Membrane," Fed. Proc. 42, 604 (1983).

20. Simchon, S., K. Jan and S. Chien, "Influence of Red Cell Deformability on Regional Blood Flow Distribution,' FASEB Abstracts 1005 (1985).

INDEX

SYMPOSIUM SERIES

ADSORPTION

96 Developments in Physical Adsorption
117 Adsorption Technology
219 Recent Advances in Adsorption and Ion Exchange
230 Adsorption and Ion Exchange—'83
233 Adsorption and Ion Exchange—Progress and Future Prospects
242 Adsorption and Ion Exchange: Recent Developments

AEROSPACE

33 Rocket and Missile Technology
52 Chemical Engineering Techniques in Aerospace

BIOENGINEERING

69 Bioengineering and Food Processing
84 The Artificial Kidney
86 Bioengineering ... Food
93 Engineering of Unconventional Protein Production
99 Mass Transfer in Biological Systems
108 Food and Bioengineering—Fundamental and Industrial Aspects
114 Advances in Bioengineering
163 Water Removal Processes: Drying and Concentration of Foods and Other Materials
172 Food, pharmaceutical and bioengineering—1976/77
181 Biochemical Engineering Renewable Sources of Energy and Chemical Feedstocks

CRYOGENICS

224 Cryogenic Processes and Equipment 1982
251 Cryogenic Properties, Processes and Applications 1986

CRYSTALLIZATION

110 Factors Influencing Size Distribution
193 Design Control and Analysis of Crystallization Processes
215 Nucleation, Growth and Impurity Effects in Crystallization Process Engineering
240 Advances in Crystallization From Solutions

DRAG REDUCTION

111 Drag Reduction
130 Drag Reduction in Polymer Solutions

ENERGY

Conversion and Transfer

5 Heat Transfer, Atlantic City
57 Heat Transfer, Boston
59 Heat Transfer, Cleveland
75 Energy Conversion Systems
79 Heat Transfer with Phase Change
87 Advances in Cryogenic Heat Transfer
113 Convective and Interfacial Heat Transfer
118 Heat Transfer—Tulsa
119 Commercial Power Generation
138 Heat Transfer—Research and Design
162 Energy and Resource Recovery from Industrial and Municipal Solid Wastes
174 Heat Transfer: Research and Application
189 Heat Transfer—San Diego 1979
202 Transport with Chemical Reactions
208 Heat Transfer—Milwaukee 1981
216 Processing of Energy and Metallic Minerals
225 Heat Transfer—Seattle 1983
236 Heat Transfer—Niagara Falls 1984

Nuclear Engineering

53 Part XIII
56 Part XIV
94 Part XX
104 Part XXI
106 Part XXII
119 Commercial Power Generation
168 Heat Transfer in Thermonuclear Power Systems
169 Developments in Uranium Enrichment
191 Nuclear Engineering Questions Power Reprocessing, Waste, Decontamination Fusion
221 Recent Developments in Uranium Enrichment

ENVIRONMENT

78 Water Reuse
97 Water—1969
115 Important Chemical Reactions in Air Pollution Control
122 Chemical Engineering Applications of Solid Waste Treatment
124 Water—1971
126 Air Pollution and its Control
133 Forest Products and the Environment
137 Recent Advances in Air Pollution Control
139 Advances In Processing and Utilization of Forest Products
144 Water—1974: I. Industrial Wastewater Treatment
145 Water—1974: II. Municipal Wastewater Treatment
146 Forest Product Residuals
147 Air: I. Pollution Control and Clean Energy
148 Air: II. Control of NO_{xx} and SO_x Emissions
149 Trace Contaminants in the Environment
151 Water—1975
156 Air Pollution Control and Clean Energy
157 New Horizons for the Chemical Engineer in Pulp and Paper Technology
165 Dispersion and Control of Atmospheric Emissions, New-Energy-Source Pollution Potential
170 Intermaterials Competition in the Management of Shrinking Resources
171 What the Filterman Needs to Know About Filtration
175 Control and Dispersion of Air Pollutants: Emphasis on NO_x and Particulate Emissions
177 Energy and Environmental Concerns in the Forest Products Industry
184 Advances in the Utilization and Processing of Forest Products
188 Control of Emissions from Stationary Combustion Sources Pollutant Detection and Behavior in the Atmosphere
195 The Role of Chemical Engineering in Utilizing the Nation's Forest Resources
196 Implications of the Clean Air Amendments of 1977 and of Energy Considerations for Air Pollution Control
198 Fundamentals and Applications of Solar Energy
200 New Process Alternatives in the Forest Products Industries
201 Emission Control from Stationary Power Sources: Technical, Economic and Environmental Assessments
207 The Use and Processing of Renewable Resources—Chemical Engineering Challenge of the Future
209 Water—1980
210 Fundamentals and Applications of Solar Energy II
211 Research Trends in Air Pollution Control: Scrubbing, Hot Gas Clean-up, Sampling and Analysis
213 Three Mile Island Cleanup
223 Advances in Production of Forest Products
232 Applications of Chemical Engineering in the Forest Products Industry
239 The Impact of Energy and Environmental Concerns on Chemical Engineering in the Forest Products Industry
243 Separation of Heavy Metals and Other Trace Contaminants
246 Advances in Process Analysis and Development in the Forest Products Industries.

FLUIDIZATION

101 Fundamental Processes in Fluidized Beds
105 Fluidization Fundamentals and Application
116 Fluidization: Fundamental Studies Solid-Fluid Reactions, and Applications
176 Fluidization Application to Coal Conversion Processes
205 Recent Advances in Fluidization and Fluid-Particle Systems
234 Fluidization and Fluid Particle Systems: Theories and Applications
241 Fluidization and Fluid Particle Systems: Recent Advances